T0073727

Social Practices as Biological Niche Construction

Social Practices as Biological Niche Construction

JOSEPH ROUSE

The University of Chicago Press
Chicago and London

The University of Chicago Press, Chicago 60637
The University of Chicago Press, Ltd., London
© 2023 by The University of Chicago
Published 2023
Printed and bound by CPI Group (UK) Ltd, Croydon, CR0 4YY

32 31 30 29 28 27 26 25 24 23   1 2 3 4 5

ISBN-13: 978-0-226-82795-7 (cloth)
ISBN-13: 978-0-226-82797-1 (paper)
ISBN-13: 978-0-226-82796-4 (e-book)
DOI: https://doi.org/10.7208/chicago/9780226827964.001.0001

Library of Congress Cataloging-in-Publication Data

Names: Rouse, Joseph, 1952- author.
Title: Social practices as biological niche construction / Joseph Rouse.
Description: Chicago : The University of Chicago Press, 2023. |
    Includes bibliographical references and index.
Identifiers: LCCN 2023001496 | ISBN 9780226827957 (cloth) |
    ISBN 9780226827971 (paperback) | ISBN 9780226827964 (ebook)
Subjects: LCSH: Human beings. | Evolutionary psychology. | Nature and nurture. |
    Philosophical anthropology.
Classification: LCC BF341 .R677 2023 | DDC 155.7—dc23/eng/20230322
LC record available at https://lccn.loc.gov/2023001496

♾ This paper meets the requirements of ANSI/NISO z39.48-1992 (Permanence of Paper).

Contents

Sociocultural Animals

Overcoming the Nature/Culture Divide

We humans are animals with distinctive anatomy, biochemical physiology, and neural architecture; we are ecologically interdependent with other organisms amid the earth's changing climatic conditions; and we evolved from ancestral social primates. Human lives and ways of life are also socially and conceptually complex, and they are shaped by diverse traditions, institutional structures, and political economies that also institute manifold social roles and identities. We are guided by sophisticated ways to conceive and reason about how we live and should live, and we are beset by widespread political domination and massive social and economic inequalities. No one seriously denies either that the people sharing these ways of life are animals or that the physiological, ecological, and cognitive lives of these animals who evolved from a common primate ancestor are now embedded in complex institutions and culturally diverse interactions and self-conceptions with often vast differences in abilities to satisfy basic needs or achieve fulfilling lives. Biological conceptions of human animality and social, psychological, or political explications of how we live together and comprehend and criticize one another have nevertheless mostly proceeded in mutual isolation.

One result of this isolation is a bifurcated and duplicated conception of humans and human ways of life. Humans have bodies composed of physical and chemical processes organized in functional biological mechanisms but are also persons with lives composed of actions, projects, social roles, and careers. Human bodies are developmental processes of growth, differentiation, functional organization, and aging; the persons inhabiting those bodies develop through acculturation, education, and maturation. Human bodies reproduce sexually with only moderate sexual dimorphism; persons take up diverse sexual interactions, gendered identities, and kinship relations.

Human bodies inhabit complex ecological environments shaped by many other living organisms; persons share social, cultural, and linguistic environments with many other persons. General acknowledgment that persons in some sense *are* their bodies accompanies often sharp differentiation of the conceptual relations that characterize people's biological lives and culturally differentiated social worlds.

This book aims to help overcome that mutual isolation and duplication. It develops an understanding of human ways of life that integrates a biological understanding of human bodies and their development, ecology, and evolution with conceptions of the practices and institutions in which people live and engage one another. The book thereby undermines conceptual separations of natural from social "worlds" and both from a so-called inner world of the mind or brain that are now deeply entrenched in the common sense of intellectual life. Those separations are also reinforced by many empirical research programs framed in their terms. Promising developments in the sciences and social theory nevertheless now enable recognizing those conceptual partitions as barriers to adequate understanding of ourselves and our prospects, and they provide resources to dismantle those barriers. The book's primary aim is to work out a constructive alternative that encompasses the richness, complexity, and fragility of human ways of life without relying on these familiar intellectual divisions.

The project has a complex relation to established research programs and theoretical frameworks. John Searle proposed what he called a "philosophy of society" as a new philosophical subfield. He writes, "I believe it will deepen our understanding of social phenomena generally and help our research in the social sciences if we get a clearer understanding of the nature and mode of existence of social reality. We need not so much a philosophy *of* the social sciences of the present and the past as we need a philosophy *for* the social sciences of the future and, indeed, for anyone who wants a deeper understanding of social phenomena" (2010, 5). Searle prefaced that proposal by claiming that such accounts must recognize human social life as part of the natural world, dependent on and consistent with the natural sciences generally and evolutionary biology in particular. His analysis did not fulfill that commitment. Searle's only subsequent mention of evolution was to insist that the evolution of language had no bearing on the conception of language underlying his proposed philosophy of society (2010, 65–66).

This book develops a philosophy of society much more centrally engaged with biology and human evolution, through what Peter Godfrey-Smith has characterized as a "philosophy of nature."

> When doing philosophy of nature in my sense, a writer comments on the overall picture of the natural world that science, and perhaps other types of inquiry, seem to be giving us. . . . The claims made by a good philosophy of nature do have to be *consistent* with the claims made by science. But the concepts employed by a good philosophy of nature do not have to be the same as those used in the relevant science, and the organization and presentation of information in the two projects can be quite different. (Godfrey-Smith 2001, 284)

The book's project also exemplifies Wilfrid Sellars's conception of philosophy as concerned with "how 'things' in the broadest sense of that term 'hang together' in the broadest sense of that term" (2007, 369). In either case, such philosophical projects draw on multiple scientific disciplines without identifying with any one but ask instead how their inquiries are mutually accountable. This conjoined *philosophy of society as integral to nature* engages ongoing research programs in the social and life sciences along Godfrey-Smith's or Sellars's proposed lines instead of Searle's. Searle suggests that a philosophy of society should set the agenda for social-scientific research. Godfrey-Smith envisions a more open-ended conversation among scientific research programs and a philosophy of nature, which draws on scientific understanding while integrating work from multiple disciplines.

A philosophical integration of human sociality and animality does not entail that the human sciences must accordingly reconstruct their domains or their methods of inquiry, data collection, and analysis. Disciplines such as economics, anthropology, psychology, and their subfields may still address intelligible domains of interlinked concepts, skills, and standards that genuinely disclose aspects of human life from within their discipline's constitutive concerns. As Godfrey-Smith (2001, 289) also emphasizes, scientific research programs may have good reasons to focus on some causal factors or emergent patterns while relegating others to background noise mostly omitted from their theoretical models or blocked from how they frame their empirical research. Research programs also often outrun and sometimes revise established conceptualizations. They need the freedom to pursue their own projects and conceptual orientations and make their case for how those projects engage others. Such pragmatic focus within a research program is consistent with a philosophy of nature's more inclusive or holistic understanding. What is disclosed within limited conceptual domains nevertheless remains answerable for its implications for what is disclosed elsewhere and vice versa. We can say of the epistemic aspirations of particular sciences what Brian Cantwell Smith said about consequential deployments of sophisticated artificial intelligence systems.

If we are to trust their deliberations, [scientific disciplines] need to be . . . able to take responsibility for the adequacy of the abstractive ontologies in terms of which they register the world. Otherwise we should use them only in situations where we are prepared to take epistemic and ontological responsibility for every registration scheme, every inferential step, and every "piece of data" that they use along the way. (2019, 80)

Scientific expertise is often developed in terms internal to a discipline's methods and conceptual development, but its character as expertise and understanding *of* the phenomena in its domain partly depends on accountability to what is registered conceptually in other partially autonomous articulations of the world.[1]

In developing an integrated philosophy of society as part of nature, this book attends to its roots in promising research programs, indicates responses to some well-known conceptual problems, and points toward better ways to understand urgent contemporary concerns. Its primary focus is nevertheless constructive development of this integrated conception rather than critical assessment of familiar alternatives or the development of specific research programs in its terms. The persistence of entrenched conceptual orientations is unassailable absent plausible alternatives. Replacing Godfrey-Smith's or Searle's locutions, I describe this reconception of people's sociocultural animality as a *philosophy of natureculture*, borrowing the term "natureculture" from Donna Haraway (2008), who also insists on the inseparability and mutual irreducibility of biological processes and cultural practices. I thereby locate human evolution and development in a more inclusive domain of natureculture rather than treating human ways of life as uniquely biosocial.[2] Biologists now often refer to diverse animal cultures as aspects of their evolution and development, and both the continuities and divergences between human and other animal cultures are important (Boyd and Richerson 1985; Avital and Jablonka 2000; Whiten et al. 1999; Jablonka and Lamb 2005).

Conceptual separations of natural and social worlds and corresponding domains of natural and human sciences did not arise by happenstance or

1. I have developed a more specific account of how disciplinary domains are accountable to other sciences, without reference to Godfrey-Smith's conception of philosophy of nature (Rouse 2015, chap. 10). This book's project does not depend on that more specific account, although I still endorse it.

2. Ingold and Palsson (2013) propose a "biosocial" reconception to integrate social and biological processes. I prefer "natureculture" to avoid confusion with Rabinow's (1996, chap. 5) genomic conception of "biosociality," and because Ingold and Palsson's term has specific disciplinary roots in biological and sociocultural anthropology. My account does have greater continuity with Ingold (2001), however.

oversight, however. Philosophers and scientists in both natural- and social-scientific disciplines have had principled reasons for conceptually separating human bodies as part of nature from human agency and cognition in complex social worlds. This bifurcation is rooted in two distinct comparisons of humans to other animals. Human anatomy, physiology, or neurological organization have extensive and far-reaching continuity with others in the primate order and more encompassing biological taxa. Those continuities expand and deepen with cellular structures, biochemical processes, or DNA sequences and their genetic expression and regulation. Evolutionary theory has compelling explanations for those continuities; the evolutionary descent of *Homo sapiens* within the lineage of the great apes and a broader history of life has not been seriously questioned for a long time.

Matters look different concerning how humans and other animals live. Research on animal behavior and social relations has yielded new recognition and appreciation of the cognitive sophistication and social complexity of many nonhuman animals. Even in this light, however, many aspects of human ways of life seem unprecedented in familiar and far-reaching ways. People undertake scientific investigations, historical reflection, artistic expression, and moral reasoning. Apparently incommensurate aspects of human ways of life also encompass the material infrastructure of our built habitats and technological systems; the scale, complexity and diversity of political institutions, economic interactions, and cultural traditions; languages and other forms of communication and education; and even how people satisfy common biological needs for food, water, sex, infant care, predation avoidance, or waste excretion. These contrasts made it seem reasonable and appropriate to say that although human social life and cognitive capacities are enabled by biologically evolved bodies and processes, they are constituted, shaped, and regulated in other ways. Those rationales also cite differences between the inexorability of natural causes or laws and the contrasting freedom and rationality of human cognition and action or the social-discursive shaping of human lives and institutions.

These and other prominent rationales for intellectual segregation of nature from human society and culture are no longer sustainable, however. Reasons to reject these considerations arise from developments in evolutionary biology, philosophy, the social and cognitive sciences, and interdisciplinary science studies. An integrated, naturecultural conception of human ways of life is reinforced by how those developments complement and partially revise one another. Further support comes from a different direction in acknowledging the range of contemporary concerns not readily separable into natural and sociocultural components: climate change and energy provision; health, disease, and health care; agricultural production and food insecurity; sex, gender and sexuality; waste disposal and environmental justice; technological change and the structure of work; and more. The recent transformations of people's "social life" to accommodate the widespread circulation of the SARS-CoV-2 virus are vivid reminders that people's social interactions have always been intimately entangled with other organisms.

Consider the most comprehensive grounds for the supposed autonomy of a human social world, stemming from familiar conceptions of "modern" scientific understanding. The supposed scientific disenchantment of nature leaves no place for a normative order, whether stemming from divine creation, cosmological order, natural law, or an essential human nature. Consequently, normative accountability could only be instituted by or within human ways of life. The natural sciences locate human *bodies* within a causally interactive or law-governed domain of inexorable flows and patterns. The explanatory overlaps between physical, chemical, and biological sciences and the evolutionary continuity of humans and other organisms undercut treatment of human biology as exceptional. Genetically, physiologically, and ecologically, human bodies are thermodynamically constrained biochemical processes, subject to inexorable natural laws or causal nexuses and resembling other animals with whom people share ancestry. Humans are nevertheless also rational *agents* or *persons* living amid biologically unprecedented cultural traditions, social institutions, and economic transactions. These higher-order patterns in human ways of life do not proceed with the causal or lawful inexorability of scientifically intelligible nature. Human social life forms meaningful patterns whose structure and transformations are mediated by reasoned intelligibility to human participants. That intelligibility is situated amid historically divergent cultural traditions and their only-partially comprehended encounters with one another. The causal inexorability of human biological functioning thus seems overlaid with a mostly autonomous space of freedom, contingency, meaning, and normative assessment.

A naturecultural account of human ways of life accords with some aspects of this familiar conception of modern scientific understanding.[3] Nature has no overarching normative order that could explain or legitimate social practices or political institutions. Human social institutions, discursive practices, and cultural differences are biologically unprecedented in ways that bear on their normative accountability. Scientific understanding of the natural world

---

3. I do not endorse a supposed seventeenth century Scientific Revolution that sharply breaks from ancient, medieval, or Renaissance cosmologies, but this disagreement over the historiography of natural philosophy does not affect my argument here.

is nevertheless not bereft of normative significance. Functional and evolutionary biology understand living organisms as processes sustained and reproduced through ongoing interaction with their surroundings. Some aspects of those surroundings thereby coalesce as meaningfully configured environments integral to those ways of life. Organisms and their lineages succeed or fail in maintaining their constitutive processes amid changing circumstances, which are thereby threats or opportunities for their ongoing maintenance and reproduction. Humans answer to many more and more variegated normative concerns than just individual survival or continuation of the lineage. I nevertheless argue that these complex forms of normative accountability and authority are evolutionary descendants of the constitutive biological normativity of organisms and their lineages. My claim is *not* that the content or authority of the normative concerns shaping human ways of life is biologically determined. After all, evolutionary phenomena typically exhibit contingency and variation rather than inexorable necessity. More important, however, the diversity and complex mutual accountability of these normative concerns are evolved outcomes. Moreover, adequately understanding these more complex, evolved forms of normative accountability requires recognizing their continuing biological significance.

A second prominent rationale for distinguishing a social-historical human world from a more encompassing nature arose from transformative scientific developments at mid-twentieth century. The modern evolutionary synthesis and molecular genetics encouraged identifying biological evolution with slow-paced processes of genetic mutation and selective adaptation to exogenously changing environments. Only genetic changes transmitted through germlines could be ultimate causes of biological form and function. The basic features of human biology thus seemed to emerge in the Pleistocene epoch with genetically, anatomically, and behaviorally modern hominins. On this conception, subsequent historical developments in human lifeways were biologically ephemeral except by partly insulating humans from further selective adaptation by diminishing environmental impacts on human populations. Meanwhile, cognitive sciences modeling human psychology on digital computation encouraged principled separation of cognitive software from the physical hardware or biological "wetware" of computers or brains. The behaviorist tradition in psychology used animal behavioral models to investigate or extrapolate how external reinforcement shaped human behavior. The cognitive sciences instead highlighted rule-governed ratiocination to analyze how human ways of life are autonomously shaped by reasoned assessment within complex traditions of linguistic meaning and sociocultural normativity. Here, the book draws on promising new directions and conceptual orientations in evolutionary biology and the cognitive sciences. These developments significantly revise those often-attributed consequences of the modern evolutionary synthesis and the cognitive revolution in psychology and linguistics.

The constructive roles of the modern evolutionary synthesis and the cognitive sciences in underwriting conceptual divisions of nature from culture were complemented by cautionary tales from failed efforts at reunification. Successive attempts to expand the modern synthesis to encompass human ways of life within a more comprehensive biological science have manifestly failed to account for the phenomena. Despite often-triumphal rhetoric, sociobiology (Wilson 1975), human behavioral ecology (Hinde 1974; Cronk, Chagnon, and Irons 2000; Winterhalder and Smith 2000), and evolutionary psychology (Barkow, Cosmides, and Tooby 1992; Buss 1999) too often trafficked in speculative storytelling with weak evidential support and debilitating inattention to cultural and individual variation. Evolutionary psychology's account of the supposed modularity of evolved minds also conflicts with subsequent work in the neurosciences (Anderson 2014) and evolutionary biology (Sterelny 2003, chaps. 10–11). From another direction, however, oversimplified biological research programs for human societies and cultures were paralleled by some programmatic proclamations of the social construction of scientific knowledge. In their problematic forms, these programs seemed to assimilate natural scientific concepts and practices to the other side of all-too-traditional dichotomies between social and natural domains.[4] Theoretical differences between this book's naturecultural orientation and earlier programs for a biological conception of human social life figure prominently in what follows. Less is said about the book's divergence from social or cultural constructivisms, which I have thoroughly addressed elsewhere (Rouse 1996b; 2015, part 2).

The failures of earlier efforts to naturalize human psychology or social life are not confined to empirical shortcomings and theoretical limitations, however. The problematic and consequential history of appeals to scientific authority to naturalize racial or sexual differences or legitimate colonial exploitation provide a more serious cautionary tale. These issues are not this book's

4. Incautious proclamations of sociocultural priority over natural-scientific understanding stood out from important, careful historical, social, and philosophical studies of scientific practices that often challenged familiar nature-culture distinctions. Incautious recapitulations of familiar dichotomies often garnered more attention than careful scholarship that challenged them. In some cases (e.g., Shapin and Schaffer 1985; Pickering 1984; Collins and Yearley 1992; or Bloor 1976) provocative but incautious claims extracted from careful scholarship became widely circulated proxies for entire projects.

primary focus, but it supports promising directions for recognizing, analyzing, and responding to these problems. A naturecultural approach to the science and politics of race, sex and gender, sexuality, or neocolonial domination converges with other work emphasizing how social practices of racial,
sexual, and neocolonial oppression are often closely entangled with the biology and biomedicine of bodies and environments and cannot be adequately
understood through oppositions of social practices to bodily processes.[5]

New Directions in Biology, Social Theory, and Cognitive Sciences

The philosophy of nature, or natureculture, proposed in this book develops
a broader philosophical orientation stemming from achievements and ongoing work in the sciences. It draws on promising conceptual developments
and empirical discoveries in multiple scientific fields and shows how their
constructive and mutually transformative engagement offers new possibilities for human self-understanding and responses to pressing contemporary
issues. Three broad research domains—in practice-based social theory; evolutionary biology; and embodied, embedded, extended, and enactive (4E)
approaches to cognition—are especially relevant. Collectively, they advance
the prospects for naturecultural integration of human biology with the social,
cultural, and cognitive complexity of human ways of life. Subsequent chapters elaborate their contributions in detail, but here I indicate their differing
roles in the book's project.

The book's main arguments revise a prominent strand in social theory
and the social sciences which identifies *practices* as the basic makeup of human ways of life. Practice-based conceptions of social life go back decades,
with prominent roles in philosophy, social theory, many social sciences, historiography, and interdisciplinary science studies.[6] The concept was introduced to show how what people do is enabled and conditioned by synchronic
and diachronic relations among prior performances and their anticipated
continuation. As is often the case in philosophy, my revised approach concep-
tualizes this domain differently. Practice-based approaches to social theory
almost invariably identify its "social" character as mostly autonomous from

5. A large literature now criticizes the conjoined scientific and moral-political failures of
naturalizations of racial or sexual differences and colonial or neocolonial exploitation. Instructive examples include Haraway (1989), Duster (2003), Reardon (2005), Anderson (2003), Hatch
(2016), Nelson (2016), Fausto-Sterling (1992), Hubbard (1990), Martin (1991), Longino (2013),
Richardson (2013), Pitts-Taylor (2016), Fine (2010), or Adas (2014).
6. Turner (2007) and I (Rouse 2007, forthcoming) both review this work, and highlight key
issues and theoretical divisions within practice-based approaches to social theory.

people's biological lives as animals. I build on this tradition but reframe its analyses of social practices within a naturecultural conception of humans as animals inhabiting and reconstructing practice-differentiated environments. Theoretical accounts of practices have struggled over difficult conceptual issues, empirical failings, and unresolved disagreements. The book shares many animating commitments that differentiate practice-based conceptions from other accounts of human social life, but it finds a common source for the conceptual and empirical difficulties preventing adequate fulfillment of these commitments. These problems arise from conceiving practices as social phenomena comprising a domain of inquiry largely independent of human biology. Chapter 1 reviews why practice-based social theory remains important and promising and what difficulties block adequate fulfillment of that promise.

This revisionist conception of practices draws on ecological-developmental biology, niche construction theory, and other work in evolutionary biology and related fields to argue that human developmental environments are composed of diverse, interdependent practices. These biological developments are thus a primary background to the book's main arguments. Many biologists contributing to this work argue, controversially, that their research coalesces as a major revision of the modern evolutionary synthesis that has guided evolutionary biology since the mid-twentieth century (Laland et al. 2015; Jablonka and Lamb 2020; Pigliucci and Müller 2010; Oyama, Griffiths, and Gray 2001; Sultan 2015; Laland et al. 2014). Defenders of a resilient modern synthesis respond that it accommodates the theoretical import of these empirical results and research projects (Wray et al. 2014), while historians of biology worry that participants in this controversy polemically oversimplify the complex history of evolutionary theorizing (Smocovitis 2020). I am more concerned with the broader import of these developments for understanding human ways of life. The book remains agnostic about their eventual consequences for evolutionary theory and the orientation of evolutionary biological research, despite sympathy with and reliance on many conceptual and empirical achievements motivating calls for an "extended synthesis." A philosophy of natureculture can incorporate these developments within its broader project without taking a controversial stand on the configuration of evolutionary biological research.

The biological developments playing especially central roles in the book begin with integration of developmental processes into evolutionary understanding (Gilbert, Opitz, and Rapp 1996; West-Eberhard 2003; Wagner 2000, 2001, 2003; Amundson 2005). The significance of these efforts is enhanced by greater recognition of phenotypic plasticity, phenotypically driven genetic

INTRODUCTION   11

accommodation, and their implications for ecological mediation of gene expression in development (Pigliucci 2010; Lewontin 2001b; Sultan 2015; Gilbert
2016). Ecological-developmental biology and the ubiquity of microbial symbiosis in eukaryotes also encourage rethinking biological individuality and
the interdependence of organism and environment (Gilbert 2017; Gilbert and
Epel 2009; Sultan 2015; Gilbert, Sapp, and Tauber 2012). The conceptual innovation featured most prominently here is the evolutionary significance of
niche construction, which recognizes causal interaction between organisms
and their developmental and selective environments as consequential in both
directions (Odling-Smee, Laland, and Feldman 2003; Lewontin 2001a, 2001b).
The import of understanding language evolution as a niche-constructive process in which languages and the hominin lineage coevolve (Dor and Jablonka
2000; Bickerton 2014; Dor 2015; Laland 2017; Tomasello 2008; Rouse 2015),
and recognition of human evolution and development as dramatically shaped
by iterated, "ratcheted" niche construction (Tomasello, Ratner, and Kruger
1993; Tomasello 2008, 2019; Sterelny 2012a, 2021; Laland 2017) are especially
central.[7] Because these biological developments frame the book's main arguments, I summarize their primary lessons for understanding human ways of
life in chapter 2 and discuss more specific implications as they arise.

   The book's naturecultural account of practices as biologically niche constructive often makes common cause with recent developments in a third
research area in the cognitive sciences and philosophy of mind which emphasizes the enactive embodiment of cognition in organisms' ongoing practical-
perceptual interaction with their surroundings.[8] The book relies less on detailed developments in embodied cognition than in ecological-developmental
biology and niche construction or the achievements and conflicts of practice-
based social theory and therefore does not review them systematically. The
book's arguments nevertheless align with the overall orientation of many of
these developments, particularly those situating cognition in the dynamics
of bodily interaction with environmental affordances (Chemero 2009; Noë
2004, 2009; O'Regan 2011; Anderson 2014; Gallagher 2017, 2020).

---

7. Tomasello, Ratner, and Kruger (1993, 495) introduced "ratcheting" as follows: "Once a
practice is begun by some member or members of a culture others acquire it relatively faithfully,
but then modify it as needed to deal with novel exigencies. . . . This accumulation of modifications over time is often called the 'ratchet-effect,' because each modification stays firmly in place
in the group until further modifications are made."

8. I use "embodied cognition" as a more compact expression encompassing work on embodied, environmentally embedded, extended, and/or enactive ("4E") approaches. Those differences are less consequential for my project, which primarily intersects theirs in relations between organismic bodies and their developmental and selective environments.

The book proceeds more in parallel than in concert with these projects in the cognitive sciences for several reasons. First, I focus on language and other conceptual capacities within this overall approach to human cognitive abilities, whereas much work on embodied cognition attends to other aspects of perception, action, and bodily skill.[9] Since I emphasize people's practical-perceptual uptake of discursive practices as integral to their developmental and selective environments, however, work in embodied cognition clearly complements my project. A second reason for less extensive engagement with details of embodied cognition research is its focus on the neurological basis for perceptual and practical interaction with an environment, whereas my primary concern is the body/environment nexus itself. Neuroscientific understanding is important, but developing both aspects of these connections would have lengthened the book beyond reasonable expectations of readers and exceeded my own competence. The complementary work of researchers in embodied cognition is nevertheless critical to the longer-term success of the project.

A third reason for reticence about the book's relation to embodied cognitive science is agnosticism about a central divide among its approaches, which I take as an unresolved empirical question. Anthony Chemero distinguishes radical embodied cognitive science from embodied cognitive science: "Radical embodied cognitive science [is] the scientific study of perception, cognition, and action as [a] necessarily embodied phenomenon, using explanatory tools that do not posit mental representations. It is cognitive science without mental gymnastics. . . . Embodied cognitive science embraces the necessity of embodiment and the value of dynamical explanation, but combines them with the computational theory of mind" (2009, 29–30). The book is mostly agnostic about this issue, because its biological orientation closely accords with that of "radical embodied cognitive science." My argument situates people's cognitive or conceptual capacities within ongoing bodily engagement with a niche-constructed developmental and selective environment. It remains an open empirical question whether understanding how people negotiate responsiveness to that environment utilizes neural representations or incorporates environmental features within an extended computational system, as proposed by less radical approaches. I agree with Chemero, however, that appeals to representational or computational processes should be empirically grounded in organism/environment interaction.

9. Di Paolo et al. (2018) incorporates language use within an enactivist conception of cognition, drawing extensively on phenomenology and post-Marxian dialectics.

The current introduction, the critical review of practice-based social theories in chapter 1, and the overview of developments in evolutionary biology in chapter 2 are background for the book's main argument. That argument aims to show how human animality is pervasive within and central to the practice-differentiated complexity of our ways of life. We can recognize social or cultural complexity as biologically evolved and sustained through developmental reproduction and transformation of practices and their mutual interactions without reducing that complexity to simpler biological needs or functions. The account thereby aims to vindicate central commitments of practice-based social theories while resolving problems confronting conceptions of practices as more narrowly social phenomena.

Practices as Biological Niche Construction

A naturecultural conception of practices as biologically niche constructive is an alternative to familiar conceptual divides between natural and social worlds. Brian Epstein recently argued that conceptions of the human social world are framed by a fundamental misconception.

> Implicit . . . in the practice of the social sciences from its earliest days is a particular analogy between the social sciences and the natural sciences. Namely, that the objects of the social sciences are built out of individual people much as an ant colony is built out of ants, or a chimpanzee community is built out of chimpanzees, or a cell is built out of organelles. When we look more closely at the social world, however, this analogy falls apart. We often think of social facts as depending on people, as being created by people, as the actions of people. We think of them as products of the mental processes, intentions, beliefs, habits, and practices of individual people. But none of this is quite right. . . . Philosophers and social scientists have an overly *anthropocentric* picture of the social world. (2015, 7)

Epstein rightly targets overly anthropocentric conceptions of a social world, but I endorse his diagnosis on different grounds. Epstein accepts the conceptual autonomy of the social world but argues that the metaphysics of social facts and social groups grounds, frames, and anchors them in considerations extending beyond facts about people. Even social groups cannot be understood without attending to facts that ground and anchor their constitution as groups which are not limited to facts about their members. I take the problematic anthropocentrism of most social ontologies to arise more fundamentally from anthropocentric conceptions of *people*.

Epstein acknowledges and works to overcome the initial strangeness of decentering the social world from people and their interactions. Challenging an anthropocentric understanding of people may seem a greater affront to prereflective common sense. A decentered conception of people and human ways of life nevertheless follows straightforwardly from aspects of being human that are often readily accepted in other contexts. Four initial insights drive this proposed reconception of practices as forms of biological niche construction.

(1) *People are organisms whose bodily capacities, practical orientations, and continuing existence are constitutively entangled with the biological environments with which the human lineage coevolved and with which individual humans codevelop in diverse ways.* Organisms are not self-contained entities with the property of being alive. A living organism is a *process* that incorporates continuing interaction of its body with its biological environment. The organism exists, *as* a living body rather than a disintegrating corpse, by actively and interactively finding and taking in resources from its environment, avoiding or blocking threats, and excreting waste products. Those interactions and exchanges enable, sustain, and transform their own continuation as an interdependent process. These sensory, metabolic, and behavioral processes not only work together to keep the organism alive. Organisms' development throughout their life span and differential reproduction as a lineage are environmentally situated and mediated. The characteristic sociality of human ways of life belongs to the environmentally mediated maintenance, development, and reproduction of living human bodies. A nonanthropocentric conception of people begins by recognizing that people are not agents with self-contained capacities and interests who happen to exercise those capacities in some settings rather than others. People's capacities, interests, and lives are shaped by lifelong codevelopment with diverse environments.

(2) *The familiar differences between human ways of life and those of other organisms arose from coevolution with the distinctive environments shaping human development in each generation.* Ecological-developmental evolutionary biology recognizes that biological environments are not equivalent to physical surroundings. The developmental and selective environment of an organism encompasses those aspects of its circumstances that matter to its way of life. Two organisms similarly located in space and time typically inhabit and encounter different biological environments. Moreover, organism and environment change together in mutually significant ways. Organisms and their lineages develop and evolve in response to their environments, but those responses also reconfigure those environments. The distinctive and characteristic features of human ways of life have been shaped by cumula-

tively niche-constructive coevolution with increasingly multifarious, differentiated environments. Not only have humans repeatedly reconstructed their surroundings into a built environment and biologically reconfigured their ecosystems through codomestication, cultivation, habitat construction or diminution, and predation of other organisms. Human behavioral and expressive innovations reshaped the developmental environments of their own subsequent generations, leading to reliable, transformative reproduction of descendant behaviors and expressive capacities.

The physical reconstruction of people's material surroundings is an important aspect of that coevolutionary history, greatly diversifying the situations people inhabit. People live in varied settings around the globe, and have developed and scaffolded abilities to move through other settings we cannot actually inhabit. These materially reconstructed sites partially buffer vulnerability to harmful environmental impacts and facilitate new kinds of performance and interaction. These transformations never stand alone, however, for other persons are salient components of human developmental and selective environments. That salience is heightened by developmental neoteny—human biological development is enabled and supported by long periods of juvenile dependence on others' care and guidance. Moreover, what is salient are the postures and orientations of those bodies, their expressive performances, and their involvement in more complex patterns of situated, practical activity. These cooperative relations among people do not constitute a relatively autonomous social world of interactions with one another but instead intensify interdependence with other aspects of their biological environments.

(3) *Diverse but interdependent practices are the characteristic shape of human developmental and selective environments.* Organisms within any lineage vary genetically, and further variation arises from development in different environments. Humans evolved a novel form of more extensive variation, however, by differentiating a range of interacting, interdependent practices. Practices on this conception are spatially dispersed, temporally extended patterns of situated bodily activity. They are shaped by the closely coupled mutual interdependence of those activities with material circumstances that incorporate comparably situated activities and capacities of others. These patterns of closely coupled, situated performance only exist through continuation and differential repetition of mutually supportive performances in sufficiently conducive circumstances. Those circumstances include other practices that enable, support, or sometimes conflict with a practice in limited ways that contrast to more intimate interdependence among the situated performances of the practice. Humans now sustain ourselves as living bodies

in a continuing lineage and develop in characteristically diverse ways by participating in many interdependent practices.

Recent efforts to track and explain hominin evolution emphasize enhanced social learning, supported by active teaching and other mutually cooperative behavior (Sterelny 2012a, 2021; Laland 2017; Tomasello 2019). Other social animals often coordinate their activities by attention and response to others' behavior, pursuing mutually responsive but individually goal-directed activities (Tomasello 2019, 2014, 2008). Hominins evolved more strongly cooperative behaviors that are reciprocally interdependent. Michael Tomasello and his colleagues characterized the evolutionary importance of these aspects of human behavior as "the interdependence hypothesis" (Tomasello et al., 2012). I endorse their formulation but reinterpret it. Where Tomasello and colleagues spoke of the interdependence of persons through shared intentionality and joint action, I emphasize the *dynamically coupled* interdependence of people's postures and performances which constitutes temporally extended practices. These formulations may converge, as Tomasello's recent work (2019) characterizes "shared" and "collective" intentionality in ways more akin to temporally extended practices and projects than to synchronic states of individuals or groups. People's situated performances in a practice depend on one another, and would not normally be done at all or in the same way except in response to and anticipation of ongoing patterns of coupled performance. People's interdependence as persons follows from the constitutive interdependence of what they do and how their capacities and needs develop in those settings.

Talk of "coupled" postures, performances, and circumstances highlights the reciprocal influence of organisms and environments that can be modeled by sets of coupled differential equations with interdependent variables (Lewontin 1983, 2001b; Odling-Smee, Laland, and Feldman 2003).[10] The coupled oscillator models often employed in radical embodied cognitive science similarly emphasize nonlinear mutual responsiveness among coupled components (Chemero 2009, 49–50, 63–64). Interactions of coupled components thereby form a single, nondecomposable system. Under different circumstances, the feedback characteristic of coupled systems can either stabilize patterns of interaction or accelerate change. Without insisting on a role for coupled oscillator models of cognition, I am claiming that the coupled dynamics of interdependent systems aptly models differences between the novel

---

10. Chapter 3 further discusses modeling organism-environment relations as sets of coupled differential equations with interdependent variables.

forms of cooperative activity in the hominin lineage and the merely coordinated individual behaviors commonly undertaken by other social animals.

An important part of the argument for this conception of practices is that
the circumstances in which their constitutive performances occur are integral
to that coupled interdependence, as both grounds and targets of participants'
practical orientations. Sustaining such coupling in changing circumstances
requires work, because patterns of coupled performance can become partially decoupled. Such partial decoupling, which I later characterize as misalignments among performances or circumstances, is consistent with people
remaining dependent on one another's performances and needing to restore
their coupled interdependence. That restoration is a consequential achievement that must be continually maintained and reproduced. Practices change
over time due to the challenges of maintaining the coupled interdependence
of ongoing performances in response to changes in their circumstances, including changes in others' performances and relations to other practices.

Performances in a practice are usually closely coupled in multiple ways,
with changes in one performance leading to compensating adjustments in
others. A practice composed of closely coupled performances may also have
more limited dependence on simpler interfaces with other groups of closely
interlinked performances. Herbert Simon's (1981) and John Haugeland's
(1998, chap. 9) analysis of hierarchically arrayed systems, "nearly decomposable" into component systems interacting at interfaces, helps understand how
performances and the practices they compose are mutually interdependent.
Their conception of nearly decomposable systems abstracts from the kinds of
systems and interactions involved to call attention to a difference in kind between interactions within a component system and those among components
at multiple hierarchical levels. The intelligibility of the system as a whole requires discerning relatively simple interfaces between components whose internal workings are more closely interdependent.

I adapt this approach to analyze relations among complex networks of interdependent practices, subpractices, and coupled performances making up
those practices. What Haugeland calls the "intimate" interdependence among
performances of a practice contrasts to simpler "interrelations" among practices at interfaces. Simon's nearly decomposable systems are often conceived
as relatively static, but Haugeland emphasizes the framework's application to
systems whose components move or change, if the interfaces among components shift accordingly. The practices making up human developmental
environments change the boundaries and interactions among their components and interfaces, often through adjustments to maintain their mutual
alignment. Accommodating or building on changes at those interfaces may

nevertheless require more complex, interlinked changes within the practices affected by them. Humans have developed highly complex, multilevel interdependences among practices, with distinctions between "intimately" coupled performances and more limited interfaces with other couplings iterated at different levels.

This influential conception of complex systems as nearly decomposable into components and interfaces has a second role in the book's argument. An implicit rationale for sustaining familiar conceptual divisions of social from natural entities, processes, or "worlds" has been that social relationships and natural processes are nearly decomposable components of human ways of life that mostly interact at relatively well-defined interfaces. I argue instead that these familiar distinctions do not demarcate distinct components of our complex ways of life. Boundaries among the practices that now make up the evolved patterns of human lives often cut across those familiar distinctions. Haugeland (1998, chap. 9) makes a parallel argument that human and other cognitive systems are best decomposed into systems in which familiar boundaries between mind/brain and body or body and world are not relevant interfaces. His argument for understanding cognition as embodied and my argument that social practices are forms of biological niche construction are not merely analogous. Establishing those connections in detail would nevertheless require a more extensive discussion of cognitive systems than this book can accommodate.

(4) *People's practice-differentiated way of life is articulated horizontally and vertically and shaped by dynamic interactions among these forms of differentiation.* Human ways of life are horizontally differentiated by partially interdependent practices. Each practice both enables and requires participants to act in ways dependent on conducive circumstances and others' supportive performances. Sustaining their interdependence allows people to participate in some practices while relying on other practices to fulfill their biological needs and provide supportive performances and circumstances. People's individual lives then become vertically differentiated by participating in multiple practices, whether synchronically or sequentially throughout their lives. Possibilities and prospects for vertical differentiation of a single life are vastly multiplied by horizontal differentiation of human ways of life to encompass an increasing variety of practices and the dispersed and often reconstructed settings where they take place.

Vertical and horizontal differentiations of human ways of life interface in two important ways. First, as noted earlier, the closely coupled interdependence of people's performances and circumstances within a practice depends in more limited but often consequential ways on alignment with other, horizontally differentiated practices. Those other practices typically involve other people interacting in differently located and equipped settings while sustaining more limited interactions with still other practices. Second, the people engaged in closely coupled activities within a practice also live vertically differentiated lives that involve them in other horizontally differentiated practices. Participation in a practice is typically affected by whether and how it accommodates involvement in other practices. The ongoing mutual adjustments between the vertical and horizontal differentiation of practice-based ways of life integrally shape the evolving dynamics of that way of life. Different practices support, conflict or supplant one another and those practices must accommodate participants' involvement in other practices. Many ongoing practices are sustained by those mutual adjustments, while others emerge, change, or disappear.

Chapters 3 and 4 develop these four initial considerations as the basic structure of a naturecultural conception of practices as the niche-constructive configuration of our developing and evolving ways of life. These ways of life evolved in the hominin lineage through iterated, often ratcheted cycles of niche-constructive coevolution. People's initial biological development as embodied agents is shaped by situated dependence on caregivers whose postured performances are practically and perceptually salient aspects of their developmental environments. People's bodily capacities and developmental trajectories have coevolved with many generations of neotenous dependence to enhance that salience. Caregiving practices and relationships now differ greatly among humans as a result. The discursive practices integral to children's developmental environments provide the most striking variations, but children's caretakers also typically engage in other practices amid and alongside caregiving. Children gradually encounter and become involved in those other practices, and throughout their developmental dependence encounter and participate in other settings and practices with varied separation from primary caretakers. As they attain physical and neurological developmental maturity, most people separate more fully from earlier caretakers and take up other practices where their capacities and characteristic performances continue to develop in closely coupled interdependence with others' situated performances.

Chapter 3 examines how people encounter and respond to their circumstances from a bodily set, stance, or posture.[11] In ordinary parlance, postures are synchronic positionings of one's body to enable some movements and

11. Kukla's (2018) related conception of embodied "stances" extends and modifies Dennett's (1987) account of explanatory stances, which I further elaborate in Rouse (2021).

impede others, while also shaping patterns of perceptual salience and occlusion. I extend the term to encompass dynamically linked synchronic postures as practical bodily orientations in and toward a situation. Bodily postures allow encountering, exploring, and responding to circumstances by conjoining perceptual, proprioceptive, and affective uptake with practical and expressive response. Alongside proponents of embodied cognition, I argue that people's postured engagement with their surroundings integrates perceptual exploration and practical activity. Postures are not taken up de novo, but are shaped instead by prior encounters and engagements. They incorporate gradually developed capacities to coordinate responsive and expressive bodily movement and engage with circumstances meaningfully configured by tasks, opportunities, threats, and other practical significance. Human bodies at birth have limited degrees of freedom of movement and coordinated control, but these abilities are developmentally scaffolded, refined, and expanded by situated activity, growth, and prosthetic support or enhancement.

Postures are not simply structured orientations of a living body itself, however, because people's postures and resulting abilities are shaped by how they are grounded, directed toward, engaged with, and sometimes prosthetically extended by surrounding circumstances. People's bodily postures are responsive to and differentiated by increasingly complex and varied biological environments resulting from iterated niche construction in our lineage. Other people are especially salient components of these developmental environments. The bodily postures and behavior of others are not separate from a shared physical environment but instead belong to one another's mutually encompassing biological environments. People do not encounter other humans *within* shared circumstances; other people and what they do are integral components *of* their environments. The varied environments in which people develop are characteristic of human ways of life and shaped by ongoing reconstruction of human behavioral patterns in situ. The distinctive features of human development resulted from ongoing coevolution of the hominin lineage with niche-constructed environments that people both depend on and reconstruct through mutually interactive life processes.

The salience of spoken and written language and other expressive performances is integral to people's postured engagement with their practice-differentiated developmental environments. Human bodies are poised to speak and listen as well as move and perceive. We develop as "linguistic bodies" (Di Paolo et al. 2018) who normally take up and contribute to the discursive practices around us as integral to our own postured, expressive responsiveness to the world. Although languages are pervasive and flexible examples of expressive uptake of human developmental environments, they coexist with other expressive practices: ostension, images, music, dramatic impersonation, dance, modeling, sport, and more. As one consequence of that expressive articulation, human developmental and selective environments are spatially dispersed, temporally extended, and imaginatively enhanced.[12] People develop abilities to track, anticipate, and respond to more wide-ranging circumstances, in part by reliance on and response to the abilities and performances of others, and by reconstructing other circumstances to facilitate those abilities.

Discussion of people's postured responsiveness to materially and behaviorally niche-constructed environments sets the stage for chapter 4 to outline the book's revisionist account of practices as the basic makeup of those developmental environments. This conception is an alternative to earlier conceptions that identified practices by dispositional regularities, shared presuppositions, or governance by common norms, and to more recent work characterizing practices as "bundles," "arrays," or "ensembles" of performances while eschewing general conceptions of how they belong together. Participants in practices need not exhibit performative regularities, or share presuppositions, commitments, or goals. What makes their situated performances part of a practice that conditions and enables those performances is instead their coupled interdependence. By relying on other performances that align with and complement their own in conducive circumstances, people accomplish many outcomes that could not be achieved singly. Humans also thereby live more differentiated lives by relying on the conjoined capacities and performances of others in appropriately reconstructed settings. People now rely on others to fulfill many basic biological needs and provide needed support or complementation for their own skills and performances. These patterns of interdependent, situated performance only exist through their ongoing, differential reproduction, but people's skills and normative orientation develop in settings conducive to sustaining and reproducing that interdependence. Several extended examples in chapter 4 illustrate how practices can be sustained over time while conjoining different kinds of performance by differently situated participants. Often they work together effectively without sharing a conception of the practice or its goals. These examples also show how practices and their constituent performances may have different and sometimes changing interdependence at multiple levels.

The remainder of the book works out further consequences of this initial reconception of practices as the niche-constructive differentiation of human

---

12. Dor (2015) argues that the core evolved function of languages was to instruct and coordinate people's imaginative grasp of distant or merely possible circumstances.

ways of life. These consequences are closely interconnected. At their root is an evolutionary transformation of the normative accountability of human ways of life. The goal-directedness of organisms' life processes toward the continuation of those processes constitutes a limited form of biological normativity. Organisms develop over their lifetimes, those developmental patterns change in response to environmental differences, and their lineages evolve through resulting redistributions of life cycles in descendant populations. The only normative issue raised by these changes, however, is whether they succeed or fail in sustaining the lineage. The contrast between this limited form of biological normativity and the diverse normative concerns shaping human ways of life has often motivated conceptual differentiation of human sociality from biological development and evolution. Chapter 5 instead shows how the normative complexity of our ways of life is a two-dimensional modification of the one-dimensional normativity of other organismic lineages. That modification not only introduces new normative concerns at issue in particular practices. The need to accommodate those practices together in their mutual interdependence complicates the normativity of human ways of life as a whole. New normative concerns—for justice, individual or community autonomy, rationality, epistemic accuracy or understanding, cultural unity or diversity, and more—arise as issues among different possible ways to sustain practices in concert. Together, the normative concerns animating particular practices and the integrative concerns for their mutual interdependence extend the normativity of human ways of life beyond whether the lineage survives to encompass *how* it continues. Recognition of the biological basis of these diverse normative concerns also provides new insight into how they acquire content and normative authority. They are shaped by future-directed concerns over issues that arise in developing and sustaining practices as mutually interdependent and over what is at stake for people's lives and the lineage in how those issues resolve.

The niche-constructive coevolution of humans with languages and other conceptual practices both exemplifies and contributes to that normative complexity. Protolinguistic communicative repertoires arose in the hominin lineage in coordinating and sustaining newly cooperative ways of life. Languages coevolved with human bodies and ways of life, as selection pressures for easier acquisition and use of communicative repertoires also enabled diverse uses, more complex linguistic forms, and further selective advantages (Dor and Jablonka 2000). Understanding languages as consequential forms of coevolutionary niche construction provides new insight into languages and the discursive practices that sustain and reproduce them. Languages are practices with their own constitutive normative concerns for communication and sense making. They also become integrated into other practices, contributing to and coordinating their performances and articulating and expressing normative concerns driving those practices. The pervasive integration of language and other expressive practices into human developmental environments is thereby materially transformative of the worlds people inhabit. Chapter 6 considers discursive practices and the languages they embody as forms of coevolutionary niche construction; chapter 7 examines the corresponding discursive articulation of human biological environments.

The horizontal differentiation of practices and the vertical differentiation of people's lives across the practices they take up does not only diversify human ways of life, however. People are situated within and amid those practices with different capacities and achievements and different constraints and limitations. Those differences in how people's lives are situated have often been analyzed by social theorists in terms of power—differences in people's power to undertake various activities or achievements and their power over one another's capacities and performances. Despite the familiarity and importance of these differences, social theorists' analyses of both the concept of power and the phenomena it describes have been both contested and fraught. Situating these issues within a naturecultural conception of practice-differentiated ways of life allows reconceiving power and the significance of power relations in people's lives. Power is closely intertwined with the more complex two-dimensional normative accountability of practice-based ways of life. What matters are *normatively* significant differences, both in what people have power to accomplish and in constraints or harms imposed on them. The concept of power is thus closely tied to the normative concerns at issue in the achievement and exercise of those powers, which result from the coupled interdependence of a practice-differentiated way of life. These concerns centrally incorporate powers to make sense of human ways of life discursively *as* open to assessment. That capacity allows both changing human ways of life to accord better with the sense they can make and changing the sense ascribed to various phenomena to accord better with how they matter to what is at issue and at stake in human ways of life. Chapter 8 explores how to situate the critical analysis of power relations within the two-dimensional normativity of people's evolved, practice-differentiated ways of life.

The book concludes by considering two aspects of the finitude of human capacities for making sense of our ways of life and bringing them into accord with the sense they can make. Kant argued for the constitutive finitude of human cognition: its dependence on and limitation by the passive deliverances of the senses and its consequent objective accountability. Earlier chapters shifted the locus of Kantian cognitive finitude from supposedly passive

sensation to the practical-perceptual interdependence of people's bodily postures with their developmental environments. One further aspect of the finitude of human niche-constructive capacities is that people's capacities for sense
making and reconstruction are shaped by already-extant practices, infrastructure, and ways of life, including the discursive practices through which critical
reflection can be undertaken. People can only introduce new practices, new
conceptualizations, and transformed ways of life by making an intelligible
place for them amid already-ongoing practice-differentiated life cycles. As John
Haugeland once noted, "to say that [human ways of life are] 'stuck with' their
current situation does not at all mean that [they] cannot move on; on the contrary, it means [they *have*] *to*—and *can only* move on *from there*" (2013, 144).
The situatedness of human capacities for sense making and transformation
then has implications for what it can mean for those capacities to be objectively
accountable.

A second further aspect of the finitude of people's capacities for critical sense
making and transformation of their ways of life is their interdependence with
other organismic ways of life. This aspect of people's practice-differentiated
ways of life is mostly suppressed in earlier chapters to foreground the integration of social life within our niche-constructive biological development and
evolution. Despite this strategic neglect, however, this aspect of human finitude
is a central consequence of an integrated naturecultural conception of how we
live. Humans and other organisms inhabit mutually overlapping environments
amid shared dependence on the physical capacities of earthly environments.
People only take up discursive communication, contestation, and reconciliation with one another, but we are not the only agents who actively affect our
possibilities and fate as living organisms in a lineage. The prospects for sustaining and ameliorating human ways of life in the face of lineage-threatening
issues we now confront depend not only on the sense we make together of how
to respond to political and climatological challenges. We must also negotiate a
shared way of life with the many other organisms with whom we cannot negotiate discursively. The *social-ecological* interdependence of human ways of life
with our many companion species and our shared earthly habitat is thus among
the most consequential conclusions of the book.

1

The Social Theory of Practices

The concept of a practice has been deployed across a wide range of disciplines to characterize the basic makeup of human social interaction. Conceptions of social practices figure prominently in social theory, many social sciences, historiography, philosophy of language, philosophy of science, and interdisciplinary science studies.[1] Stephen Turner noted that "a large family of terms [are] used interchangeably with 'practices,' among them . . . some of the most widely used terms in philosophy and the humanities such as tradition, tacit knowledge, *Weltanschauung*, paradigm, ideology, framework, and presupposition" (1994, 2). Related concepts of performativity (Butler 1990, 1993) or enactment (Mol 2002) are prominent in feminist theory and science studies. My integrated naturecultural reconception of human ways of life endorses some central concerns that motivate practice-based social theories. Despite these insightful shared concerns, however, practice-based approaches also confront serious conceptual and empirical difficulties. Those problems have a common source in taking practices as a *social* domain largely autonomous from human biology and the natural world. Understanding practices as structuring organism/environment relations in the human lineage thus aims

1. Prominent examples include, in social philosophy, Brandom (1979), MacIntyre (1981, chap. 13), Taylor (1985b), Schatzki (1996, 2002, 2019), Risjord (2014), Haslanger (2018); in sociology and anthropology, Giddens (1984), Bourdieu (1977, 1990), Ortner (1984, 2006), Reckwitz (2002), Shove, Pantzar, and Watson (2012); in historiography, Polyakov (2012), de Certeau (1984); in management studies, Nicolini (2012), Tengblad (2012); in philosophy of language, Brandom (1994, 2000), Taylor (2016), Ebbs (2009); in philosophy of science and interdisciplinary science studies, attention to scientific practices is now pervasive. Many anthologies focus on the concept, notably Schatzki, Knorr-Cetina, and von Savigny (2001), Pickering (1992), Soler et al. (2014), Hui, Schatzki, and Shove (2017), and Buch and Schatzki (2019).

to build on central contributions of practice-based social theories while systematically addressing and overcoming their limitations and conflicts.

Practices: The Very Idea

The term "practices" is sometimes used atheoretically to refer to domains of organized activity without further analysis of how those domains are organized or bounded.[2] Early philosophical accounts were also oriented away from a more general conception of social life. Rawls's influential "Two Concepts of Rules" exemplified this orientation. According to Rawls, he "use[s] the term 'practice' throughout as a sort of technical term meaning any form of activity specified by a system of rules which defines offices, roles, moves, penalties, defenses, and so on, and which gives the activity its structure. As examples, one might think of games and rituals, trials and parliaments" (1955, 3n1). Searle extended Rawls's conception by distinguishing regulative from constitutive rules. Constitutive rules do not specify how to perform antecedently recognizable activities, but institute a domain of practices and performances stipulatively governed by those rules (1969, sect. 2.5). Rawls distinguished justification of performances of a practice by appeal to its rules from justification of the practice as a whole on other grounds. His use of the term thereby emphasized the normative autonomy of particular domains constituted by systems of rules, rather than suggesting a general conception of social life.

Three closely related concerns guide revision and expansion of the concept of a practice to characterize the basic makeup of social interactions and institutions. A first concern arose from reflection on Ludwig Wittgenstein's (1953) discussion of rule-following and Martin Heidegger (1962) on understanding and interpretation. Wittgenstein and Heidegger complicate accounts of social life and understanding that assign constitutive or governing roles to rules, norms, conventions, or meanings, including Rawls's or Searle's accounts of practice-constitutive rules. Identifying normative governance with imposition of and obedience to rules traces back at least to Kant. Kant's (1993) distinction of inexorable causal laws from authoritative norms that obligate but do not determine actions influentially shaped subsequent divides between supposed "realms" of nature and of human activity. Wittgenstein initially pointed out that rules are not self-interpreting. Given only a rule, it remains possible to follow the rule in divergent ways. One might specify how

---

2. Gallagher (2020, chap. 9–10) exemplifies a resolutely atheoretical use independent of practice-based social theory.

to interpret the rule, but the interpretation would be another rule open to divergent applications. Wittgenstein drew the following complex conclusion:

> This was our paradox: no course of action could be determined by a rule, because every course of action can be made out to accord with the rule. The answer was: if everything can be made out to accord with the rule, then it can also be made out to conflict with it. And so there would be neither accord nor conflict here.
>
> It can be seen that there is a misunderstanding here from the mere fact that in the course of our argument we give one interpretation after another.... What this shows is that there is a way of grasping a rule which is *not* an *interpretation*, but which is exhibited in what we call "obeying the rule" and "going against it" in actual cases. (1953, 1:201)

Practice theorists have drawn similar lessons from Heidegger's (1962) claim that interpretation draws on a more basic understanding or competence that is not explicitly articulated. The challenge taken up in practice-based social theories is to characterize this grasp of rules that is more basic than interpretations, and understand how it is "exhibited in actual cases." The practice—how people take up the relevant concepts and concerns in what they do—is not specified or determined by the rule, but instead has a constitutive role in understanding what the rule specifies.

Why have these considerations from Wittgenstein and Heidegger mattered for practice-based conceptions of social life? These social theories remain within the post-Kantian tradition that identifies human activities and ways of life as *normatively* ordered. Wittgenstein and Heidegger challenge conceptions of normative ordering as governance by explicit rules or norms. If to act in light of norms is to follow a rule, and rule-following can be done correctly or incorrectly, a vicious regress of rules would render normative governance impossible. Saul Kripke (1982) notoriously starts from skeptical concerns about rules or norms in his provocative interpretation of Wittgenstein. Most practice-based social theories instead construe their criticisms as constructive. They highlight a level or dimension of normative ordering and responsiveness expressed *in* what people do that is more fundamental than explicit interpretation *of* what they do. They invoke the concept of a practice for this background orientation, understanding, or competence that enables following rules, obeying norms, and articulating and grasping meanings as concretely embedded in interrelations among its performances or enactments. Practice theorists develop Wittgenstein's enigmatic claim that rules and rule-following draw on "agreement in forms of life" (1953, 1:241), and Heidegger's more elaborated exposition of everyday human being as articulated by "what one does." In both cases, how a performance is responsive and accountable to others "in actual cases" or in "what one does" constitutes practices as temporally extended patterns of constitutively interrelated performances comprising an ongoing, normatively ordered "form of life."

A second concern guiding efforts to discern meaning and normative accountability "in practice" is for an alternative to shared commitments underlying the opposing positions of behaviorism and the cognitivism that often supplanted it.[3] Behaviorists dominated psychology and many "behavioral sciences" in the United States at mid-twentieth century. Their suspicions against mental or intentional concepts led them to redirect the human sciences toward studying behavior conceived as publicly observable, "thinly" describable movements. This orientation was reductive or eliminativist: behavior was described in nonintentional, nonnormative terms congenial to a strict empiricist. Charles Taylor characterizes this empiricist/behaviorist orientation as an aspiration to describe human life through "features which can supposedly be identified in abstraction from our understanding or not understanding experiential meaning, [in] brute data identifications" (1985, 28). The difficulty is that objectively similar movements could perform quite different actions or no action in particular, while psychologically or socially similar activities could be accomplished by very different publicly observable behaviors. Behavioral descriptions were psychologically or socially gerrymandered.

In both psychology and philosophy, the dominant response to behaviorist failings was a cognitivist turn that explains people's actions by appeal to psychological states such as representational beliefs and motivational desires. Digital computers were a powerful model for distinguishing the physical hardware, or "wetware," of brains from a cognitive software of representational states and computational transformations. Rule-governed computations over internal representations provided conceptual classifications of what people do that behaviorists could not express in thinly descriptive, behavioral terms. Cognitivists nevertheless implicitly shared behaviorists' conception of manifest, publicly accessible behavior as only thinly describable in physical or causal terms. Because others' performances and situations are not already meaningful, one must appeal to mental representations and cognitive processing to explain meaningful responses to circumstances. For thoroughgoing cognitivists, any significance accruing to actions and circumstances apart from their physical lawfulness must come from meanings conferred on

---

3. Taylor's early work on social practices addressed behaviorist approaches explicitly (1964, 1985b, chap. 1). Other practice-based approaches appearing after the eclipse of behaviorism said less about that contrast.

them by agents' psychological states and the rules or procedures of reasoning governing interrelations among those representations and attitudes.

Practice-based accounts respond differently. In characterizing practices as indispensable background to explicit articulations or interpretations of people's situated actions, they retain behaviorists' focus on publicly situated performances rather than mental events or states. They do not avoid intentional or normative locutions, however, but aim to make them accessible and comprehensible. While attending primarily to "outward" performance rather than "inner" belief or desire, practice-based accounts characterize such performances in what Ryle (1971) and Geertz (1973) characterized as "thick" descriptive terms rather than behaviorists' extremely thin language. Practices are meaningful configurations *of* what people do in meaningful situations rather than only having meaning imposed on or infused within them by animating beliefs, desires, and intentions. Rules, norms, and concepts get their content and normative authority and force from their embodied configuration in publicly accessible activity. Taylor offers a characteristic account of this move: "The situation we have here is one in which the vocabulary of a given social dimension is grounded in the shape of social practice in this dimension; that is, the vocabulary would not make sense, could not be applied sensibly, where this range of practices did not prevail. And yet this range of practices could not exist without the prevalence of this or some related vocabulary" (1985b, 33–34). Language use does not publicly express inner psychological states but comprises a larger pattern of publicly accessible interaction intelligible as such.

A third concern motivating practice-based approaches is to circumvent alternative assignments of explanatory priority to individual agency or social or cultural formations.[4] At issue in these debates is whether social theory and the social sciences can and should refer to and understand social wholes (institutions, cultures, social structure, traditions, shared worldviews, etc.) not decomposable into actions by or states of individual agents.[5] The autonomy of anthropology, sociology, history, social philosophy, or other disciplines

4. Zahle and Collin (2014) review the current state of this controversy. Epstein (2015) also frames his social metaphysics to circumvent this perennial controversy without explicitly discussing practice-based alternatives.

5. Philosophy of social sciences uses "holism" in two ways. One sense denies that social or culture entities ("wholes") are analyzable as actions or states of individual agents. In another sense, prominent in reflection on psychological states and linguistic meanings, properties are holistic if one thing cannot have the property unless many other things also have it. Orthographic ambiguities allow distinguishing acceptance of supraindividual entities as "wholism" and interdependent property ascriptions as "holism."

studying social domains of inquiry might seem enhanced if irreducibly social or cultural structures are their proper object. Critics of social or cultural
wholism nevertheless question what a social or cultural "structure" could be
except composites of individual agents or their actions and abilities. They
pose methodological and epistemological doubts that knowledge of irreducibly social wholes could be grounded in evidence. Wholists respond in turn
that individual actions are often only intelligible in social or cultural contexts.
As individual actions apart from supraindividual settings, such familiar activities as voting, paying money for goods, performing religious or political
rituals, or speaking a language might make no sense. Larger social contexts
may seem indispensable to distinguish voting from raising a hand or marking
papers, uttering words from making noises, or recognizing the interchange-
able significance of cash, credit cards, electronic transfers, or other forms of
monetary exchange. Individual actions and agents may only make sense as
components of a larger culture or society.

Practice-based approaches respond to these disputes by acknowledging
that each side grasps something important. At one level, practices are composed of individual performances.[6] These performances nevertheless only
take place intelligibly against a more or less stable background of other interrelated performances. Practices constitute the background replacing earlier wholist appeals to culture or social structure. The relevant social structures and cultural backgrounds are understood dynamically through their
continuing reproduction in practice, including uptake by new practitioners.
While there is no more to the practice than its ongoing performative reproduction, these performances cannot be properly characterized or understood
apart from belonging to or participating in a practice sustained over time
by interrelations among multiple participants and their performances. The
anthropologist Sherry Ortner concluded that, "The modern versions of practice theory appear unique in accepting all three sides of the . . . triangle: that
society is a system, that the system is powerfully constraining, and yet that
the system can be made and unmade through human action and interaction"
(1984, 159).

6. Most practice theorists refer to actions of individual agents, but some acknowledge "dialogical" actions not decomposable into individual performances (Taylor 1991, 310). Heidegger-
ians attribute everyday activities to anonymous, undifferentiated agents rather than individuated
selves. Foucauldians take individuation as constituted rather than presupposed by performances, exemplified by Judith Butler's claim: "gender is always a doing, though not a doing
by a subject who might be said to preexist the deed. . . . There is no gender identity behind the
expressions of gender; that identity is performatively constituted by the very 'expressions' that
are said to be its results" (1989, 25).

Emphasis on dynamic social interactions as governing or constraining individual performances gives practice-based conceptions of social life a strongly historical orientation. Such dynamics also recognize "cultural" background as not monolithic or uncontested, an especially important consideration in recent anthropology. Anthropologists long worked with a conception of culture as unified and systematic, typified by Clyde Kroeber and A. L. Kluckhohn's classic formulation: "Culture consists of patterns, explicit and implicit, of and for behavior acquired and transmitted by symbols, constituting the distinctive achievement of human groups, including their embodiments in artifacts; the essential core of culture consists of traditional (i.e., historically derived and selected) ideas and especially their attached values; culture systems may, on the one hand, be considered as products of action, on the other as conditioning elements of further action" (1963, 181). Instead of positing a unified culture, practice-based accounts recognize alternative practices within a cultural milieu, different positionings within and conceptions of those practices, and contestation of their governing norms. Moreover, practice-based conceptions enable different treatment of cross-cultural interaction resulting from migration, political domination, or trade relations. Instead of treating cultural interaction as translation between whole cultural systems, practice theorists recognize localized patterns of partial interpretation and exchange distinguishable from other practices and meanings within overlapping fields of cultural practice.[7] Acknowledging cultural dissonance also recognizes different uses and meanings of cross-cultural interaction within intracultural politics (Traweek 1992, 1996).

Most practice-based conceptions of social life take up these three concerns. Practice-based social theories characteristically locate normative governance and accountability in particular practices rather than general norms of rationality, and "in practice" rather than constitutive rules or explicit interpretations. Social interactions occur in meaningful situations that are continually reconfigured by interrelations among previous performances and people's responsive uptake and continuation or revision of those patterns. What individual agents do is meaningfully enabled and constrained by their situations, while also reproducing and reconfiguring those situations for subsequent performances. The challenge has been to work out a coherent and reasonably comprehensive account of practices, how they are sustained and reproduced, how they both enable and condition their constitutive performances, and how practice-constituted performances interact with other

7. Galison (1996, chap. 9) treats such partially comprehended interactions and exchanges as "local coordination" rather than systematic translation and understanding.

practices. This book aims to vindicate these core concerns of practice-based social theories by responding more adequately to their difficulties.

Challenges and Conflicts

Conceptions of practices and their constitutive performances as the basic makeup of human social life have confronted at least three prominent challenges. I first consider how best to analyze interrelations among participants in practices and the material circumstances of their performances. I then take up a central issue for practice-based conceptions: how performances belong to and are enabled or conditioned by a practice, and how that belonging together is sustained in and through subsequent performances. A third consideration is how practices constitute a publicly accessible locus of meaning and understanding. This last consideration is central to practice-based challenges to a cognitivism that locates meaning, understanding, and critical assessment in individual thought and experience and to associated analyses of social life that start from individual agency. Practice-based accounts of social life have pursued opposing strategies on this issue, yielding different conceptions of practices. These alternative conceptions highlight different aspects of social life, and each strategy has difficulties accounting for others' exemplary cases. One strategy emphasizes flexible bodily skills for coping with situations, including participants' embodied responses to one another's performances. Alternative strategies take different approaches to language use as a public domain that enables making sense of and responding to one another's performances in partially shared circumstances.

SOCIAL PRACTICES, MATERIAL SETTINGS, AND EMBODIED PERSONS

A first consideration for conceptions of practices as basic constituents of social life concerns their scope. Do practices only involve what people say and do and relations among those performances, or do they also encompass the material settings within which those doings and sayings take place? Practice theories originated as conceptions of social life as distinct from the natural world that incorporates human organisms. The material world is a natural-scientifically intelligible causal nexus. Emphasizing the autonomy of a social world encourages thinking of practices as composed of what people say and do, how those sayings and doings respond to one another, and the social relations and relationships constituted by mutual responsiveness. People nevertheless act *with* equipment, *on* things, *in* a more or less conducive setting.

Absent these objects or settings, these actions would be very different. Much of what people say is also about shared circumstances, often via ostensive or other indexical indications that only make sense in their surroundings. Both the sayings and doings are perceivable, causally efficacious performances, and their conjoined effective and receptive significance is integral to their public accessibility as a situation in the world rather than a so-called inner domain of mental states.

Most practice-based social theories now acknowledge that social practices are always materially situated, while *analytically* distinguishing worlds of so-cial relations from causally efficacious natural objects. Thus, Sally Haslanger talks about "a mutual dependence between the social and the material world. Objects have normative significance in the context of our practices, and our practices are, in turn, shaped by the objects we interact with" (2018, 243). Haslanger implicitly suggests that social and material worlds are distinct but inseparable components of *the* world which interact at relatively simple and well-defined interfaces. Theodore Schatzki (2002) made this suggestion ex-plicit, distinguishing social practices composed solely of people's sayings and doings from more encompassing social "orders" or "arrangements" that incor-porate their material settings. While recognizing that "practices are intrinsi-cally connected to and interwoven with objects" and that practices and orders are "not distinct ontological regions" (2002, 106), Schatzki ascribes analytical or explanatory priority to practices in the narrower sense: "Activities and ob-jects are not equals here. The character of social existence is, in the end, much more the responsibility of practices than orders" (Schatzki 2002, 117). Assign-ing such comparative responsibilities presupposes the separate intelligibility of activities ("sayings and doings") apart from their circumstances, and of ar-rangements of objects whose determinacy and interrelations are independent of those activities, which then come together to constitute "social existence."

Robert Brandom suggested a more integrated conception of discursive practices as material in claiming that "standard discursive practices . . . are solid (even lumpy) in that they involve actual objects and states of affairs" (1994, 632). On closer inspection, however, he still separates social-normative and causal domains.[8] The supposed incorporation of actual objects within discursive practices turns out only to "involve" them via scorekeeping *judg-ments* of speakers' perceptual and practical reliability (1994, chap. 4). Those judgments are thoroughly ensconced within a social-normative domain of abstract normative scores. Norms only govern judgments and causes only affect things.

---

8. Chapter 6 below revises Brandom's account in this respect.

Quill Kukla and Mark Lance reject Brandom's abstraction of discursive scorekeeping judgments from people's bodily engagement with the material world. They nevertheless still assign explanatory priority to how people take up their material circumstances interactively and thereby confer normative significance on material things: "The material features of rain constrain what sorts of world-involving normative practices can be developed in relation to it, and once these are developed, rain has concrete normative significance from inside these practices. The rain need not 'tell us' anything or 'hold us' to anything. *We* are the ones who institute, maintain, and practice the norms of vinification, baseball, fashion, and so forth. But we cannot do so except as embodied beings who engage with rain and its absence; within such engagements, rain has specific normative meanings and consequences" (Kukla and Lance 2014, 26). Only people do the telling, holding, instituting, or practicing that confers normative significance, but they do so as embodied beings in "concrete" relations to material surroundings.

Understanding material settings as conceptually distinct but inseparable from the practices comprising a social world often turns on distinctions between objects' causal capacities and their social-normative significance within a nexus of practices. One difficulty is that the social-normative significance of objects is closely interdependent with their causal capacities. The boundaries of objects are in turn closely tied to their social-normative significance, while assigning causal capacities *to objects* depends on the boundaries between object and context.[9] These issues become even more acute when attention turns to the people whose performances and interactions confer social-normative significance on objects and to their bodies as natural objects that also thereby acquire such significance. As Kukla notes in a different context, "One's material body, in all its detail, situated in its material environment, fully determines one's capacities. Everything I can do or be, I can do or be because of the body I have, in the context it is in. This includes my capacity for reasoning, my trained skills, my temperament, and so on" (Kukla 2008, 519–20).

The challenge for a *social* theory of practices is then to understand both the identity of and differences between a social body and a biological body to account for conferral of normative significance on otherwise anormative objects taken up within people's concretely embodied engagement with their surroundings. A biological understanding of how bodies are environmentally situated challenges even residual conceptions of social practices as a nearly

9. Smith (1996, 2019), Cartwright (2003, 2019), and Haugeland (1998, chap. 10) acutely analyze close interdependence of object boundaries, causal relations, and normative significance in different contexts.

decomposable aspect of humans' place in the world, and thus also requires a different conception of how practices reconfigure the normative significance of human environments.

BELONGING TO A PRACTICE

A second consideration in developing the concept of a practice is to understand how situated performances belong to a practice, and how practices are held together both synchronically and in their ongoing reproduction. This issue is complicated by recognition that practices are nested—some practices incorporate other practices as components—and they intersect, overlap, or interrelate with other practices that nevertheless remain distinct. As Haslanger recently noted:

> Social structures consist of interconnected practices. Structures are of different sizes and can be found at different levels. For example, a high school involves not only practices of teaching and learning, but also wage labor, accreditation, food distribution, sporting events, fundraising, security systems and policing, etc. The particular school is part of a structure of public education in the school district, and across the country. It connects the local practices with broader ones, and educational practices with labor practices. (2018, 232–33)

Agents also participate in multiple, vertically differentiated practices within any extensive temporal span, and the same performance can belong to different practices, perhaps under different descriptions. Identification of practices is thus not sharply bounded, and may also be affected by interpreters' cultural expectations and theoretical interests without thereby implying that differently bounded practices manifest from different interpretive orientations are not "real patterns" (Dennett 1991b). Practices need not admit of anything akin to necessary and sufficient conditions for performances as their constituents.

The question of how practices are constituted and sustained is nevertheless important in two ways. The first reason emerges in contrast to Rawls's early concept of practices. On Rawls's conception, a practice is defined by its constitutive rules, whose taken-for-granted authority within the practice is established and justified on other grounds, such as general utilitarian reasons for practices of punishment (Rawls 1955, 4–8). In taking practices as more basic than rules, however, one must identify the unity of a practice on a different basis arising from within the practice itself. The second reason for the importance of how practices are constituted arises from the spatial dispersion and temporal extension of practices to incorporate participants in disparate regions over successive generations. The links or other relations among those performances supposedly *enable* them to occur in this form, account for their *intelligibility*, and thereby *sustain* a practice. Analysis of how performances belong together as a practice thus cannot group them arbitrarily but must provide a basis for understanding their constituent performances.

In what Stephen Turner (2007) calls "classical" theories of practices, some commonality among their performances identifies and sustains the unity of a practice. Specifying the common features that could define a practice, showing how they could constitute normative authority, and discerning them empirically has been difficult. Any common feature postulated or discerned among performances of a practice would have to be complex for multiple reasons. Individual participants or their performances may have different roles in a practice and consequently take differing forms; the relevant "sameness" may be in relations among performances rather than the performances themselves. Doing the same thing in different circumstances may require doing it differently. Moreover, social practices can presumably incorporate and accommodate errors in or resistance to conformity to what others do, and some "noise" that partially obscures the underlying regularity (Dennett 1991b). Finally, whatever common considerations constitute a practice would have to incorporate the different performances that train novice participants or correct deviant or erroneous performances, along with responses to training or correction.

Classical practice theories strategically split by conceiving this common ground according to alternative readings of Wittgenstein's claim that "'obeying a rule' is a practice" (1953, 1:202). As Taylor noted:

> There are two broad schools of interpretation of Wittgenstein, . . . two ways of understanding the phenomenon of the unarticulated background [to rule-following]. The first would interpret . . . the connections that form our background [as] just de facto links, not susceptible of any justification. For instance, they are imposed by our society; we are conditioned to make them. . . . The second interpretation takes the background as really incorporating understanding; that is, a grasp on things which although quite unarticulated may allow us to formulate reasons and explanations when challenged. (Taylor 1995, 167–68)

The former strategy identifies a practice by an exhibited or dispositional regularity in performance. The inspiration for regularist accounts often stems either from Heidegger's appeal to anonymous conformity to what "one" (*das Man*) does, or Wittgenstein's remark that, "If I have exhausted the justifications [for following the rule in the way I do], I have reached bedrock, and my spade is turned. Then I am inclined to say, 'This is simply what I do'"

(1953, 1:217). The latter strategy instead appeals to commitments implicit in or presupposed by the performances that make up a practice, whose commonality is open to further interpretation but not determined by it. The difference between the two strategies concerns whether a practice is unified and sustained causally or normatively. On the former strategy, practices supposedly introduce normatively ordered domains of activity via causally shaped dispositions and performances, "baking a normative cake with non-normative ingredients" (Brandom 1994, 41). The latter strategy treats that order as irreducibly normative: aspiring participants can succeed or fail in "catching on" to the practice, and the practice can succeed or fail in attracting new participants or sustaining and developing their competence.

Taylor cited Kripke's (1982) influential book on Wittgenstein to exemplify the first strategy, but Kripke was hardly alone in ascribing socialization into shared practices through imitation, training, and sanctions that enable and enforce the continuity of practices by straightforwardly causal means. Bourdieu most prominently exemplifies this strategy among practice theorists in the social sciences, claiming that, "The objective homogenizing of group or class *habitus* that results from homogeneity of conditions of existence is what enables practices to be objectively harmonized without any calculation or conscious reference to a norm and mutually adjusted in the absence of any direct interaction or . . . explicit coordination" (Bourdieu 1990, 58–59). Michel Foucault (1977, 1978) also often emphasized continuities in performance enforced or self-imposed by surveillance, normalization, examination, or confession amid changing interpretations and justifications of those continuities.

Taylor defended the second approach for which continuation of a practice involves *understanding* previous performances: "We have to think of [humans] as self-interpreting [animals]. [We are] necessarily so, for there is no such thing as the structure of meanings for [us] independently of [our] interpretation of them; for one is woven into the other. . . . Already to be a living agent is to experience one's situation in terms of certain meanings; and this in a sense can be thought of as a sort of proto-'interpretation'" (Taylor 1985b, 26–27). Practitioners learn a practice from others' performances, including responses that correct improper performances by oneself or others. Taking up a practice is not merely imitating others' movements, or training and discipline by causal means, but requires appropriate uptake. In developing that uptake, "our aim is to replace [a] confused, incomplete, partially erroneous self-interpretation by a correct one, and in doing this we look not only to the self-interpretation but to the stream of behavior in which it is set" (Taylor 1985b, 26).

Thomas Kuhn's well-known account of scientific research practice exemplifies Taylor's second strategy. Kuhn argued that shared paradigms—"accepted examples of actual scientific practice . . . which include law, theory, instrumentation, and application together" (1970, 10)—were more fundamental than any explicit rules derived from them in guiding subsequent work in a scientific community. According to Kuhn, scientists can "agree in their *identification* of a paradigm without agreeing on, or even attempting to produce, a full *interpretation* or *rationalization* of it. Lack of a standard interpretation or of an agreed reduction to rules will not prevent a paradigm from guiding research" (1970, 44). Advocates of this second strategy take practices as held together and guided by acceptance, presupposition, or recognition of some common basis for competent participation in the practice without needing or even permitting its explicitly accepted articulation as constitutive rules. Attributions of implicit or background competence take varied forms: reliance on models of exemplary performance in Kuhn; "intersubjective" or "metabiological" meanings of performances and vocabularies (Taylor 1985b; 2016, 6); flexible skills (Dreyfus 1979, 2014; Polanyi 1958); norms implicit in community responses to performances of a practice as correct or incorrect (Brandom 1979); or appeal to goods or excellences internal to practices that "can only be identified or recognized by the experience of participating in the practice in question" (MacIntyre 1981, 188–89).

Stephen Turner (1994) mounted a withering, disjunctive critique of these strategies, focused on the continuity and effectiveness of postulated practice-constitutive commonalities. He argued that advocates of the first strategy failed to make good on the causal arguments needed to show how regularities in past performance or training and sanctions deployed to instill and enforce them effectively sustain those regularities in subsequent performances. A deeper problem also undercut the first strategy: a finite set of performances exhibits indefinitely many regularities. One can identify performances as instances of the same practice in multiple ways, with no grounds to identify the relevant practice with any one of them. Alternative conceptions of the underlying regularity and its reference class would provide differing verdicts on which performances belong to a practice and whether subsequent performances were in accord, but subsequent reconceptions informed by these outcomes would remain underdetermined even by additional evidence, since the gerrymandering problem recurs (Brandom 1994, 18–46). Turner then argued that proponents of the second strategy have not shown that the shared presuppositions or normative aspirations they ascribe to practices actually guide their performances or carry over to subsequent performances. Turner's arguments are reinforced by parallels to the gerrymandering argument familiar from W. v. O. Quine (1960, 1969) or Donald Davidson (1984, 2005b) on interpretive indeterminacy. Theorists typically ascribe holistically interlocking presuppositions to determine which performances belong to a practice. Any ascribed presuppositions sufficient to account for those performances would also allow for alternative attributions that systematically vary some presuppositions to compensate for changes in others. Since these attributions have no basis other than explanations of how performances belong together and make sense, alternative attributions of presuppositions undercut the credentials of any to explain the identity or continuity of a practice. In effect, the second strategy could only rationally reconstruct a practice rather than explain how its performances actually occur. Turner's conclusion was uncompromising: "The idea of 'practice' and its cognates has this odd kind of promissory utility. They promise that they can be turned into something more precise. But the value of the concepts is destroyed when they are pushed in the direction of meeting their promise" (Turner 1994, 116).

Few advocates of practice-based social theories endorse or acknowledge Turner's criticisms or the related problems of gerrymandering or indeterminacy, but subsequent accounts mostly take a different tack. "Postclassical" practice-based accounts of social life circumvent attributions of common elements or shared implicit background to performances of a practice by instead identifying amorphous "sets," "nexuses," "bundles," "blocks," "pools," "arrays," or other organized groupings of situated sayings and doings, while remaining reticent about how they belong together.[10] These groupings are not arbitrary, although postclassical accounts do generally acknowledge practices' loose boundaries and open-ended possibilities for continuation and revision. Schatzki suggests that "the organization of a practice determines which doings or sayings belong to it" (2019, 130), but "organization" here refers not to some organizing activity, but only to interrelations among the performances themselves as somehow organized. The implication is that their binding into groups is local and particular: "In doing things like driving, walking, or cooking, people (as practitioners) actively combine the elements of which these practices are made. . . . Practices emerge, persist, shift, and disappear when *connections* between elements of these three types [materials, competences and meanings] are made, sustained, or broken" (Shove, Pantzar, and Watson 2012, 14–15). These accounts usually emphasize that multiple components are united in various, changing ways in or by performances themselves, without appealing to or characterizing generally the kinds of connections established or how activities effect the combinations. These unifying performances then

10. The quoted terms are drawn from Schatzki (2002, 2010, 2019), Reckwitz (2002), and Shove, Pantzar, and Watson (2012) to exemplify Turner's "postclassical" category. Turner includes Rouse (2002), but discussion below highlights important differences Turner overlooked.

supposedly continue prior patterns by differential repetition of their composition: "Practices are open sets of activities. . . . A practice persists whenever
an additional practice-composing action is performed. Whenever such an
action is performed, it turns out that the practice had persisted from the time
of the previous practice-composing action to the time of this one" (Schatzki
2019, 28). The question of what makes the action practice-composing is nevertheless circumspectly avoided except in particular cases. The postclassical
theorists *start* with an organized nexus of activities—driving or skateboard-
ing (Shove, Pantzar, and Watson 2012), hospital life (Blue and Spurling 2017),
or making and selling Kentucky bourbon (Schatzki 2019) or Shaker furniture
(Schatzki 2002)—and examine how those activities develop and change without explicating why those groupings or others like them in unspecified respects are the basic makeup of social life. That strategy provides a reasonable
basis for empirical research but leaves important issues open for understanding the character and significance of human ways of life.

BODIES AND THEIR SKILLS AND AFFECTS

A third consideration plays an important role in working out practice-based
conceptions of social life. A central concern has been accounting for how
people understand, evaluate, and respond to material and social circumstances as meaningful. Practice-based approaches take these capacities for
understanding and evaluation to be deployed and sustained in practices as a
shared, publicly accessible domain of social life. That aspiration has nevertheless mostly been pursued from two alternative directions. One route emphasizes that practices are composed of *bodily* performances and affects. Understanding involves flexible skills or competence to encounter and respond to
objects and others' situated performances as significant. An alternative route
starts with language use and what is explicable in public language as locus of
a socially interactive domain of conceptual understanding and assessment.
People are both bodily agents and language users, but practice-based conceptions of social life have struggled to understand bodily skills and discursive
articulation together.

Making bodily skills and affective responsiveness central to a publicly
accessible, practice-differentiated space of meaning and its uptake counters
intellectualist conceptions of culture and social life. Attending to bodily comportment also initially seems to aid two important themes of practice-based
approaches. These approaches aspire to reconcile natural-causal and social-
normative dimensions of human activity to account for how complex forms
of social life are interactive outcomes of situated individual agency. Human

bodies seem initially promising sites for understanding how to accommodate these supposedly exclusive conceptual registers together. Bodies are causally effective and affected natural entities with internally unified capacities for self-directed movement and expression. Their mutual interaction with other bodies are also both causally constraining and normatively constitutive. The resulting patterns of interactive performance might thereby implement forms of meaning and normative accountability not already found in their natural-causal capacities. These practice theorists understand human bodies as both the locus of effective agency, affective response and cultural expression, and vulnerable targets of discipline, normalization, and other power relations. This dual character of bodies, or at least their intelligibility within these two distinct registers, seems initially well suited to capture seemingly opposed considerations. The challenge is to show how these dual conceptions or conceptual registers work together in human bodily interaction with other bodies and a partially shared environment. Understanding them together is critical, since capacities for spontaneous, self-directed activity are often disciplined or channeled by training, while people can also resist efforts at disciplinary control. The danger is that appeals to the body's role in social life merely name the coincidence of causal and normative or individual and social conceptual registers while obscuring their lack of reconciliation.

This danger is manifest in how Taylor's two strategies for reading Wittgenstein have counterparts in practice-based conceptions of human embodiment. Some theorists characterize bodily dispositions or habits as the locus of continuity in social practices. A practice can be sustained over time, because it is inculcated in ongoing dispositions or habits of individual agents. For example, Bourdieu explicates his influential conception of the *habitus* by claiming that "the dispositions durably inculcated by the possibilities and impossibilities, freedoms and necessities, opportunities and prohibitions inscribed in the objective conditions . . . generate dispositions objectively compatible with these conditions and in a sense pre-adapted to their demands" (1990, 54). Such conceptions exemplify the idea that causally implemented patterns of behavior provide background that enables rule-following and other complex normative activity. Their appeals to imitation, repetition, imprinting, training, sanctions, and other "objective conditions" make the human body a crucial intermediary in transmitting, acquiring, and reproducing social practices. Expressive bodily capacities help reproduce these acquired patterns in new circumstances. This first strategy nevertheless offers recurrent temptations to equivocate between resolutely causal analyses of acquired habits or dispositions and accounts of their more richly expressive and flexible exercise. Bourdieu, for example, went on to characterize *habitus* as both "the product of a particular class of objective regularities" and also a form of "spontaneity without consciousness or will" (1990, 55, 56).

More commonly, practice theorists pursue a second strategy that finds continuous background to discontinuous performances of a practice in bodily skills and intentional directedness rather than dispositions or habits. This strategy discerns an intermediate domain between two seemingly exhaustive characterizations of how bodies interact with their surroundings. Bodies as natural objects are sites of causally induced or constrained movements and processes. They have mass, occupy space, encounter forceful impacts and chemically reactive materials, undergo injury and pain, and process nutrients and wastes. Bodies are also understood as more or less transparent vehicles for consciously reflective action. Much of what people do can be explained by attributing beliefs, desires, and practical reasoning, and inferring what they do from what they rationally should do, without needing to ask how the resulting intentions are causally implemented.[11]

Practice-based conceptions of bodily skills and capacities often appeal to human performances that resist both approaches. Polanyi, for example, cites an expert pianist's touch and an ordinary bicyclist's balance among countervailing forces as cases intelligible neither as simply causally determined nor as outcomes of practical reasoning. Here, he claims, "rules of art can be useful, but do not determine the practice of an art; they are maxims, which can serve as a guide to an art only if they can be integrated into the practical knowledge of the art [and] cannot replace this knowledge" (1958, 50). The body is the locus of "practical knowledge" or "skillful coping" (Dreyfus 1979, 2014) as neither merely causally conditioned nor consciously articulable and reasoned. With such examples in hand, practice theorists often then extend the domain of skillfully active and responsive bodily capacities to make inroads into cases previously understood causally or rational-reflectively.[12] While exemplary cases are usually skillful encounters with objects, equipment, or situations, social relations of expression and uptake are also prominent in how practices exert normative authority over what people do, most strikingly in Judith Butler's (1990) account of gender performativity.

---

11. Dennett's (1987, chap. 1) thought experiment of Martian physicists who can predict all behavior causally shows the importance of by-passing causal implementation; they fail to grasp a "real pattern" in the world by not recognizing the autonomy of intentional description from causal realization (Dennett 1991).

12. Dreyfus (1979, 2014) presents expert chess play, a paradigmatically intellectual, rationally reflective activity, as instead exemplifying practical skills of pattern recognition manifest in meaningful perceptual gestalts and bodily solicitations toward some chess configurations rather than others.

The most developed accounts treat skillful bodily activity as intentional directedness toward and responsiveness to the world.[13] That directedness is grounded in practical unification of one's own body, an implicit practical "I can" analogous to the Kantian "I think" tacitly accompanying conceptual judgments. Unlike objects, whose motions are decomposable into movements of their parts, bodies work together in unified skillful movements. Bodily movement is intentionally directed toward a meaningfully configured situation without intentional intermediaries of linguistic meanings or spatial representations.[14] This unmediated bodily engagement is proposed to account for how people can pick up on and respond to expressive movements of others without having to infer their intentions or articulate their meaning discursively. That recognition transforms Turner's concern about how practices are transmitted to others. Implicit in transmission is that a performance present and complete in one embodied agent is then imparted to another in equally self-contained form. Advocates of somatic intentionality deny that bodily activity is like that. When skillful responsiveness is sufficiently involved with others' bodily performances and their differently positioned and postured orientation toward shared surroundings, the result is not transmission of a skill from one agent to another, but "dialogical" shaping of action "effected by an integrated, nonindividual agent" (Taylor 1991, 310).

Emphasis on flexible bodily skills contrasts skillful movement to ingrained habits or other causally induced repetitions or dispositions. Skills do not repeat movements or connections between environmental cues and bodily responses, but instead permit flexible response to changing circumstances. In learning a bodily skill one acquires effective bodily orientation toward and responsiveness to task-situations that are meaningful as affordances whose realizations require appropriately varied performance under varying conditions. Dreyfus adds that explicit rule-following and merely habitual motions do contribute to skilled performance, but as characteristic of novices rather than experts (Dreyfus and Dreyfus 1986). When first learning a skill, people "go through the motions" in awkward but often explicitly specified ways or by imitation that has yet to grasp just *what* it is imitating. As movement becomes more familiar and integrated, however, people pick up on relevant

13. Merleau-Ponty's (2012) classic account of somatic intentionality is variously developed by Gibson (1979), Todes (2001), and Dreyfus (2014), and also by work in radical embodied and enactive cognitive science (Chemero 2009; Noë 2004, 2009; Anderson 2014; Gallagher 2017).

14. Accounts of bodily intentionality differ over whether bodies are oriented toward objects or a situation and its affordances. Chapters 4, 7, and 8 below reinterpret bodily orientation toward objects as directed toward affordances of a discursively articulated environment, as a special case of organisms' constitutive intra-action with biological environments.

patterns, leaving rules behind and even violating them in successful, skillful performances.

Practice-oriented emphasis on bodily agency, skill, intentionality, expressiveness, and affective response might initially seem to rest uneasily with roles for social-normative constraint. Practice-based accounts nevertheless insist that individual actions are shaped by social practices and their embedded norms, and often recognize bodies as targets of social normalization and constraining or productive forms of power (Foucault 1977). Can the spontaneous, expressive body, and the docile, normalized body inhabit the same embodied life and even the same bodily performances and skills? Perhaps surprisingly, this combination is often conceived not just as a compatible coexistence of opposing body-world vectors, but as mutually reinforcing.[15] Foucault in particular identified *power* relations as distinct from merely causal impositions of force, and insisted on speaking of power as "including an important element: freedom. Power is exercised only over free subjects, and only insofar as they are free. By this we mean individual or collective subjects who are faced with a field of possibilities in which several ways of behaving, several reactions and diverse comportments may be realized" (Foucault 1982, 221). Foucault is hardly alone. Kant and especially Hegel emphasized freedom and normative constraint as connected; social practices institute meanings, possibilities, goods, and norms allowing humans to understand themselves and act for reasons, in part by constraining their sense of available possibilities. In Brandom's succinct formulation, "The self-cultivation of an individual consists in the exercise and expansion of expressive freedom by subjecting oneself to the novel discipline of a set of social practices" (1979, 195). Practice theories distinctively contribute by locating both constraining discipline and expressive freedom in coordinated bodily engagement with the world.

Practice theorists rarely discuss *how* bodily spontaneity and constraint function together, however. Kantian themes of freedom as self-imposed constraint by norms do not readily mesh with how training, disciplining, or sanctioning bodily performances produces new flexible skills. Kantian norms are only constraining through agents' understanding and acceptance of them as binding, thereby subjecting themselves to internally generated and imposed constraints. By contrast, discipline and training are externally imposed as material constraints, even if the actions they constrain are taken up and appropriated within a bodily schema. The apparently ineliminable causal-material aspect of discipline and training seems lacking in post-Kantian discussions

15. Chapter 9 below discusses the entanglement of constraint with resistance and power with normativity.

of normative constraint. Nor does this misalignment apply only to Kant-inspired, broadly cognitive or discursive conceptions of unconstrained rational freedom. Social-practice-oriented versions of the Merleau-Pontian "I can" often seem abstract and immaterial in leaving no room for residual inflexibility, misrecognition, infelicitous response, injury and pain, or other forms of "I cannot" pervading bodily grip on the world, except as deficient modes of idealized, smooth, seamless functionality.[16]

Failure to grasp how bodily capabilities and causal or normative constraints function together is pervasive, even in human bodies' structure and orientation. Bodies do have an evolved structure: a bilaterally symmetrical body plan with an asymmetrical dorsal/ventral axis, a resulting vertical balance and forward orientation for mobility and manipulation, functionally differentiated and asymmetrically "handed" quadroped limbs, distinct but interactive sensory modalities, moderate and sometimes indeterminate sexual dimorphism, and other characteristic features and ranges of variation. These bodily capacities and organization are also modulated, interpreted, extended, and refined in individually and culturally varied ways. Many variations in bodily habits, styles, skills, and ways of life are central to practices as patterns of bodily performance. Somatically oriented practice theories are nevertheless reticent about how trained and expressive social bodies inhabit evolved and developing biological bodies. Feminist theorists explicitly considered one prominent manifestation by distinguishing biological sex from cultural constructions of gender, but they have similarly encountered greater difficulty in understanding how they belong together as mutually implicating or conflicting. Butler's (1990, 1993) work on gender performativity troubles the distinction, but some critics respond that it only addresses "how *discourse* comes to matter" while "fail[ing] to understand how *matter* comes to matter" (Barad 2007, 192). Other practice-based accounts are typically less attentive to how to integrate people's biological makeup and earthly environment with how bodies are made (developed) and "made up" (Hacking 2002, chap. 6) through bodily-expressive, skill-developing social practices.

LINGUISTIC PRESUPPOSITIONS AND DISCURSIVE PRACTICES

Language's role within practices or language itself as a practice has been a strategic alternative to bodily skills and affects for situating meaning and

---

16. Merleau-Ponty (2012) emphasized disease or injury as meaningful deformations of an "intentional arc," but this concern has not carried over into accounts of skillful social practices.

normativity within publicly accessible social practices. What calls for a practice-based account in lieu of appeals to individual psychological states is the discursive articulation of social life in public language rather than its flexibly skillful embodiment. A shared language or linguistically expressible background then unites the disparate performances of a practice. Wittgenstein again succinctly suggested how practices enable the intelligibility of their constitutive performances by sharing a language: "It is what human beings *say* that is true and false; and they agree in the *language* they use. That is not agreement in opinions but in form of life" (1953, 1:241). Emphasis on the interdependence of language use and practices or ways of life has nevertheless invoked two different senses of language and of practices as social phenomena.

The most widespread but also problematic approach emphasizes an articulable but tacit background understanding uniting some community, culture, or other participants in shared practices. These presuppositions may be tacit not because they cannot be said but because they are so obvious to everyone engaged in a practice that they need not be said, or because participants cannot recognize serious alternatives. They may function as norms of material inference that express conceptual contents already implicit in a vocabulary; someone objecting to these presuppositions may seem not to disagree but only to fail to understand what was said.[17] Proponents of practices as constituted by shared presuppositions also cite philosophical accounts of how available descriptions can either shape actions, intentions, and ways of life or are mutually interdependent (Anscombe 1957; Davidson 1980; Taylor 1985b, chap. 1; 2016), and how new possible descriptions can constitute new kinds of person (Foucault 1971, Hacking 2002, chap. 6). Community-constitutive presuppositions are invoked to account for how adherents' participation in a practice makes sense to the participants themselves. It is one thing, however, to recognize how people talk about their activities and institutions as integral to how they understand and engage in social interaction, or appeal to reliance on similar words, concepts, or inferential relations. That also does not yet show that they understand these characterizations in the same ways or that a shared understanding constitutively sustains or reproduces a practice. To that extent, the decisive challenge to accounts of practices as constituted

17. Sellars (2007, chaps. 1–2) and Brandom (1994, 2000) most clearly develop Hegel's notion of conceptual content expressed in material inferences. Both conceptions of how language and social practice are conjoined deploy Sellarsian conceptions of material inferences as constitutive of conceptual contents. I use Sellars and Brandom here to discuss tacitly shared presuppositions even though they reject the resulting conception of practices.

by shared presuppositions still comes from Turner's criticisms of "classical" practice theory.

These approaches nevertheless do capture an insight often overlooked by their critics. This insight is well expressed in one of Taylor's examples of linguistically constituted social practices.

> The vision of society as a large-scale enterprise of production in which widely different functions are integrated into interdependence . . . is not just a set of ideas in people's heads, but is an important aspect of the reality in which we live in modern society. And at the same time, these ideas are embedded in this matrix in that they are constitutive of it; that is, we would not be able to live in this type of society unless we were imbued with these ideas or some others which could call forth the discipline and voluntary coordination needed to operate this kind of economy. (Taylor 1985, 46)

Apart from claims to shared semantic content underlying these discursive articulations of economic life, Taylor insists that the language articulating those activities belongs to a public space of meaning and normative accountability integral to the activities themselves. Those concepts and inferential patterns are part of "social reality." Understanding and engaging in those activities and understanding and using the language work together. Despite effectively criticizing classical theories of social practices, Stephen Turner ironically provides an example for this line of argument in misdescribing it. Turner concluded that, "Together with such concepts as ideology, structures of knowledge, *Weltanschauungen* and a host of other similar usages, the idea that there is something cognitive or quasi-cognitive that is 'behind' or prior to that which is explicit and publicly uttered that is implicit and unuttered became the common currency of sociologists of knowledge, historians of ideas, political theorists, anthropologists, and others" (1994, 29). Turner rightly does not limit this strategy to explicit talk of practices; theorists characterize similar conceptions in other terms. To talk about "cognition" here nevertheless expresses a question-begging commitment to the explanatory priority of psychological states which these accounts challenge. They do not posit something cognitive or "quasi-cognitive" presupposed by linguistically constituted practices; they argue that conceptually articulated cognition instead presupposes social practices where language use and other activities are mutually constitutive.

An alternative conception of practices as linguistic or discursive does not appeal to linguistically expressible presuppositions as shared commitments unifying practices and communities of participants. Language use itself is analyzed as a practice in which agents participate via publicly accessible utterances—audible noises, visible marks, discrete gestures—that can be intelligibly grasped by themselves and others. Conceptions of language as social practice have been modeled as "radical translation" (Quine 1960), "the game of giving and asking for reasons" (Sellars 2007; Brandom 1994), or "radical interpretation" (Davidson 1984). On these accounts, language use is strongly holistic: a noise or mark is only a linguistic expression if interpreters can systematically interconnect it with many other such expressions. Interpreters must be able to make sense of these expressions in the circumstances of both speaker and interpreter. There would be no constraints on admissible interpretations and no basis for evaluating interpretations except by treating utterances as "about" and assessable with respect to accessible circumstances. The interpretation must itself be expressible "in a language"—that is, by correlating utterances in one expressive repertoire with utterances one ought to make or accept in another repertoire in those circumstances. In its classic versions, speakers' utterances, actions, and attributed beliefs and desires are interpreted together as mostly rational under those circumstances: as interpreted, their beliefs and utterances are mostly true and their actions reasonable in light of their beliefs and desires. Interpretations are undertaken in light of interpreters' own beliefs, which are nevertheless only locally privileged and open to reciprocal interpretation and assessment in other terms. Although any interpretation depends on commitment to massively common ground among participants as mostly rational truth-speakers, they need not agree on common content. They share the *practice* of tracking one another's utterances and actions in relation to their own and vice versa. Language as a social practice thus becomes the basis for ascribing and assessing beliefs and desires rather than the reverse.

These two versions of discursive practices start from different conceptions of language and sociality. One attributes "*a language*" or other shared semantic content to a community and identifies participants by implicit acceptance of its constitutive presuppositions. Its "sociality" arises from how participants belong to discursive communities. The other starts with "*language*" as a practice shared by anyone whose public expressions are interpretable as making sense, regardless of whether they share specific beliefs, desires, dispositions, or linguistic structure. It is "social" as interactions among individuals rather than with a group. The two conceptions are nevertheless independent, strictly speaking. One could treat social practices as activities governed by shared presuppositions, whether one treats language as an interpretive social practice or accepts a very different conception. One could likewise treat language as a social practice of mutual interpretation, whether or not one thinks that its participants also take up practices defined by shared presuppositions.

A DUALISM OF BODIES AND LANGUAGE

These strategic treatments of language as situating meaning and normativity in publicly accessible practices coexist uneasily with accounts of practices as involving bodily skills or dispositions, even though some theorists attempt to straddle that divide.[18] More commonly, somatic and discursive conceptions of practices place one another's primary concerns on opposing sides of their own constitutive oppositions. Somatic approaches often identify a preconceptual "stratum" of understanding as more basic to practices than conceptual, theoretical, or scientific understanding, and they often concede the latter phenomena to cognitivists. Linguistic accounts instead place bodily skills on the causal side of a constitutive distinction between causal interaction and social-discursive normativity. Taken together, practice-based accounts of social life incorporate a dualism, where "a distinction becomes a dualism when its components are distinguished in ways that make their characteristic relations to one another ultimately unintelligible" (Brandom 1994, 615). Brandom cites Descartes on body and mind as exemplary, but many practice-based accounts of social life transpose Cartesian dualism into divisions of bodily skill from discursive conceptualization. Their mutual disconnection is manifest from both directions.

Somatically oriented practice theories are replete with reference to what can be shown but not said or competently enacted only when freed from verbal mediation. These theorists secure the primacy of practices as meaningful performances by how their bodily character outruns verbal articulation. Bodily skills are irreducible to conceptually articulated representations in language or thought. Marcel Mauss (1979) on French and American styles of walking or Clifford Geertz (1973, chap. 15) on how Balinese villagers run, squat, stroke the feathers of a fighting cock, or avoid bodily acknowledgment of others are influential accounts of nonlinguistic practices. If crucial components of a practice are skills, dispositions, habits, or other performances, then

18. Foucault (1977, 1978) and Taylor (1985a, 1985b, 1995, 2016) emphasize both somatic and discursive aspects of social practices. Neither explains *how* the disciplined spontaneity of bodies functions together with semantically articulated performances in discursive practices. Taylor's (2016) attempt only shifts the boundary between the somatic and the discursive to move aspects of linguistic activity to the somatic side.

descriptions of them cannot have the constitutive role some theorists ascribe
to implicit presuppositions or other semantic commitments. One might well
describe these embodied styles, as Mauss or Geertz do, but there is no semantic content to how one walks, runs, squats, strokes an animal's feathers, or
avoids eye contact, even though Geertz treats such kinesthetic performances
as texts to be "read." Reading them and performing them remain altogether
different domains of activity.

   Bourdieu, Dreyfus, Taylor, Polanyi, and other somatically oriented practice theorists thus emphasize a level of meaning and understanding which, if
not utterly inaccessible to language, is nevertheless more a matter of skillful
performance and perceptual uptake and recognition. The skillful know-how
underlying social practices supposedly bypasses verbal expression, even or
perhaps especially in its acquisition or transmission, which supposedly leaves
rules and descriptions behind. To sustain this distinction, however, they need
an account of language use that upholds the contrast. Sometimes this distinction is drawn between linguistic or cognitive representation and bodily practice, as in Dreyfus (1979, 2014). Bourdieu (1990, 1991) instead distinguishes
between the logic or standpoint of practice and what is constituted within a
scientific field or linguistic market. Taylor splits the difference with a bifurcated conception of language, in which people both "inform ourselves about
the world of self-standing objects" and "also build landscapes of meanings,
both human meanings and footings . . . [in which language] has a constructive, or constitutive function" (2016, 332–33). These two aspects of language
then sit uneasily side by side in his account. Taylor concludes that "the linguistic capacity is essentially more than an intellectual one; it is embodied . . .
so our language straddles the boundary between 'mind' and 'body'; also that
between dialogical and monological" (2016, 333), but he leaves that boundary
intact and does not attend to any "straddling" across it.

   The ability to speak and understand language, after all, also requires practical, bodily know-how. Linguistic fluency might have been a telling example of
Dreyfus's distinction between expert skill and incompetent, rule-guided action. The difference between the halting, uneven, and sometimes-unintelligible
speech of a language learner, and the smooth, rapid, usually-coherent flow of
a fluent speaker would exemplify embodied skillfulness, as would comparable
differences in the perceptual skills needed to discriminate and understand
others' utterances. Those who must rely on phonetic and grammatical rules
or defined word-meanings do not yet speak or understand a language; they
work with it rather than in it. The difficulty in recognizing language-learning
as bodily skill-acquisition comes from apparent opposition between supposedly tacit or inarticulate bodily skills and semantic contents expressed in lan-

guage use. Philosophers who acknowledge the importance of bodily skill in language use often invoke stratigraphic metaphors. Practical and perceptual aspects of language use supposedly comprise one "level" of linguistic competence, while semantics and pragmatics are understood separately.[19] The difficulty with these metaphors is that supposedly different levels of linguistic understanding and competence are realized in the same performances. One cannot exercise semantic competence without also exercising practical/perceptual bodily skills of speaking and listening or reading and writing, for they are the same performances.[20]

Parallel problems arise for conceptions of social practices sustained by shared presuppositions, conceptual frameworks, or vocabularies, and of language itself as discursive practice. Each linguistically-based conception of social practices strategically distinguishes causal domains of bodily interaction from normative relations among linguistic expressions, and implicitly presumes some transducer or other interface between rationally processed semantic contents and causal interactions of speaking and hearing bodies.[21] Somatically based accounts of practices have difficulty reconciling the biological formation of bodies as given causal structures with the spontaneously skillful responsiveness of bodily practice; conceptions of practices as linguistically constituted assimilate both within the causal domain while confronting similar difficulty connecting a body's causal capacities with its discursive skillfulness. Ian Hacking, for example, discusses how new kinds of people and new patterns of behavior emerge as historically "made up," by noting that "microbes' possibilities are delimited by nature, not by words [while] what is curious about human action is that by and large what I am deliberately doing depends on the possibilities of description" (2002, 108). Butler (1990, 1993) locates a comparably sharp distinction differently when emphasizing the discursive construction of sex and discursive materializations of bodies, without connecting that discussion to those bodies' biological development.[22]

19. Dreyfus (2002, 313–22) explicitly uses stratigraphic metaphors for different "levels" of linguistic competence. Ebbs (2009, chaps. 4–5) attributes to Davidson's anomalous monism a mistaken stratigraphy distinguishing a practical-perceptual orthographic or phonetic "level" of word-forms from interpretive grasp of their semantic significance.

20. I have critically assessed Dreyfus's and Todes's stratigraphic conceptions of somatic intentionality and conceptual understanding elsewhere (Rouse 2000, 2005a, 2013).

21. Haugeland (1998, chap. 9) criticizes assumptions of a transducing interface between rational semantic relations and bodies' causal interactions with an external world, or between brains and bodies.

22. I have critically assessed Hacking's distinction elsewhere (Rouse 1987, chap. 6). Barad (2007, chap. 5) has a nuanced critical discussion of Butler.

Inability to explicate adequately how semantic normativity relates to causal interactions among material entities has been a deeper problem for conceptions of language as discursive practice. Accounts of the intralinguistic, rational determination of semantic contents and their inferential or other holistic interrelations ultimately depend on showing how those relationships are accountable to causal capacities and interactions of worldly objects. Each account—Quine on holistic adjustments of theories at the "tribunal" of sensory surface irritations, Sellars on integrating the manifest and scientific images of humanity-in-the-world, Davidson on token identity of mental and physical events, John McDowell on relations between law-governed first nature and conceptual capacities inculcated by second nature, or Brandom on judgments of practical and perceptual reliability—fails.[23] The underlying difficulty is their effort to separate rational, normative relations among semantic contents from their realization by humans as living organisms who evolved and developed in discursively articulated environments. This book aims to circumvent that difficulty.

The Roots of Discontent

Practice-based social theories have many sophisticated versions, and I draw on their many insights. I endorse the common concerns laid out at the beginning of the chapter and reiterate the importance of people's skillful, affective bodily engagement with surroundings *and* our discursive articulation of "social" practices in language and other conceptual repertoires. The challenges and conceptual difficulties discussed above and others specific to some accounts of social practices have a common root in conceptions of practices as social interactions whose intelligibility is largely independent of biology. The remainder of this chapter initially indicates how human biology is implicated in these conceptual blockages.

The best work in practice-based social theories now mostly recognizes that social practices cannot be understood apart from their material settings, but their accounts are almost invariably "interrelationist" (Haugeland 1998, chap. 9). They treat social practices as people's interactions *with* one another *in* a material setting, with *relations* among practices and settings importantly

---

23. McDowell (1994) and I (Rouse 2002, chap. 2, 3, 6) have both developed extended criticisms of Quine's and Davidson's dualisms; I have addressed Sellars on the manifest and scientific images (Rouse 2015); I have analyzed McDowell's own difficulties (Rouse 2002, chap. 2; 2015, chap. 5); and I have taken on Brandom (Rouse 2002, chaps. 6–7; 2015, chap. 5; and below, chap. 6).

shaping both. They nevertheless still maintain principled distinctions between practices and settings which often turn on differences between causal relations among situated performances and the meaning and other normative significance that practices confer on material things. Haugeland's contrast class to interrelationist conceptions is the "intimacy" of practices' "embodiment and embeddedness in the world, [where] the term 'intimacy' is meant to suggest more than just necessary interrelation or interdependence but a kind of *commingling* or *integralness* . . . [that] undermines their very distinctness" (Haugeland 1998, 208). Many characteristic problems of practice-based social theories arise from not recognizing participants in practices as animals whose interactions are aspects of the intimate interdependence of living bodies with their biological environments.

Material settings of social practices are unintelligible apart from what people do there. Even when people are absent, the seats in empty classrooms, clothes in closets, parking lots for closed offices, or blazed trails through the woods are configured by how people use them. More importantly, we cannot understand people's performances without recognizing what they do and the bodily capacities they develop as intimately responsive to how those performances are materially grounded, directed, and equipmentally mediated. Bodies do not move in empty space; they normally move on and against a solid surface, through air, in a gravitational field, oriented by other surroundings that configure a task at hand. The social-practical significance of what people do is also commonly differentiated by where it occurs: at home, in the office, in a car, on the farm, in the bathroom or kitchen, on the subway, and so forth. These settings were not only constructed for those purposes but are continually maintained; abandoned settings quickly change and decay. Moreover, spatial relationships among diverse practical settings are closely shaped by the proximity and transport needed for timely movements among the practices that vertically differentiate people's lives. The intimate interdependence with which practices incorporate material settings results from the niche-constructive coevolution of human bodies with the environments where they develop and sustain themselves biologically.

Whether practices are understood as intimately situated or not, however, a bigger conceptual challenge confronts practice-based accounts of social life. They need to understand how a practice both conditions and enables its dispersed and temporally extended performances. Similar problems arise for how different practices belong together and condition one another and their constituent performances. The effort to identify practices by constitutive common features among their performances—performative or dispositional regularities, implicit commitment to shared presuppositions, or authoritative governance by implicit norms—has manifestly failed. In retrospect, their failure should not be surprising. The performances we might identify pretheoretically as belonging to the same practice often enact different roles, undertake similar roles differently, have participants at different stages of experience with different degrees of commitment or involvement, and often encompass disagreement about the aims or norms of what participants nevertheless do together.[24] Empirically, anthropological practice theorists are right to identify cultural practices as sites of difference, contestation, and power in lieu of commitments to shared cultures. Postclassical practice theorists usually start from empirical examination of pretheoretically identified "arrays," "nexuses," or "bundles" of performances, without a more general sense of how a performative nexus is established and sustained, or why these particular arrays are the basic makeup of human social life rather than merely patterns of contingent correlations.

This book proposes that the root difficulty with conceptions of social practices arises from theorists' effort to understand the unity and constitutive significance of practices abstracted from their biological significance for human ways of life. The challenge in understanding practices has two aspects. First, the social-theoretic concept requires a nonarbitrary basis for identifying temporally extended and spatially dispersed performances as performances *of* a practice—why these performances and not others, and why they belong to this practice rather than others. Second, that identification has to disclose the normative authority with which practices both enable and condition their own constituent performances. The book argues that what resolves these two difficulties is biological, both in continuities of human biology with other organisms and in changes wrought by human evolution that include more complex forms of normative accountability.

A compact summary of this resolution can be expressed in two claims. First, the normative authority with which practices enable and condition their subsequent performances arises from the constitutive neediness of humans as organisms. A living body is not a self-contained entity, but a dynamic, vulnerable process of maintaining itself individually and within a lineage by extracting needed resources from its environment and exporting disruptive wastes. People act to sustain and reproduce practices as part of their ongoing biological dependence on their environment as living organisms. Evolutionary biology helps understand how and why such interactive processes exist,

24. Arguably, the nexuses, arrays, pools, or bundles of activity that "postclassical" practice theories take as sites for empirical investigation are roughly coextensive with this pretheoretical conception of practices.

how they are sustained and reproduced over time, and how they can fail and go extinct. Second, humans have gradually evolved a distinctive, practice-differentiated way of life. People no longer sustain themselves in more or less the same ways, but instead take up different activities, in concert with mostly supportively aligned performances by others in appropriately conducive circumstances. Those practices mostly work together to sustain and revise their own supportive interconnections. Evolutionary theorizing long presumed that a central problem for understanding human evolution is understanding the genetic basis for altruistic behavior that benefits others as much or more than oneself or one's genetic kin. Underlying that conception has been an implicit presumption of baseline individual self-sufficiency. More recent work has rejected that presumption and reconfigured the problem as the evolution of interdependent performance and the development of relevant capacities for interdependent performance in each generation (Sterelny 2012a, 2021; Tomasello et al. 2012; Tomasello 2019). People act in alignment with others' situated performances to accomplish things neither could readily accomplish by themselves or even make sense of as individual goals. Moreover, those practices—ongoing patterns of interdependent performance—similarly depend in more limited ways on supportive alignment with other practices. Such mutual interdependence takes ongoing effort and adjustment to sustain—an environment of mutually interdependent practices is maintained under changing conditions, in distinctive forms of what evolutionary biologists now call niche construction. The practice-differentiated character of human ways of life can nevertheless be understood as conjoined outcomes of natural selection and iterated cycles of that niche-constructive differentiation and articulation of mutually interdependent practices.

The ensuing chapters develop the biological background for this conception of human evolution, and work out this reconception of practices as the configuration of organism-environment interactions in the human lineage. The evolution and development of human bodies, skills, and languages are best understood as coevolving forms of niche construction. The resulting normative accountability of performances of a practice to one another and to other interdependent practices enables and conditions those performances. It also accounts for the diversity, mutability, and intertwining of the normative concerns arising in human ways of life. The authority those concerns carry and the possibilities for critical assessment and transformation of human ways of life in their light arises from their temporality. All biological lives and lineages are temporally open patterns whose success or failure is shaped by their development and evolution. The normative diversity of people's practice-differentiated way of life enables a more complex temporality, in which the normative concerns internal to practices, other normative concerns arising from their interdependence, and the continuation of a practice-differentiated way of life are at issue and at stake in *how* that interdependence is sustained and adjudicated. The biologically based environmental dependence of that way of life then also allows further recognition of how that normative accountability is not simply anthropomorphic. For now, the promissory conclusion to this chapter is that practice-based conceptions of human sociality have encountered their characteristic difficulties due to efforts to analyze a social world as intelligible apart from the biological processes of human bodies and ways of life. The constructive corollary to that conclusion is to vindicate the concerns that originally motivated practice-based social theories and many of the insights attained in their development, but only when resituated within an evolutionary biological understanding of human animals and our ways of life.

This reconception indicates how a practice-differentiated way of life positions people differently amid those practices as differences in power—in abilities to accomplish ends, avoid harms, and affect the performances of others, with different constraints on what they can do.[25] It also resolves practice theorists' difficulties in understanding how bodily skills and affects work together with the discursive articulation of language and other conceptual repertoires. Those difficulties resulted from neglecting biological processes of environmentally situated bodily development that include the evolution and development of language in discursive environments.

25. Brown (2017) highlights the lack of adequate understanding of power within practice-based social theories.

2

Ecological-Developmental Niche Construction

Practice-based conceptions of social life and culture and most other social theories share commitment to the conceptual autonomy of a social world, which blocks attention to human biology. Social theorists accept broadly Darwinian conceptions of human evolution while presuming that language, culture, and complex patterns of social life mostly insulate modern human populations from biological selection pressures. They take evolved features of human bodies and behaviors and biological variation for granted in the social sciences and social theory. The presumed independence of social practices from biological influence or explanation has been reinforced by critical assessment of Darwinian accounts of contemporary human social life. Programs in human sociobiology (Wilson 1975), human behavioral ecology (Hinde 1974; Cronk, Chagnon, and Irons 2000; Winterhalder and Smith 2000), or evolutionary psychology (Barkow, Cosmides, and Tooby 1992) encountered extensive, detailed criticism from philosophers and social scientists and widespread dismissal within social theory and the social sciences.[1]

These programs for evolutionary understanding of human behavior and social life have appealed to the modern evolutionary synthesis, dominant in biology since the mid-twentieth century. Recent developments in many biological subfields enable more constructive rapprochement between practice-based conceptions of social life and human biology. Biologists now disagree

---

1. Kitcher (1985) and Lewontin, Rose, and Kamin (1984) on sociobiology and Buller (2005) and Lloyd and Feldman (2002) on evolutionary psychology offer prominent criticisms of these evolutionary accounts of human sociality. Evolutionary psychologists (Symons 1989; Cosmides and Tooby 1987) have been programmatic critics of human behavioral ecology. Laland and Brown (2011) provide sympathetic comparative assessment of each approach.

whether these developments require extending or reconstructing central te-
nets of the modern synthesis or research programs in evolutionary biology.[2]
Their significance for biological understanding of human ways of life is un-
questionable, however. Inadequacies in earlier biological programs for understanding of social life are thus insufficient reason to discount the relevance
of evolutionary and developmental biology for social theory. This book primarily draws on developmental evolution, ecological-developmental biology,
developmental systems theory, niche construction, new work on animal behavior, and some aspects of cultural evolution from among the many programmatic developments in contemporary evolutionary biology but does not
attempt a comprehensive overview.[3] This chapter instead thematically introduces biological initiatives relevant to practice-based approaches to human
social life. It begins with the constitutively normative character of biological
processes grounded in relations between organisms and their environments.
That conception requires a constructionist conception of biological development, which works together with the evolutionary significance of niche construction. Those considerations then clarify the implications of recent work
on early hominin evolution for understanding our subsequent ways of life.

Biological Normativity

Practice-based conceptions embed the normative ordering of human life in
practices and their interrelations rather than norms whose authority is established on different grounds. I argue that this complex, practice-articulated
normative order is an evolved modification of the constitutive normativity of
biological processes. Justifying this claim begins by spelling out how contemporary evolutionary theory allows for biological normativity.

Biology concerns the structure, functioning, development, interaction,
and evolution of organisms. Understanding organisms functionally suggests
how natural processes could have normative import. A functional system can
succeed or fail in fulfilling its function and has that function even if it fails;
otherwise, talk of malfunction or failure would not make sense. The difficulty is understanding how entities have a function. Artifacts are functional
because their designers or users assign them a function, often building them
for that purpose. Evolutionary theory takes analogies to designed artifacts

---

2. Smocovitis (2020) emphasizes the conceptual complexity of the modern synthesis, arguing that debates over its revision are framed by only partial recognition of this complexity.

3. Laland et al. (2015), Jablonka and Lamb (2020), Pigliucci and Müller (2010), and Sultan
(2015) review many of these programmatic developments.

seriously while dispensing with designers.[4] Natural selection is often taken as a surrogate for designers, but that will not quite do. What evolutionary theory does is explain aspects of organisms' processes or behavior as fulfilling functional roles in the way of life of a population which contribute to its comparative reproductive success ("adaptive fitness").

Susan Mills and John Beatty (1979) emphasize that adaptive fitness is not the *actual* reproductive success of individuals, populations, or phenotypic traits. Success might result from factors other than fitness, and equating fitness with success would also trivialize its explanatory power. Fitness is instead a *propensity* to produce offspring which is normally assessed with optimization models (Maynard Smith 1978). Key assumptions in these models are the range of phenotypic variation, the aim or goal to maximize, the population structure, and its mode of inheritance (Maynard Smith 1978, 52). What is maximized is not itself a functional role, but the continuation of the life of an organism and its reproduction in subsequent generations.[5] Functional explanations in evolutionary biology thus presuppose a goal for the functional process or subsystem.

This teleological basis for biological normativity has recognized historical antecedents, notably Aristotle's (1941) conception of organisms. Aristotle distinguished activities whose goal was not part of the activity (as a house is separate from the building of that house) from activities whose goal is to continue or reproduce that activity. Living organisms exemplify this latter sense of activities as energeia: they can succeed or fail in maintaining and reproducing those processes under changing conditions and thereby acquire normative significance. Aristotle regarded organismic processes as stable, but his conception of organisms as energeia is independent of that presumption. Mark Okrent noted that, "Darwin was the greatest Aristotelian of the nineteenth century. Darwin agrees with Aristotle . . . on the central issue of whether [biological processes] are evaluable in a non-arbitrary fashion even if they were not made by some rational creator. . . . For Darwin and Darwinians, living organisms are those [processes] that carry the principle of nonarbitrary normative evaluability in themselves" (2007, 68). Darwin's evolutionary Aristotelianism nevertheless complicates specifying the goal-directedness of those organismic processes.

4. Lewens (2004) is a philosophically sophisticated discussion of concepts of design and function and analogies between artifacts and organisms in evolutionary biology.
5. Models often address a component of fitness, such as rates of energy intake while foraging, which might itself be understood functionally (Maynard Smith 1978, 52). The modeled component of fitness is still functional *because* it contributes positively to maintenance and reproduction of the lives of organisms and their descendants.

Biological processes of physiology, behavior, and reproduction develop and evolve, and those changes are evaluable for whether they sustain those processes through changes over time. Organisms die and their lineages go extinct. Other organisms and lineages persist through long sequences of life cycles. Identifying *what* perishes or persists is nevertheless complex and important. Organisms are material entities, but their material components change continuously and turn over multiple times throughout their life spans. Moreover, material components of an organism persist and dissipate after the organism's death and often become constituents of other organisms. For Aristotle, what supposedly persists and is reproduced is not a material entity but its supposedly constant organismic form. Organismic forms nevertheless also change continuously. Organisms grow, develop, and die. Metamorphic changes also make development markedly discontinuous. The dead metaphor of "development" suggests teleological culmination in mature, adult form, but aging continues the developmental process until death. Organisms reproduce to sustain a lineage, but reproduction is descent with modification that results in evolutionary change in the reproducing populations.

The goal of an organism's physiology and behavior is not external to its functioning, however, but is instead the ongoing reproduction of that process. The key outcome of Darwin's Aristotelianism is that this constitutive goal is not describable abstractly, since it is irreducibly deictic. The goal is to continue *this* process, as a differential reproduction. Organisms are temporally extended, open-ended patterns of developmental and evolutionary change. This revisionist Aristotelian conception of organisms as evolving energeia crucially differs from *externally* goal-directed processes. Consider cybernetic conceptions of teleology as negative feedback relations between a system's input-output relations and a prespecified goal. The classic paper on cybernetics by Arturo Rosenblueth, Norbert Wiener, and Julian Bigelow (1943) characterizes purposiveness as external: "The term purposeful is meant to denote that the act or behavior may be interpreted as directed to the attainment of a goal—i.e., to a final condition in which the behaving object reaches a definite correlation in time or in space with respect to another object or event" (1943, 18). By contrast, the goals of organisms as energeia are not determinately specifiable, because they are continuations of changing processes. Organismic goals have no final condition; their "finality" is *failure* to sustain their life processes.

Organism and Environment

Recognition that organismic "forms" vary within populations and evolve through changes in their variance requires further revision to Aristotelian conceptions of biological teleology and normativity. An ambiguity in Okrent's (2007, chaps. 2–4) account of Darwin's Aristotelianism brings out the difficulty.

> For Darwin and Darwinians, *living* organisms are those individuals that carry the principle of nonarbitrary normative evaluability in themselves. Nonarbitrary standards for evaluating goal-directed events are borrowed from nonarbitrary standards for evaluating the entities in which they occur. . . . Since to be alive is to do something that allows it to continue to live, it is life itself that provides the standard against which a particular living thing is to be evaluated. The end or goal of the process is the very process itself, for the end of the process is that the process should continue. (2007, 68–69)

Okrent alternately assigns biological purposiveness to living organisms as *entities* subject to nonarbitrary normative assessment and then to the life *process*. The life process is not the same as individual organisms, however. As life processes, organisms are not physical objects bounded by an outer integument. The integument that marks the bounds of an organismic entity begins to disintegrate when the life processes transpiring *across* that boundary cease; its apparent boundedness is a dependent component of that larger process.[6] Nor are organismic processes equivalent to the systematic organization, functioning, and differential reproduction of an internally complex system. Organismic *bodies* only remain alive and develop through ongoing interaction with their environments: taking in resources, utilizing tools, responding to signals, avoiding or overcoming threats, seeking or exploiting opportunities, and excreting wastes. *Organisms* are situated, interactive processes whose functioning is constitutively interdependent with their environments.

An alternative approach identifies the life process with organisms as entities by treating them as *agents* of that process. The boundary between internal processes and external environments is dynamically maintained through processes of *self*-constitution and *self*-maintenance, importing resources needed to sustain those processes, blocking or avoiding external threats, and exporting waste products that otherwise threaten their persistence from within. Those processes change over time through development and environmental interactions. What supposedly remains constant throughout the life span is the organism's agency in adjusting its internal processes and responding to changed external circumstances to maintain the constitutive boundary

---

6. Recognizing that organismic processes extend beyond organismic bodies does not block classification of organisms by similarities among bodily forms. For the disciplinary purposes of systematics, similar forms are useful markers for similar processes (Lange 2007). Thanks to Mark Okrent for raising this possible objection.

between organism and environment, partly by orchestrating two-way traffic across the boundary.

This "autopoietic" tradition that locates the agency of organismic processes on one side of the organism/environment boundary faltered in turn. One reason for its failure is that biologists now recognize that eukaryotic organisms are holobionts, symbiotic and endosymbiotic communities rather than discrete individuals (Gilbert 2017; Gilbert, Sapp, and Tauber 2012; McFall-Ngai et al. 2013). Scott Gilbert recently concluded from conjoining the ecological-developmental turn in evolutionary biology with recognition of eukaryotic organisms as symbiotic holobionts that, "If developmental symbioses represent the rule and not merely the exceptional case, then the entire notion of 'autopoiesis' must be abandoned. We are not adults entering into symbiotic relationships with other adults or microbes. Rather, the processes that made us adults are already the interactions between us and our microbes. . . . Upon reflecting on the data accumulated by McFall-Ngai, we may need to conclude that 'All development is co-development'" (Gilbert 2002, 213). Understanding organisms as aspects of interdependent processes extending into their environments makes them the loci of organism-environmental coconstitution, codevelopment, and coevolution rather than self-maintaining agents.

If the goal-directedness of life processes cannot be identified with physically bounded entities or self-organizing agency, however, so the environmental component of life processes cannot be equated with organisms' physical surroundings. Different organisms in the same physical vicinity encounter and inhabit different environments, because different aspects of their physical surroundings matter to their life processes. Organismic physiology, development, and evolution are *phenomena*, patterns of multifaceted *intra-action* (Barad 2007). "Intra-action" differs from interaction, which implicitly takes interactants as specifiable apart from their mutual relations and interactions. Intra-active patterns are more basic than their components, and determine their boundaries and characteristics. Features of an organism's physical surroundings only belong to its environment through involvement in its way of life as process. Richard Lewontin incisively captured this point about organism/environment relations.

> Every element in [an ornithologist's] specification of the environment [of a bird species] is a description of activities of the bird. As a consequence of the properties of an animal's sense organs, nervous system, metabolism, and shape, there is a spatial and temporal juxtaposition of bits and pieces of the world that produces a surrounding for the organism that is relevant to it. . . . It is, in general, not possible to understand the geographical and temporal distribution of species if the environment is characterized as a property of the

physical region, rather than of the space defined by the activities of the organism itself. (Lewontin 2001b, 52–53)

Populations of the fruit fly *Drosophila pseudoobscura* found in deserts in the western United States and Mexico compellingly exemplify this point. One might expect these flies to be adapted to low-moisture environments. They are instead *more* hydrophilic than other fruit flies. That fact is less surprising on recognizing that these flies do not live in deserts; they live in high-moisture environments within a millimeter of plant leaves.

The environments relevant to evolution are also entangled with populations of organisms in a second important way. Robert Brandon and Janis Antonovics note:

> One might think that external environmental heterogeneity was a necessary condition for ecological environmental heterogeneity. But whether this is so depends on whether or not we include conspecific organisms as part of the external environment. . . . For any individual organism, its conspecifics can be influential factors in its development and reproduction, and systematically relevant to selection within and of the population. Consequently, the selective environment for a population must incorporate interactions among conspecifics in partially shared circumstances as a social-ecological environment in cases of density-dependent and frequency-dependent selection. (1996, 165)

As Lewontin noted, "if the fitness of a genotype is dependent upon the mix of other genotypes in the population, then nothing has been learned by measuring the 'fitnesses' of the genotypes in isolation" (2001b, 57). This aspect of organism-environment intra-action is especially central to the book's argument: social relations are integral to human biological environments.

Recognizing that organisms do not live in so-called external environments of physical surroundings but in social-ecological environments relevant to their life processes has obvious implications for evolution. Natural selection is a process whereby heritable variation in organisms' fitness within a population— their propensity to produce offspring in an environment—changes the population's composition over time. Evolutionary theory long regarded selective environments as organism-independent causes of differential reproductive success. Recent evolutionary biology recognizes that this environmental role is not specifiable independently of organisms' life processes. Evolution is not organismic adaptation to given environments, but coevolution of organisms' bodily morphology, physiology, and development with the configuration of their selective environments (Brandon and Antonovics 1996).

Similar considerations revise understanding of organisms that intra-act with their environments in selectively significant ways and populations that coevolve with those environments. This aspect of organism/environment coupling is less often remarked, but Lewontin's characterization of avian environments is instructive. Every biologically significant trait he mentioned involves responsive intra-action with some aspect or aspects of the organism's environment. An organism's sense organs, nervous system, and metabolism are sensory, nervous, or metabolic through interconnected responsiveness to that environment. An organism's life process, including its constituent traits and behaviors, *is* that larger intra-active pattern.

The locus of biological goal-directedness and normativity is thus the temporal and causal continuity of environmentally intra-active processes of physiology and behavior and their development, reproduction, and evolution. The autopoietic tradition is almost right. It correctly treats development and evolution as dynamic processes whose later stages are produced from earlier ones without determination by continuously present components or by causal predetermination. The membranes that partially segregate organisms' bodies from their surroundings as far-from-equilibrium thermodynamic systems are nevertheless products of more encompassing processes rather than organismic self-maintenance. Developmental systems theorists identify that process with life cycles encompassing organismic bodies and the resources for their ongoing development and reproduction (Griffiths and Gray 2001; Griffiths and Stotz 2018). Those resources include other symbiotic or commensal organisms as well as persistent environmental resources and products of earlier life-cycle stages. Evolution is differential reproduction or extinction of these developmental cycles. The "agents" propelling that process are the coupled components of the life cycles. An evolutionary individual is "a life cycle whose components cannot reconstruct themselves when decoupled from the larger cycle" (Griffiths and Gray 2001, 213). Organisms are not individuals but instead dependent components of more encompassing life processes.

The biological normativity of purposive Aristotelian energeia thus concerns intra-active life cycles rather than organismic bodies. What establishes their goal-directedness is the interdependence of their component events and sequences; each takes its form in support of and supported by how other events and sequences proceed. As Griffiths and Stotz concluded, "A series of developmental events is a single process because those events serve a common evolutionary goal, namely to maximize the representation of cycles descended from them in future generations vis-à-vis the representation of the variant cycles with which they compete. . . . [More precisely,] an individual life cycle is a token of a life-history strategy, and that strategy is its telos and its principle of genidentity" (2018, 239). That goal does not exist before the emergence of a developmental sequence. The contingent establishment of recognizable developmental and reproductive cycling that incorporates and accommodates changing conditions retroactively constitutes and projects its own goal-directedness.[7] The goal-directedness of biological systems as energeia is thus temporally constituted and becomes successful only in retrospect.

Evolutionary emphasis on a unitary goal of survival and reproduction of organisms and lineages seems to mark an important contrast between the biological normativity of other organisms and humans' more complex and plural goal-directedness and normative accountability. What is at stake in life-history strategies and governs assessment of their success or failure is only whether they produce descendant cycles as a lineage, and not whether or how those descendants differ from their predecessors. Some life-cycle sequences are remarkably stable over thousands of generations, while others dramatically shift bodily forms, development, behavior, environmental resources and challenges, and hence their goal-directed way of life. Both routes exemplify biological success, differing only in how it is achieved. Success is never irrevocable, because life processes and lineages remain vulnerable to death and extinction. Biological normativity arises from that vulnerability, which makes the continuation of lineages intelligible as *success* through their resistance to that ever-present possibility.

Human ways of life are also open to nonarbitrary assessment on those grounds. One hominin lineage, *Homo sapiens*, has succeeded so far, while all others have been extinguished. Human ways of life nevertheless also answer to an extraordinary range of other normative concerns, which collectively address not just whether those ways of life persist but how they proceed and what they become. That difference is indeed important, but we first need to understand how these more complex and variegated normative concerns relate to the characteristic biological normativity of people's lives and the human lineage. Only then can we recognize human normative plurality and complexity as an evolved form of biological normativity.

Development as Construction

A long, influential tradition in biology responds to the processual character of development and evolution by seeking a stable, underlying locus of

7. Appeal to recognizability as constitutive of goal-directed developmental life cycles echoes Dennett (1987) on predictive/explanatory stances as norm-constitutive disclosures of "real patterns" (1991) and Heidegger's (1962, 1982) insistence that the being of entities (but not entities) depends on an understanding of being. Brandom (2002, 50) characterizes such dependence on *intelligibility* from a stance or ability-to-be as "sense-dependence" rather than "reference-dependence": A is sense-dependent on B if understanding or recognizing A requires understanding or recognizing B, but A's existence need not depend on B's existence. Okrent (2018, 39–42) shows how this issue bears on the goal-directedness of organismic ways of life.

organismic identity and continuity. August Weismann's distinction of organisms' stable hereditary germplasm from changing somatic tissues initiated modern, post-Darwinian versions. Subsequent examples include identifying character-determining Mendelian factors or "genes" with chromosomal locations, Erwin Schrödinger's (1946) influential suggestion that a stable, "aperiodic crystal" was the persistent material basis of life, the reconception of genes as coding and regulatory sequences of DNA base pairs, and eventually a shift from material genes to immaterial information successively encoded in base sequences of DNA and RNA, amino acid sequences of proteins, and their phenotypic expression.

In evolutionary theory, one result was the notorious omission of embryology or developmental biology from the modern synthesis. John Maynard Smith neatly summarized this strategy: "One consequence of Weismann's concept of the separation of the germline and soma was to make it possible to understand genetics, and hence evolution, without understanding development" (1982, 6). Only the germline is replicated in reproduction of an organismic lineage, so developmental processes through which germ cells become an organism are literally segregated from the evolutionary process. This strategy is encapsulated in metaphors of a genotype as repository of "information" as the ultimate cause of an organismic "phenotype."[8]

The metaphorical slippage in these concepts is manifest from either direction. Genotypes acquire their typology from a genetic sequence specifiable biochemically by base pairs in DNA molecules. This biochemical specification is not yet *genetic* specification, however, except via developmental processes in which transcription of those sequences is regulated and translated into organismic form. What information about organismic form was transmitted via the DNA sequence is not determined by the sequence alone, but by developmental processes in genomic, cellular, organismic, and ecological environments.[9] From the other direction, phenotypes, the organismic forms supposedly determined by genotypes, are not general patterns or forms, but life processes that coconstitute organism and biological environment from

8. Metaphors of genotype, phenotype, and information transfer from one to the other illustrate the difference between a philosophy of nature, or natureculture, and specific research programs in biology. These metaphors do important work in many research programs, but their assumptions and occlusions require critical assessment of their broader import.

9. Susan Oyama (1985) thus emphasized the *ontogeny* of information. Organismic form and activity is "informed" not by predetermined genotypes, but by developmental intra-action of the organism-environment complex.

birth to death. Development is not maturation into final adult forms supposedly specified by genes, but the entire life process. The concept of a pheno*type*
implicitly generalizes from sexual forms of reproduction in concert with the
germline/soma distinction. The organism continues to develop after reproducing, but subsequent development supposedly does not affect its offspring
or its lineage. Hence, the phenotype would be the adult form at sexual maturity. That conception neglects the many forms of asexual reproduction, but
also erases the developing adult organisms' contribution to the environments
in which their offspring develop.

Even "phenotypic plasticity," introduced to describe the partial independence of development from genetic determination, retains vestiges of this
metaphor of organismic preformation. Instead of identifying the phenotype
as a single form prespecified by a genotype, it recognizes bounded ranges of
variation across different environments as "norms of reaction" predetermined
by genotypes. Norms of reaction, however, are only measured components of
developmental responsiveness to specific environmental variations. Convert-
ing norms of reaction expressing localized measures of developmental plasticity of a genetic sequence into a phenotype would require integrating them
subjunctively across the range of actual or possible developmental outcomes
in relevant possible environments.

The metaphor of genotypic information prespecifying phenotypic forms
worked together with the supposed externality of organisms' selective environments to identify evolution with changing frequencies of genetic alleles
within populations as adaptations to those environments. "Allele" invokes
similar metaphorical slippage as "genotype," from specific DNA sequence
fragments to environmentally intra-active variations in life processes. That
metaphoric complex is complemented by an implicit sense of development as
step-by-step, "internal" unfolding of a genetically specified ("programmed")
phenotypic form. This conception of organismic development and evolution clashes with the environmental intra-activity of every stage and level
of the life process, including genomes in cellular environments, cells in somatic environments, and bodies in social-ecological environments. Lewontin
(1982, 1983) influentially suggested that the metaphorical complex of development and evolution as the unfolding and adaptation of genetic information
should be replaced by understanding development as construction. Organismic forms and ways of life are constructed over time by multiple intra-
acting components. Gene sequences, epigenetic markers, cellular processes,
bodily functions and activities, and shifting developmental environments
produce, maintain, and reproduce organisms and their characteristic patterns

of environmental affordance.[10] These processes maintain recognizable continuity within taxa and across generations due to the reliable availability of
some of their components and recurring reconstruction of many others.

Genocentric conceptions of evolution were sustained by presumptions that
only genetic components of developmental processes are reliably passed on
through multiple generations. That presumption is now widely recognized as
false. Replication of a DNA sequence would not be a genetic phenomenon unless it were accompanied by and encapsulated in cellular structures that enable
and regulate transcription and translation and supplied with conducive surroundings and sufficient accessible nutrients to drive that process energetically.
In eukaryotic organisms, many key developmental processes are initiated or
regulated by microbial symbionts acquired from parents or other conspecifics pre- or postnatally (Gilbert, Sapp, and Tauber 2012; McFall-Ngai et al. 2013;
Gilbert 2017). Some indispensable components of the developmental process,
from diurnal and seasonal variations of solar radiation to critical features of organismic habitats, are reliably sustained or reproduced independent of the organism. Others are cyclically reconstructed through reliable provision of their
requisite components and circumstances. What is sustained and differentially
reproduced over generations is not just organisms and their genetic sequences
but common and relevantly variant aspects of developmental life cycles and
processes of organism-environment intra-action (Griffiths and Gray 2001). The
developmental life-cycle components and patterns for individual organisms
vary across populations, and those differences affect which cycles reproduce
in subsequent generations. The resulting redistribution of life-cycle patterns in
descendant populations is the evolutionary process.

Niche Construction and Coevolution

A constitutive feature of ecological-developmental conceptions of evolution is the reciprocal causal intra-action among components and patterns of

---

10. Epigenetic markers help regulate gene expression, and people's social situation and interactions can lead to epigenetic changes, some heritable across generations. These considerations
have encouraged treating molecular epigenetics as an interface between human biology and the
"social world" (Landecker and Panofsky 2013; Warin, Moore, and Davies 2016; Meloni et al. 2018,
parts 2, 4, and 5). Such work is valuable, but still neglects the broader context where molecular
epigenetics is only one component of the integration of evolution and development. It thereby
retains misleading conceptual separation between social phenomena and a genocentric biological domain. A naturecultural conception of practices recognizes human molecular epigenetics
as one aspect of how humans develop in practice-differentiated environments that are also integral to evolution of the lineage.

developing life processes (Laland, Odling-Smee, and Gilbert 2008). Whereas evolutionary theory once treated causal relations between genotypes and phenotypes or environments and populations as unidirectional, ecological-developmental reconceptions recognize that the causal relations go in both directions. Not only is development a reciprocal intra-action among genes, epigenetic cellular processes, and cellular, organismic, and social-ecological environments. Organisms actively construct or reconstruct developmental and selective environments. These activities affect their own subsequent development and change the selection pressures they encounter along with those affecting their conspecifics and companion species. This aspect of the reciprocal causality of biological development is niche construction, "the process whereby organisms, through their metabolism, their activities and their choices, modify their own and/or each other's niches" (Odling-Smee, Laland, and Feldman 2003, 419). Niche construction is the key process that both allows and requires incorporating social practices within a more expansive ecological-developmental conception of human development and evolution.

Earlier conceptions of evolution are "externalist" (Godfrey-Smith 1996) in the sense that evolutionary change is supposedly driven by unidirectional causal effects on the distribution of phenotypic variants within later generations of a population due to organism-independent features of the population's local environments. The sources of heritable phenotypic variation within those populations—genetic mutation, sexual recombination, migration, and lateral gene transfer in microorganisms—were treated as independent of environmental "selection" among those variants that causes differences in their subsequent survival and reproduction.[11] Organisms in subsequent generations are better adapted to their environments only in the sense that those variants comparatively more successful in surviving and reproducing in those environments increased their numbers or proportion within the population. This conception of adaptive evolution does not constitute biological progress because the environments to which organismic populations adapt are a moving target. Environmental change would then be exogenously driven, so changes that were once adaptive can later become maladaptive and vice versa.

Externalist conceptions rightly emphasize the importance of natural selection in the evolution of organismic populations across multiple generations,

---

11. Strictly speaking, migration is not an environment-independent source of variation. Some environments are basins of attraction that transform occasional movements into more permanent migration. For theoretical purposes, however, migration can be analyzed as variation due to features of environments from which migrants came, and thus effectively independent of local selective environments.

and many sources of selectively relevant environmental change are exogenous. Niche construction theory nevertheless demonstrates that earlier models omit a second important cause of changes in distributions of organisms and ways of life. Natural selection, ecologically mediated development, and niche construction together constitute reciprocal rather than unidirectional causal relations among populations of organisms and their environments. The coauthors who built niche construction theory from its beginnings in Lewontin's (1982, 1983) classic publications conclude that "niche construction should be regarded, after natural selection, as a second major participant in evolution" (Odling-Smee, Laland, and Feldman 2003, 2). Earlier evolutionary theorists did not deny that organisms change their environments. They only claimed that those changes are not reliably heritable and sufficiently cumulative on an evolutionary time scale to make discernible or significant differences in evolutionary outcomes.

When niche construction is properly understood, however, its evolutionary significance cannot be ignored. We already saw that organisms are niche-*constitutive*. The developmentally and selectively relevant features of environments are not independent properties of physical locations but are instead specified by how the organism makes a living. This specification is accomplished causally by organisms themselves through transformative interactions with some aspects of their surroundings and not others. Niche construction is thus a pervasive, obligate aspect of all life, not an occasional or isolated phenomenon. As John Odling-Smee, Kevin Laland, and Marcus Feldman emphasize, "Niche construction is not the exclusive prerogative of large populations, keystone species or clever animals; it is a fact of life. All living organisms take in materials for growth and maintenance, and excrete waste products. It follows that, merely by existing, organisms must change their environments to some degree" (2001, 116). These effects cannot be dismissed as heterogeneous, mutually counteractive noise, since the evolved similarity of organisms comprising a species or larger taxon contributes to cumulative effects. Kevin Laland, John Odling-Smee, and Scott Gilbert remind us that, "Collectively, developing niche-constructing organisms in a population act as unidirectional 'biological pumps' in their environments, provided that they constantly do the same things, to the same environmental components, generation after generation" (2008, 557). This cumulative character is especially apparent in weed organisms, whose niche-constructive life processes render their immediate surroundings uninhabitable for subsequent generations. Weeds are not driven to extinction by their niche-destructive life process but become mobile lineages whose selective environments regularly change through migration and habitat colonization.

Despite cumulative, unidirectional niche-constructive activities of organisms sharing a habitat and a way of life, most species are not obligate-migrational weeds. They can collectively modify shared habitats in a single direction because niche construction is a multispecies phenomenon.[12] Organisms' niche-constructive effects are not confined to how they change selection pressures on their own lineages, which is a further reason to reject autopoietic conceptions of organisms as individual, self-constituting agents. Niche construction does change those selection pressures, but also transforms environmental conditions for other organisms, which may affect them in turn. The gaseous composition of earth's atmosphere is an instructive example. The current composition of approximately 20 percent oxygen resulted from cumulative niche construction by cyanobacteria whose photosynthetic processes utilized available carbon dioxide and excreted O_2 as a waste product. Molecular oxygen is toxic to many life forms, and gradual oxygenation of the atmosphere drove many anaerobic bacteria to extinction and others to more isolated environments. It also created favorable conditions for the Cambrian "explosion" of animal life forms, whose niche-constructive CO_2 emissions in turn enhanced prospects for CO_2-consuming plants. Odling-Smee, Laland, and Feldman characterize such interactive niche-constructive relations among species as "environmentally mediated genotypic associations (EMGA's)" (2003, 217–24), although this chapter's reasoning suggests regarding them as life-cycle entanglements rather than genotypic associations.

Organisms act on or move to an environment in one way as "counteractive niche construction" that minimizes or redirects changes in selection pressures already underway (Odling-Smee, Laland, and Feldman 2003, 46). Humans provide salient examples in the buildings and clothing that buffer seasonal and geographical temperature differences to expand their range of habitable environments. Birds' seasonal migrations not only change their climatic conditions, but also reconstruct the environments of predators to counteract predation pressures on their own lineages. Their migration reduces sustainable populations of their predators by removing an important food source part of the year. Turner (2000) is rife with examples. Earthworms

12. In the introduction I mentioned an important consequence of a naturecultural account that the practices making up human developmental environments are multispecies, social-ecological phenomena. This theme is mostly suppressed until the conclusion of chapter 9. Other organisms' effects on human ways of life are often regarded as external influences on an autonomous social world; *initial* appeal to their roles in human development and evolution would be too easily dismissed in this way. Symbiotic, codomesticated and commensal organisms are only recognizably integral to social practices after establishing human ways of life as naturecultural on other grounds.

maintained their adaptation to aquatic environments through fifty million years of terrestrial living with tunneling and burrowing that discompacts soil and reduces its matrix potentials to let them draw more water into their bodies (Turner 2000, 100–19). The extraordinary diversity of insect nests or mounds similarly mitigate the effects of changes in ambient humidity and temperature (Turner 2000, 186–200). Instead of adapting to new selection pressures from altered environments, organisms often act on those conditions to mitigate those pressures.

Niche-constructive buffering is complemented by "inceptive niche construction" in which organisms both introduce and adapt to changes in developmental and selective environments (Odling-Smee, Laland, and Feldman 2003, 45). They exploit new resources, generating selection pressures for their more effective utilization and sometimes creating the resources they then adaptively utilize. Australian eucalypts shed oily leaves and bark that contribute to the ignition and spread of brush fires, transforming the soil and light conditions for their seeds. Brush fires sometimes become more integral components of their life cycles by releasing eucalypt seeds from protective capsules (Gray 1992, 197). Many organisms develop tools, techniques, or habitats whose uses spread by social learning (Laland 2017, chaps. 2–4). They migrate to new circumstances with different opportunities and threats or limit their movements to circumscribe the conditions they encounter. Inceptive and counteractive niche construction often function together, since novel environments or new opportunities also bring challenges that can then be counteractively buffered.

Classifying niche-constructive activities by how they occur is especially instructive for understanding practices as integral to human developmental and selective environments. Organism-environment intra-actions are niche constructive in three ways.[13] The canonical form of niche construction is "perturbation," physical transformation of an organism's abiotic surroundings (Odling-Smee, Laland, and Feldman 2003, 44). Perturbative niche construction incorporates animal architecture (nests, mounds, dams, burrows, pupas, webs, etc.) (Hansell 1984, 2007; Turner 2000), chemical secretions and defenses, waste products (excrement, detritus, respiratory outputs, heat, corpses), resource depletion, and bodily interventions (blocking or chan-

13. This tripartite classification tracks Odling-Smee, Laland, and Feldman's (2003, 44–45) distinction between perturbation and relocation as forms of niche construction, adding behavioral niche construction in rough correspondence to distinctions between biological and cultural processes. Reasons for replacing "cultural processes" with behavioral niche construction extend throughout the remainder of the book.

neling currents, winds, rainfall runoff, or sunlight; trampling vegetation; crumbling rocks and discompacting soil; multiple forms of parasitism and symbiosis). The biosphere has been built up and continually reconfigured by collective forms of perturbative niche construction.[14]

A second pervasive and important form of niche construction is well-known, but less readily recognized as niche *construction*. Organisms change selection pressures on their lives and lineage by moving to different environments. These movements are recognizably niche constructive, because the developmental, ecological, or selective environment of an organism or lineage is not just its physical surroundings but consists in affordances for its way of life, including the threats and challenges afforded by different circumstances. Constructing a niche intra-actively configures or reconfigures patterns of environmental resources, opportunities, challenges, and threats that matter to an organism within what thereby becomes a way of life.

Grasping the full niche-constructive significance of migration requires construing it expansively. Animal species can permanently relocate to change the selection pressures they confront, and plants can shift their geographic range and local habitats. Early hominins notoriously shifted their primary habitat from diminishing patches of tropical forest to savannas covering larger parts of east Africa, and also responded to rapidly oscillating environmental changes that were distinctive features of the region amid the generally desiccating climatic years of the Pleistocene (Potts 1996; Bickerton 2009). Plants can most obviously migrate by changing their methods and patterns of seed distribution, but they also differentially grow or spread stems and leaves toward photon-rich environments and extend roots toward different levels of moisture or other nutrient density (Sultan 2015, chaps. 3–5). Many fish species preferentially inhabit changed currents, temperatures, or concealment provided by kelp beds, coral reefs, or bridge and dock pilings, which each resulted from other species' niche-constructive activities. Migration need not involve permanent relocation, however, but can instead consist in different movement patterns among settings. Migratory birds, territorial or catchment scavengers, predators or prey tracking one another's movements, suburban commuters, or animals changing habitats when breeding significantly change the environmental conditions and selection pressures to which they or their offspring are exposed.

---

14. McMenamin and McMenamin (1994) argue that terrestrial organisms collectively constitute a "hypersea," extending life's original oceanic environment in simultaneously counteractive and perturbative niche construction, exemplified by multiple levels of parasitism characteristic of terrestrial but not oceanic flora and fauna.

Behavioral niche construction is even less widely recognized. Behavioral patterns can have niche-constructive evolutionary significance if their presence in developmental environments of subsequent generations creates selection pressures favoring descendant behavioral patterns in their own or other lineages. Failure to recognize the niche-constructive significance of behavior has multiple sources. The common conflation of biological environments with physical surroundings is sustained in part by overlooking the pervasive developmental and selective effects of conspecifics and companion species. Organisms are then implicitly understood as bodies inhabiting an abiotic environment alongside others living in similar or overlapping environments. Environments are thereby implicitly limited to abiotic circumstances as external sources of selection pressures, independent of the bodies inserted within them and the ongoing perturbative niche construction that shaped those abiotic circumstances.[15] The need to overcome implicit distinction of physical environments from organisms that inhabit them is partly why Brandon and Antonovics (1996) emphasized frequency- and density-dependent selection. One cannot understand the environments that exert selection pressures on lineages unless other organisms, including conspecifics, are recognized as integral to developmental and selective environments, a point central to the argument of this book.

A second reason for overlooking behavioral niche construction is neglect of the evolutionary significance of development and especially the role of social learning in development. Organisms' development across a wide range of taxa is scaffolded by the behavior of conspecifics, and these patterns of social learning are often reproduced in subsequent generations (Avital and Jablonka 2000; Laland 2017). Social learning is also entangled with other selectively relevant traits, illustrated by Laland's (2017, chap. 4) group's experiments with closely related species of stickleback fish. One species, whose bony armor-plating mostly frees them from predation, exhibits minimal social learning because their relative safety reduces the costs of individual exploration. Another species more vulnerable to predators reduces their own exposure by attending to behavior of conspecifics and its outcomes. While much social learning is simply observation of others' behavior and outcomes, animal communication actively produces behavioral patterns sustained by their selective significance (Hauser 1996, Avital and Jablonka 2000). Nonhuman animal

---

15. A variant strategy incorporates *other* biological phyla within animal environments, implicitly treating them as abiotic circumstances rather than organisms co-developing and under intra-active selection pressures with the animals whose morphology, physiology and behavior are being considered.

communication mostly exemplifies "natural" rather than "non-natural" meaning (Grice 1988).[16] These patterns of behavior are significant components of those organisms' developmental and selective environments, contributing to selection pressures to reproduce those patterns in subsequent generations. In a wide range of taxa and behavioral repertoires, organisms' behavioral patterns are both selectively significant and sustained over multiple generations.

A third and perhaps most important reason for overlooking behavioral niche construction is the widespread assumption that behavior is ephemeral and only reliably affects selection pressures if genetically controlled. The extent to which behavioral patterns are sustained over multiple generations is an empirical question, with considerable evidence of long-lived patterns of behavior with no signs of genetic determination (Avital and Jablonka 2000; Jablonka and Lamb 2005, chaps. 5 and 8). Even behaviors with a genetic basis may nevertheless result from behavioral niche construction. Behavior is phenotypically plastic, but behavioral patterns that produce significant advantages also produce selection pressures for their more reliable reproduction and easier acquisition (Jablonka and Lamb 2005, chap. 7; West-Eberhard 2003, 147–55). Genetically regulated behaviors can result from behavioral niche construction if the behavior resulted from genetic assimilation of behavioral innovations or "stretch-and-assimilate" patterns of behavioral and genetic interaction that further develop those patterns (Dor and Jablonka 2000). If language is indeed a product of behavioral niche construction, and human brains and speech and auditory capacities coevolved with those behaviors (Bickerton 2009, 2014; Dor and Jablonka 2000; Rouse 2015, chaps. 3–5), then claims that behavior is too ephemeral to have evolutionary significance confront a salient, powerful counterexample.

Recognizing the prevalence and evolutionary importance of niche construction changes the overall conception of evolution. Earlier conceptions emphasized evolution as organisms' selective adaptation to exogenously driven environmental changes. Organisms vary phenotypically due to randomized changes in genotypes via mutation, sexual recombination, or lateral gene transfer. Subsequent survival and reproduction of variants within a population would then be determined unidirectionally by their comparative reproductive fitness in that population in those environments.[17] Niche construction changes

16. Although Grice's distinction is well established in philosophy of language and mind, I am arguing that "non-natural" meaning is a natural, biological phenomenon. The distinction is important, but its canonical expression is misleading and question-begging.

17. As previously noted, fitness determines propensities to reproduce successfully rather than actual reproductive success. Differences between comparative fitness values and actual

that model from evolutionary adaptation of organismic populations to their environments to intra-active coevolution of organisms and environments. Environments do have significant selective effects that change the composition of populations over multiple generations, but organisms have powerful constructive effects on their environments and those of conspecifics and companion species. Recognizing the resulting nonlinear coevolution of organisms and their environments provides an instructive basis for understanding human social life in biological terms.

Early Hominin Evolution

Against the background of these developments in evolutionary biology, recent scholarship on early hominin evolution importantly contributes to understanding our practice-based way of life.[18] Taken together, this work provides rich, detailed, multifactorial accounts of evolution of the hominin lineage as a temporally extended process shaped by positive feedback rather than a punctualist conception of the emergence of "behaviorally modern" humans.[19] Related work in psychology and animal behavior explores how some relevant considerations are now manifest in comparative ontogenetic studies of human children and great apes.[20] Work on hominin evolution is complementary to this book's project, which aims to understand contemporary human ways of life rather than human origins. It nevertheless matters that the account be intelligible as an outcome of early evolutionary transformations; its charac-

---

reproductive outcomes is determined by "drift" due to random events not directly connected to fitness.

18. Laland (2017), Sterelny (2012a, 2021), (Tomasello 2008, 2014), Tomlinson (2015, 2018), and Richerson and Boyd (2005) extensively discuss this literature.

19. Sterelny (2012a, 2021) is especially clear and informative on both the *multifactorial* character of early hominin evolution and the *gradual* emergence of distinctive forms of hominin life, in contrast to efforts to identify a dominant causal factor or critical period of differentiation. Sterelny (2012a, 75) specifically contrasts this multifactorial approach to Wrangham (2009, 1999) on the control of fire and resulting effects of cooking food or Hrdy (2009) on the social significance of alloparenting. I agree, but think Wrangham and Hrdy rightly point to relevant factors, although not with the dominance they suggest. Bickerton's (2009, 2014) more detailed account of language evolution is comparable on both counts, although Dor and Jablonka (2000), Sterelny (2012b), Laland (2017) and others importantly contribute to understanding that critical phenomenon. Wimsatt (2019) provides a more general taxonomy of factors that played a role.

20. Tomasello (2014, 2019) and Heyes (2012, 2018) offer empirically well-informed surveys of relevant cognitive or social developmental differences in human children and juvenile apes.

ter as naturecultural requires taking human evolution seriously and situating human ways of life within evolutionary processes. Moreover, continuity with work on evolution of characteristic features of hominin development and lifeways renders some features of contemporary practices more salient.

The underlying hypothesis uniting much recent work on early hominin evolution emphasizes a fundamental shift in ways of life due to ecological changes in east Africa that dramatically increased environmental variability, diminished resources available to great apes, and heightened vulnerability to predation.[21] Kim Sterelny incisively summarizes a view widely shared in broad outline: "Hominin social complexity has certainly increased. But there has also been a transformation in the ways hominins interact with, and extract resources from, their environment. . . . I take these shifts in ecological role to be a clear historical signal of the invention and establishment of a new lifeway, built around a new mode of foraging. By two hundred thousand years ago, and most probably much earlier, hominins had evolved into social foragers. Such foragers depend on harvesting high-value but heavily defended resources" (Sterelny 2012a, 10–11). That shift led to a gradual coevolutionary process sustained by exploitation of many high-value energetic and instrumental resources. Coevolution of hominins with socially complex, materially transformed environments was driven by positive feedback among increasingly obligate social interdependence, material and behavioral niche construction, changing developmental trajectories, and some changes in genes or genetic regulation, partly through genetic accommodation of advantageous phenotypic plasticity.[22]

This work often begins with social and neurological parallels among hominins, other great apes, and some corvids and cetaceans as species with capacities for flexible, intelligent behavior, social learning, and "cultural drive" (Wilson 1985, 1991). Its primary focus, however, is distinctive features of early hominin ways of life against that common background. The latter include enhanced capacities for imitation and social learning, conjoined attention (Tomasello 2019; Tomasello et al. 2005), deliberate teaching (Laland 2017; Sterelny 2012a), and consequently more extensive capacities for "high-fidelity copying" of behaviors and skills (Laland 2017, chap. 7). These abilities exemplify behavioral niche construction in which the situated behavior of one

21. Potts (1996, 2013) compellingly summarizes relevant changes in hominin physical environments during an extended period of speciation.

22. Tomasello (2008, 2014), Sterelny (2012a), Tomasello et al. (2012), Laland (2017), Bickerton (2009, 2014), and Tomlinson (2018) synthesize a much more extensive technical literature.

generation of participants changes the developmental environment of subsequent generations to produce descendant patterns of behavior and skill over extended periods. Such abilities both maintain skills and knowledge across generations and revise and improve them. Sufficient continuity in practices allows "ratcheting" these learned capacities to new levels (Tomasello, Kruger, and Ratner 1993; Tomasello 1994) while also stretching the cognitive and bodily-skilled capacities of subsequent generations through development in different behavioral environments. These capacities for reliable transmission of complex skills and information are exemplified by the long-term stability of complex Oldowan and Acheulian stone-knapping technologies. Each technical repertoire produced distinctive, similarly patterned stone tool-faces over periods of hundreds of thousands of years (Stringer and Andrews 2012, 208–11). Recent experiments show the importance of gestural and verbal teaching in passing on these complex skills; mere imitation or emulation was not sufficient (Morgan et al. 2015). The stability and geographic dispersion of these artifacts may conceal considerable behavioral diversity and cultural differentiation not preserved in the fossil record, but the need for teaching to maintain stable outcomes is clear.

Evolution of extensive cooperative interdependence among hominins, distinct from the mostly coordinated social interactions among other apes, has been widely recognized as distinctive to the lineage (Tomasello 2014, 2019; Tomasello et al. 2012; Sterelny 2012a, 2021; Laland 2017; Hrdy 2009). Related phenomena have been discussed in evolutionary theory since the 1960s in terms of the evolution of altruism and detection of "cheaters" and "free-riders."[23] Sterelny (2012a, chap. 5; 2021, chap. 2) and Tomasello et al. (2012) critically assess and challenge this formulation of the issue for early hominin evolution. Tomasello and his colleagues often characterize these forms of cooperation as development of shared and collective intentionality, but this use of terms from Searle (1995, 2010) suggests a misleading interpretation of their experimental and anthropological work. Tomasello (2019) now keeps those terms, but glosses them more adequately as akin to ongoing *practices* of communicative and cooperative interaction. Consider his initial discussion of the evolutionary emergence of "joint intentionality" through "strong and active social selection for cooperatively competent and motivated individuals."

> The radically new psychological process that emerged at this time was what we may call joint intentionality based on joint agency. A joint agent comprises two individuals who have a joint goal, structured by joint attention, each of whom has at the same time her own individual role and perspective. This

23. Hammerstein (2003) thoroughly reviews this discussion.

may be called the dual-level structure: simultaneous sharedness and individuality. The partners in joint agency relate to one another dyadically, second-
personally, in face-to-face interaction; over time they create with one another
shared experiences, the common ground on which their collaborative efforts
may rely. (2019, 15)

What Tomasello is describing is not a shared mental state but a temporally
extended, interactive practice of coordinating activities recursively in mutually encompassing environments.[24]

I draw two primary lessons from these discussions of the evolution of
cooperative interdependence. The first lesson is that abstractly defined problems of how individual agents can trust others and circumvent free-riding
on collective accomplishments were significantly resolved by how attention
to the evolution of hominin development transforms the problem. An *interdependent* way of life, such as the cooperative social foraging now widely ascribed to early hominins and the more intensive and distributed interdependence now characteristic of human lives, transforms the agents themselves.
Those transformations undermine standard descriptions of the "problem."
The question is not how individuals pursuing selfish interests manage to cooperate. People only grow up and become who they are through neotenous
development amid ongoing activities that are constitutively interdependent.
People thus only *have* significant possibilities for action and self-development,
and interests in how those possibilities are realized, through mutual support
and sustenance. Sterelny nicely summarizes the first lesson's import.

> One of the most distinctive features of human social worlds is our dependence
> on intricate networks of cooperation and the division of labor. No living hu-
> mans gather the resources needed for a successful life by their own efforts.
> None do so even with the help only of their immediate family. . . . Cheats exist.
> But the costs imposed by cheating have been kept low enough to make coop-
> erating a successful strategy. (2012a, 101)

This redescription of the problem situates detection and punishment of occasional cheaters and free-riders as only one of many issues to manage in
sustaining cooperative interdependence, rather than a defining problem for
a cooperative way of life. Moreover, it indicates why this issue's prominence
recedes through positive, niche-constructive feedback as cooperative interdependence became established. As people's activities become increasingly

24. Chapter 5 below argues that the "joint goals" of such practices can only be adequately
characterized anaphorically as what is at issue and at stake in how such coupled performances
develop over time.

coupled with and dependent on what others do, their interpersonal and institutional relationships become more extended and interrelated, enhancing
both direct advantages and long-term selective significance of effective cooperation. As a result, the issues raised by cheating and free-riding became more
localized and specific rather than a constitutive problem for cooperation.[25]

The second lesson to draw from the literature on the evolution of human cooperation is that the complexity of achieving *effective* cooperative
interdependence is easily underestimated. Evolved dispositions to cooperate with others are not sufficient to sustain effective collaboration. Even the
now-extensive developmental scaffolding of cooperative abilities, affects, and
dispositions, both through guidance by adult caretakers and enhanced opportunities for constructive interaction with peers, is not yet sufficient. These
are each important aspects of human psychological ontogeny, informatively
analyzed and brought together in recent work by Tomasello (2019). The question of how people with these evolved and developed psychological capacities
and dispositions create and sustain cooperatively independent ways of life
nevertheless only *starts* from work like Tomasello's on human psychological ontogeny. Sustaining patterns of interdependent performance is difficult
even when people are strongly motivated to cooperate. These considerations
will be especially important when we explore the diverse normative concerns
through which cooperation in and among practices is assessed.

The evolution of language from gesture, ostension, and vocal calls and
responses into protolinguistic expressions and eventually language has usually been understood as both initially enabled by and then exemplifying this
broader range of cooperative capacities. Language is then taken to play increasingly prominent coevolutionary roles in enhancing cooperation, social
learning and teaching, cultural diversity, mutual normative accountability,
and more sophisticated material culture. The development of verbal abilities both benefits from attentive emulation and cooperative teaching and significantly enhances both. Moreover, evolution of language clearly depended
on building more complex linguistic capacities through ratcheted evolution
(Bickerton 2014; Dor and Jablonka 2000; Tomasello 2008). The recent literature on early hominin evolution remains undecided in dating the emergence
of early protolanguages and their transition to more complex linguistic forms

---

25. The later emergence of more extensive, impersonal cooperation in spatially and socially
distanced practices introduced new possibilities for cheating and free-riding, but also enable
institutionalized practices for uncovering them and limiting their consequences. Daston (1992)
tellingly shows how normative concerns guiding scientific practices changed to enable reliable
interdependence among spatially and socially distanced participants.

in comparison to more well-established dates for complex artifactual skills. It is nevertheless clear that language and complex artifactual traditions are mutually supportive once both are in place. Sterelny tellingly concludes this body of work: "the distinctive character of human social life depends upon the accumulation, preservation, and intergenerational transmission of cognitive capital" (2012a, 65).

Social Practices as Niche Constructive

All organisms are niche constructive, both in their basic metabolic life processes and in characteristic activities that sustain those processes. Niche construction has occurred both locally and on planetary scales for a long time. Global niche construction is exemplified by gradual oxygenation of earth's atmosphere and transformation of much of earth's land surface from rock to soil by plants, fungi, and other microorganisms. Humans have nevertheless been distinctively and influentially niche constructive in our relatively brief presence on the planet by the standards of geological deep time. Recent designation of the current geological epoch as the Anthropocene requires care in making that claim. Arguably the niche-constructive significance of cyanobacteria in oxygenating the atmosphere exceeds the transformative effects of human-induced perturbations. Chloroplasts and their plant hosts absorbing atmospheric CO_2, and multiple phyla breaking down rocky surfaces into soil are also geophysically important. Apart from the scale of cumulative geophysical and ecological effects, however, humans have been distinctive for the range and diversity of our niche-constructive activities, the extent of their evolutionary transformation of human bodies and life prospects, their demographic effects on our own and other animal and plant populations, and perhaps above all for diversifying human ways of life.

Even in physically perturbing their environments, human ways of life are remarkable for their variety and scale. People and their codomesticated organisms have cleared, plowed, planted, and harvested vast land areas for agriculture and timber removal. Built habitats long since expanded to comprise cities and towns, now usually with infrastructure for water provision, sewage removal, electronic communication, and transport of persons and goods. Extractive mining and combustion of fossil fuels notoriously added CO_2, methane, nitrogen and sulfur oxides, and other gases and particulates to the atmosphere, with rapidly increasing effects on climate and ocean acidity. Other forms of mining and manufacture, chemical synthesis, and radiation bombardment produced myriad novel substances and objects. Much of the earth's surface is crisscrossed by electrical grids, data/communication networks, and transport systems by water, air, road, and rail. Water flows have been shifted by dams, levees, paving, dredging, drainage, and erosion as by-products of other perturbations. These initial perturbations have been matched in scale by disposal of their waste products.

Physical perturbation on these scales was preceded and accompanied by migrations from humans' early east African locales across seven continents and myriad islands with significantly different elevations and climatic conditions. These more or less permanent migrations are accompanied by ongoing patterns of movement among practically and spatially differentiated built habitats: homes, offices, factories, fields, laboratories, schools, play spaces, museums, worship sites, and residual "open spaces" and "wilderness areas." These latter movement patterns are forms of behavioral niche construction, but self-sustaining patterns of human behavior extend further. Building on earlier arguments, I show why the diverse forms of conceptual and symbolic performance pervading human ways of life conjoin material and behavioral niche construction.[26] Languages, images, musical and mimetic performances, ostensive and expressive gestures, models, equipmental complexes, and associated skills are integral components of human developmental environments that only persist through their scaffolded reconstruction across generations.[27] The resulting capacities for conceptual articulation enable further differentiation and reproduction of social roles, relations, capacities, and cultural patterns. These in turn significantly transformed human bodies and their functioning and movement.

Human niche construction also includes many "environmentally mediated genotype associations" (EMGA's; Odling-Smee, Laland and Feldman 2003), which I redescribe as life-cycle entanglements with diverse companion species (Haraway 2008). Much of the planet's biomass consists of codomesticated and cultivated plants and animals integral to human ways of life.[28] Many

26. I have argued (Rouse 2015) for understanding language, equipmental complexes, sciences, and the cooperative behaviors they enable as conjoined forms of material and behavioral niche construction, drawing extensively on Deacon (1997), Bickerton (2009, 2014), and Dor and Jablonka (2000, 2001, 2004). The evolutionary novelty and significance of the practice-based diversification of human developmental environments is a central theme.

27. Tomlinson (2015) argues that musical performance and its uptake coevolved with language and Dor and Jablonka (2000) intriguingly suggest how the coevolution of multiple forms of expressive practice contributed to the shaping of the "semantic envelope" of language.

28. Recent estimates of proportional distribution of the earth's biomass suggest the entire animal kingdom provides only a small proportion of that biomass (~2 of 550 gigatons of carbon), but the ~0.16 gT of carbon taken up by humans and codomesticated livestock now dwarf the ~0.009 gT of carbon of wild mammals and birds (Bar-On, Phillips, and Milo 2018).

other organisms are human commensals whose life cycles for are closely entangled with ours, including many animals and plants classified as wildlife: many rodents, deer, skunks, pigeons, insects and arachnids, weeds and other plants exploiting human perturbations, and myriad microorganisms.[29] Humans parallel other eukaryotes in our extensive microbial symbioses and endosymbiosis, but our other forms of perturbative, migratory, and behavioral niche construction also introduce and sustain changing forms of disease, parasitism, and partially disabling difference within human populations.

The extent of niche-constructive effects of human ways of life and the significance of conjoined material and behavioral niche construction in hominin evolution does not by itself indicate that our practice-differentiated ways of life should now be understood *as* forms of evolutionary and developmental niche construction. Making that case turns on showing how such reconceptions provide new insights. Efforts to retain sharp conceptual boundaries between nature and a human social world required placing people's cooperative and discursive interactions with one another on opposite sides of those boundaries from coevolutionary interdependence with domesticated or commensal populations and with microbial symbionts and pathogens. Emphasizing instead the constitutive interdependence of humans' practice-differentiated way of life with a diverse array of other companion species is an important aspect of a naturecultural conception of practices. That issue cannot be addressed explicitly until the end of the book, however, once this reconception of human ways of life is otherwise mostly in place. The ensuing chapters work out the basic structure of that account by showing how and why to reconstruct a social theory of practices as the naturecultural structuring of organism-environment intra-action in the human lineage.

29. I know of no estimates of what proportion of plants' 80 percent of the earth's biomass (~450 gT) comes from codomesticated plants, but other plant biomass whose presence and distribution was affected by land clearance and other human niche constructions increases that proportion significantly.

3

Postures

The preceding chapters reviewed central motivations for and challenges to practice-based conceptions of social life and introduced important recent developments in evolutionary biology. We can now draw on this background to work out an ecological-developmental, niche-constructive conception of human ways of life as a nexus of interdependent practices. This conception undertakes several important tasks together. It aims to integrate our self-understandings as living organisms, socially situated and accountable agents, and persons who perceive, conceive, and critically assess our situations, capacities, and lives. In giving up aspirations to the autonomous intelligibility of a social world, it seeks to resolve difficulties confronting earlier accounts of social practices while still satisfying their primary motivating concerns. These concerns remain compelling despite the challenges of developing an adequate, practice-based conception of human ways of life. A final consideration is an aspiration to bring clearer focus and new critical resources to pressing contemporary concerns that cut across familiar conceptual divisions of social life from natural processes and environments.

Practice-based conceptions of social life aimed to steer between individualist interpretations of action that appeal to agents' psychological states or genetic-environmental interactions, and social-wholist reifications of cultures, social structures, or traditions posited to explain individual behavior. Practices are composed of situated performances and capacities of human agents, conditioned by synchronic and diachronic relations among those performances and their circumstances. Those interrelations are more basic than explicit norms or rules, not because the latter are unimportant but because they are shaped by practices whose orientation and concerns they partially express. Those individual performances respond to encompassing patterns of practice and are oriented toward continuation of those practices through how those performances depend on, align with, act on, and are affected by what others do or could do in partially shared circumstances.

Understanding practices as naturecultural initially transforms this basic conception in two ways. First, it emphasizes that the agents who participate in practices are living bodies whose involvement in practices exemplifies the codevelopmental interdependence of organisms and their environments. Second, it recognizes practices as the characteristic shape of human environments, as outcomes of a longer history of niche-constructive coevolution of human bodies with practice-differentiated developmental and selective environments. Practices are relations of cooperative interdependence among people's temporally extended, spatially dispersed, and materially situated activities. An ongoing practice both enables distinctive human capacities, concerns, and achievements and imposes constraints. Not all the agents involved in practices are human—codomesticated, commensal, and symbiotic organisms participate in the practices making up human environments, along with predators and prey—but we will only consider the role of nonhuman organisms more explicitly at the end of the book. We first must assimilate human social relations within this larger biological context. Human ways of life are horizontally differentiated by the many practices whose mutual support enables their participants to live very different lives. Those lives are also vertically differentiated by participation in multiple practices that require them to coordinate those involvements within a single life. This chapter examines people's bodily engagement with practice-differentiated environments. Chapter 4 then works out an initial conception of practices as the characteristic shape of those human developmental and selective environments.

Bodily Postures

Many practice-based social theories rightly emphasize bodily skills, affects, vulnerabilities, and constraints but do not look closely enough at how those bodily capacities are developed, sustained, and exercised. Other practice-based approaches give central place to language and discursive interaction but do not look closely enough at how people encounter and take up those discursive articulations perceptually and practically. The commitment to understand practices as phenomena that are autonomously social has been especially debilitating. Looking more closely at practices as patterns of bodily performance not only requires attending to human biology. It requires overcoming a truncated conception of biology that marginalizes development and ecology. In this respect, the covert alignment of theories of social life with internalist conceptions of bodily physiology and narrowly genocentric conceptions of development and evolution blocks adequate understanding. A more adequately *naturecultural* conception begins with bodily processes and performances as both responsive to and partially constitutive of their developmental environments. Understanding what practices are and how they enable the intelligibility and accountability of their performances requires understanding how the performances themselves are embodied and situated, and how bodily capacities develop and evolve.

Practices are composed of bodily capacities and performances situated and exercised in publicly accessible settings integral to those practices. Practice-based accounts reject behaviorist conceptions of what people do, which restrict behavior to relatively thinly described bodily movements. They also oppose those cognitivist responses to behaviorism for which the informational or representational content of internal psychological states animates behavior and enriches its characterization.[1] Whether practice-based approaches should join with radical embodied cognitive science (Chemero 2009) in eschewing explanatory appeal to internal representational content altogether is an open empirical question. At a minimum, however, practice-based approaches insist that bodily engagement with one's surroundings takes up a thickly describable, inhabited situation whose meaningfulness is not reducible to how agents represent it cognitively or discursively.

Understanding human bodily orientation toward meaningful situations requires better understanding of the bodies of living organisms. Conceptions of bodies as discrete entities, bounded and contained by skin or other integuments, overlook how maintaining the boundary between organisms and their environs requires constant effort and ongoing intra-action. Living bodies only maintain their integrity by taking in resources from their surroundings and excreting waste products. Bodies, especially animal bodies, may seem to be self-moving entities, but that sense of autonomy is illusory. Not only does bodily movement require import of energetic and other resources and export of their remnants. Bodies only move within and in response to their environs.

---

1. As Godfrey-Smith (2002) notes, cognitivist accounts of the representational content of psychological states include both realist and interpretivist approaches—the former take relevant informational content as embedded in neurological "wiring-and-connection" (Sterelny 2003, 4), and the latter as explanatory attributions of more global patterns in organisms' responsiveness to circumstances. Practice-based approaches are typically more sympathetic to interpretivist conceptions, but also take meaningful bodily responsiveness to publicly accessible settings to extend beyond propositionally contentful states. Whether they attribute "nonconceptual content" is more contested (Schear [2013] usefully discusses the latter issue, including longer exposition of my reasons to reject nonconceptual content, briefly considered in chapter 6 below).

Whether an animal body moves on and against the earth's surface or through water, air, or the bodies of other organisms, bodily movement is responsive to and dependent on those "grounds." Bodies are *processes* that are constitutively interdependent with their developmental environments. Once that process ceases, a body is no more.[2] A corpse is not the continuing existence of a body after its death, but the process and remnants of its physical dissolution.

To understand the bodily performances whose interactions constitute practices, we need to begin with the bodily *posture*, "set," or stance from which people encounter and take up their surroundings as a situation. I use the adverbial form "bodily" postures rather than using the noun "body" adjectivally, because postures in this sense are not relations of one object, a body, to its environment or context. Postures are instead aspects of *how* organisms' bodies and their environments actively intertwine.[3] We should not think of people having a body which then takes up a posture. Bodies *are* postured orientations within a situation which both depend on and respond to how they are situated. "Posture" as a verb has a misleading colloquial sense in English as self-promoting bodily pretense, but my use of the noun should still be understood as more like a verb. Posture is how a bodily agent actively sets and orients itself practically and perceptually, and is thus something an organism continually *does* rather than a state or condition *of* its body at a time. The term foregrounds how an organism moves, balances, orients, and expresses itself and thereby also prepares itself for further movement and perceptual exploration.

Postures in this sense are not static: they include how people move within or through a situation as also ways of setting themselves to move, including perceptual and expressive responsiveness as well as locomotion. Postures are not just synchronic bodily orientations; synchronic positionings are abstracted from bodily repertoires for movement and exploration. As Quill Kukla put a similar point, postures in my sense encompass "embodied strategies for coping and coordinating . . . made up of a wide and counterfactually flexible repertoire of bodily positions" (Kukla 2018, 8–9). Bodily postures are also affective: moods and emotions are integral to how people encounter and respond to their circumstances.[4] Postures also incorporate

---

2. Dupré (2012, 2021) and Dupré and Nicolson (2018) work out processual conceptions of organisms.

3. I talk about postures rather than stances to avoid confusion with Dennett's (1987) well-known use of "stance." Kukla (2018) develops a related conception as constructive elaboration of Dennett (1987) on interpretive stances, although for somewhat different purposes.

4. I mostly endorse Prinz (2004) on emotions as organisms' perceptual openness and responsiveness to patterned changes in their bodies, although the details do not figure prominently here.

bodily nourishment and health or illness as well; these aspects of bodily life condition how people set themselves within and toward a situation. Human bodies are born already postured. Although infants have comparatively limited bodily capacities and degrees of freedom, their bodies have a bilaterally structured and forward-directed orientation with "handed" and dorsal/ventral asymmetries. They are also perceptually responsive and needy in specific ways. Babies are born with capacities to move limbs, head, eyes, mouth, and torso, which are supported by internal processes of digestion, respiration, circulation, growth, and neural responsiveness. Those capacities then develop over time through the effects of bodily growth and aging conjoined with the reinforcement, refinement, and occasional atrophy of those capacities for movement as they are exercised in particular environments.[5]

Postures are not simply properties or arrangements of a bounded entity, because they are always world-involving. This recognition reiterates that organisms are processes that constitute, engage, and depend on the environments that sustain their living. Bodily capacities for perception and movement evolved under selection pressures for effective coping with relevant surroundings, and those capacities then develop in more refined ways through ongoing responses to circumstances.[6] Organisms are embodied strategies for sustaining and developing their own capacities by coping with environmental opportunities for and threats to bodily functioning. In responding to and coping with their environments, these embodied coping strategies are world-involving in three general ways. Considering human bodies in particular, people can only be effectively postured if balanced on some "ground" (typically the earth's surface, or something grounded on that surface, such as a floor or a chair), in and against some forces, for which the earth's gravitational attraction is the most familiar and paradigmatic example. The dependence of postures on their grounding is evident from the different postures and movements available on different grounds: in water; in free fall as a parachutist, diver, or basketball player; on snow or ice with or without skis or skates; in a wheelchair; on a bicycle; or ascending or descending stairs. World-involving grounding shows itself differently when momentarily lost, as in mistakenly stepping toward a nonexistent or broken stair, tripping on an obstacle, or falling off or tipping over in a chair. Such moments incur disconnected thrashing

---

5. Ultimately, understanding how peripheral and central neural connections are formed, reinforced, and sometimes atrophy will need to be incorporated in accounts of bodily motility and perception, but this topic cannot be addressed here.

6. Bodies grow and age as they also develop and refine skills for perceptual exploration and movement, but this aspect of bodily postures cannot be addressed here.

about and momentary loss of perceptual registration, in search of restored practical grip on one's own body as balanced and effectively oriented within and on a local setting.

Bodily postures are differently world-involving when they incorporate items of equipment within a more or less unified posture. When writing with a pen, walking with a cane, typing on a keyboard, driving a car, or throwing a ball, we do not integrate these tools into the body itself, unlike an artificial limb or organ, but more or less effectively integrate them into an active posture. Moreover, once integration has been effectively achieved, it typically remains resilient even when those tools are set aside. People need not relearn to write, drive, or throw when the objects integral to these postures are no longer currently deployed. World-incorporating postures also include the important, more complicated special cases of coordination with others exemplified by dancing, deploying multiuser tools, conversing, or holding onto and maneuvering a large object together through a tight space. Here one's bodily posture is not only responsive to the heft, orientation, and causal capacities of another entity, but also to how others' postured orientation changes in response to one's own efforts to accommodate or resist their movements.

Recognition that equipment can be temporarily integrated within an effective, dynamic posture is highlighted by contrast to a third way bodily postures are world-involving. Postures are actively directed toward and oriented by some aspects or configurations of their circumstances, enabling effective bodily exploration of and active response to some things and not others. Postures do not merely render some patterns of movement easy and fluid and others more difficult. Different foot placement and weight balance are indeed only conducive to movements in some directions, and similarly for ways to direct the hands and arms or tilt the head. Orientation toward and responsiveness to circumstances are nevertheless integral to a posture rather than a consequence. In reaching to grasp a delicate teacup, we set ourselves to move and respond differently than in reaching to lift a heavy weight, pull oneself up from a handhold, or detach a firmly attached object. Differences are even clearer between those world-involving postures and rehearsals or imitations of them in the absence of their objects, as in pretending to dribble a basketball, shadowboxing, or visibly miming other activities. Imitative movements proceed from different postures configured to replace contributions from objects' resistance or accommodation. Bodily directedness *toward* a situation is integral to people's postures, but that differs from how things can be incorporated within or ground those comportments.

Understanding bodily postures as both world-encompassing and world-directed also highlights an indissoluble unity to perceiving and acting. The dominant approaches in philosophy and psychology mostly treat perception and action as discrete processes mediated cognitively. They treat perception as receptive uptake of sensory input, even if the resulting appearances incorporate subsequent cognitive processing. Further cognition then supposedly leads to action by conjoining cognitive instructions with bodily habits and skills to act spontaneously within the circumstances manifested perceptually.[7] Important work in embodied or enactive cognitive science suggests that this conceptual separation misconstrues and distorts both perception and action. Perception actively explores the world, with a visual field emerging from an appropriate set and controlled movement of the eyes, head, and body (Noë 2004; O'Regan 2011). Auditory discrimination involves setting oneself to listen and hear,[8] and tactile, olfactory, and gustatory encounters respond to relevant exploratory interactions.[9] Action in turn is perceptually responsive to the surroundings that both enable and are engaged by that activity. Effective action requires maintaining appropriately postured balance through proprioceptive and perceptual orientation, while acting on or working with objects requires ongoing adjustment to their perceptual presence. Grasping and manipulating an object or moving toward or away from it is a simultaneously perceptual and active response. Strictly, therefore, we should not speak of perception or action separately but only of practical-perceptual interaction with circumstances.[10]

The grounding and orientation of bodily postures configure their surroundings as a meaningful situation. Postures are shaped by conjoined directedness toward and responsiveness to what is manifest and salient within that situation. Some aspects of circumstances become perceptually and practically salient through postures open to some receptive encounters and oriented toward only some responses. Other possible encounters are perceptually occluded or backgrounded. Capacities to respond to some aspects of a

7. Talk of receptivity and spontaneity traces to Kant (1998), with renewed life in philosophy of language and mind through McDowell's (1994) influential interpretation.

8. Auditory perception often intrudes unwillingly, but Todes (2001) notes how being startled by unanticipated sounds requires resetting oneself to recognize them.

9. Merleau-Ponty (2012) and Gibson (1979) are classic sources for this account of perception as bodily activity, with Todes (2001), Noë (2004, 2009), Chemero (2009), O'Regan (2011), and Gallagher (2017) as important recent elaborations.

10. As chapter 2 noted, organisms' bodily responsiveness to their environment is not interaction between discretely bounded entities, but intra-action (Barad 2007) among entities defined by their situated interdependence. Since all bodily interaction with environments is intra-active, I retain the familiar term while rejecting the implication that interactants are discrete and independent.

situation are thereby constrained or made awkward without first resetting one's posture. Such occlusions and practical blockages are never total, however. The postured configuration of a field of perceptual salience is dynamic, soliciting some further explorations, open to disruption and involuntary shifts provoked by what intrudes when unprepared, and providing a basis for self-directed shifts of orientation, awareness, and activity.[11] Posture and situation are mutually coconstitutive as dynamic patterns of open responsiveness to what circumstances both afford for and demand or solicit from people's developed bodily capacities and affects. Moreover, those capacities and affects are outcomes of ongoing, lifelong reinforcement or diminution through ecologically mediated biological development.

Living a human life involves developed practical-perceptual involvement in and responsiveness to surroundings. We can distinguish but cannot separate active and receptive components of people's performances, whether in using equipment, looking for things or looking out for them, moving through partially occupied spaces, engaging in conversation, or engaging in a relaxed way with familiar settings. Emphasizing the unified practical-perceptual character of worldly involvement does not deny roles for cognitive processing or explicit interpretation but reconstrues them as integral to bodily comportment rather than intermediary between perception and action. Self-consciously reflective thought and conceptual explication also take up bodily postures that selectively engage a meaningfully configured environment, in part by occluding responsiveness to immediate circumstances. Contemplation is thereby sensitive to environmental disruption, whether by inadequately supportive grounds (such as an uncomfortable chair) or by intrusive interventions from the occluded background. Natural languages and other conceptual repertoires such as images, gestures, models, music, signs, or equipmental complexes are environmental configurations that people take up with and within bodily postures that enable their effective perceptual recognition and expressive deployment.

All living bodies are postured to let them encounter some aspects of their surroundings, partially occlude others, and enable or constrain particular responses. In this respect, human bodies and those of other animals do not differ in kind despite the many differences in detail. People's bodily postures do differ significantly from those of other organisms, however, in their grounding in and directedness toward environments that are products of their own

---

11. I have argued (Rouse 2021) that Kukla's (2018) reconception of Dennettian stances requires that they incorporate unthematic openness to and orientation toward possible stance-shifts occasioned by encounters.

species' massive, iterated niche construction. Human bodily postures and capacities coevolved with and develop in response to those distinctive environmental circumstances. Humans now live in and move among multiple purpo-
sively reconstructed spaces built and equipped to facilitate specific forms of
activity and interaction, and we continue to reconstruct those environments.
Other people within those settings also normally take up bodily postures that
are responsive to and partially constitutive of the practical orientation of the
space and the activities and orientation of others within them.

People do not merely inhabit those settings alongside and responsive
to others; we take them up *with* others in coordinated or antagonistic responses to one another as oriented toward the encompassing situation. Kukla
(2018) points out that people normally adopt different postures toward other
persons than toward other aspects of their environments—for example, by
walking around them or asking them to move rather than picking them up
and moving them out of the way. We make eye contact and listen to others
rather than staring at them, touching other parts of their bodies, or focusing
on inanimate things in their vicinity. Such characteristic postured orientations toward and expectations of one another do have familiar exceptions.
Those exceptional postures often stand out as problematic, however, and
typically evoke very different postured responses. People also engage with
things around them as significant for one another, including things others
own, use, care about, or have talked about. Kukla characterizes such ongoing,
postured interactions with one's situation as "micronegotiations": "People accommodate spaces that fit them imperfectly to their needs, partly by tinker-
ing with these spaces via small micronegotiations that emerge and coalesce
organically. . . . Plenty of micronegotiations have no representational content:
sitting on a fence; queuing up for a bus; edging forward to make a left turn at
an intersection; avoiding contact with a homeless man on a corner; tossing
money in a busker's hat" (Kukla 2021a, 8).

What is salient within and partially constitutive of environments in and
toward which postures are grounded and directed is not just others' bodies
or even their postures. People respond to others' postured responsiveness *to*
meaningful surroundings that encompass their own meaningfully configur-
ing postures. These ongoing, situated micronegotiations reconfigure spaces
and how people inhabit them. As Kukla notes in an example:

> The [boxing] gym is also divided into territories; the side of the ring may be
> the space where the insiders who have fought out of the gym for a long time
> hang out, whereas the benches by the bags may be where the casual gym us-
> ers who just come to get a cardio workout hang out. These territories may be

marked and maintained through micronegotiations: where people leave their
gym bag and equipment and how much space they take up doing it; where
they sit and their posture as they do; who they greet; how long they hang out
after they work out. . . . [Territories] depend for their existence on the entire
ecology, which is made up of the material and institutional space of the gym,
the dwellers in the gym, and the stances and place ballets that articulate the
space. (Kukla 2021a, 27)

Postures and situations dynamically configure one another by how people
set themselves to respond to a situation as partially reconfigured by others'
postured involvements.

People's mutually interactive responsiveness to mutually encompassing
situations both requires and enables distinctive postures toward one another. The vocative and recognitive capacities and performances that Kukla
and Lance (2009) highlight within discursive practice are especially consequential forms of bodily postured, micronegotiated niche construction. In
hailing others; ostensively redirecting one another's postures, saliences, and
affordances (Kukla 2017); and recognizing and responding to hails and other
calls, people engage in complex patterns of iterated, interactive and collective niche construction. Human environments are replete with articulated
linguistic utterances and other signifying performances, whether those are
spoken words, images, models, written signs or texts, musical performances,
or dramatic gestures and impersonations. These performances are salient features of people's current circumstances, but they also enable postured engagement with situations that are spatially, temporally or modally displaced from
current environs.[12] People direct themselves and one another from within
current situations toward such displaced settings, whether by talking about
them, enacting them performatively, working to bring them about, depict-
ing, diagramming, or modeling them. This dual character to such expressive
performances, as both practically-perceptually salient within immediate circumstances and also oriented and directed beyond them, will be discussed
more extensively later.

Adopting and adjusting a posture reconfigures the affordances and occlusions of those constructed environments.[13] Moreover, postures and the
capacities they enable are developmentally constructed through interaction

12. *Modal* displacement from spatiotemporal surroundings and present action-situations is
directedness toward what is merely possible, necessary, counterfactual, or even impossible, or
obligatory, permissible, forbidden, encouraged, and so forth.

13. Chemero (2009, chap. 7) discusses how to construe affordances as meaningful, organism-
involving configurations of a practical-perceptual situation.

with those environments, as people move among them and direct or reset themselves toward and in response to them. Other organisms are also niche constructive, including forms of behavioral niche construction,[14] and they often respond flexibly to conflicting solicitations for what to do. Some animals adjust their postures and activities to seasonal and migratory environmental variation. In many cases they are sensitively attuned to the affective and practical-perceptual postures of conspecifics or companion species. Human ways of life are nevertheless distinctive because our developmental environments are shaped by cooperative interdependence. Those differences start with the horizontal and vertical differentiation of practices and of people's involvement in them. People's postures and postured abilities differ from one another because they encounter and take up different practices. They nevertheless also encounter and learn to cope with the postures and performances of others in contexts that they themselves do not take up or understand in the same ways, because their own situations and performances depend on what is done elsewhere. The need to coordinate closely with others in different settings shaped by different concerns also requires adjusting or shifting what one does in one setting to accommodate the demands and possibilities of others. We need to recognize both the underlying continuity of humans with other organisms in bodily attunements to life-relevant circumstances, and the distinctive, cumulative niche-constructive transformations of human developmental environments with correlated changes to human bodies and ways of life.

Human Environments and Human Development

Since the postures people take up are directed toward, responsive to, and partly shaped by environmental circumstances, we need to understand characteristic features of those circumstances. A key starting point is that humans, like many other primates, are social animals. People live in close proximity, and are perceptually and practically salient features of one another's environments. People not only differentially attend to and notice others around them but their own postures and performances are responsive to the presence, postures, and behavior of others. Humans are neotenous, in ways that further highlight the salience and significance of other persons in their developmental environ-

---

14. Dominance hierarchies, parental supervision, and animal communication systems exemplify behavioral niche construction in other lineages.

ments.[15] Developmental psychologists have long noted that newborn infants are preferentially attentive to human faces and expressions, and readily recognize and respond to different faces. That orientation undoubtedly has a partly genetic basis, but any such predisposition is an obvious candidate for origins in genetic assimilation of a selectively advantageous behavioral pattern. Tactile and vocal/auditory interactions with others are also integral and salient features of human environments and human development. Infants deprived of extensive tactile contact with caregivers develop differently, and these patterns of development through responsive interaction with others continue throughout human life. Tactile responsiveness to one another's touch and the differential maintenance of intimacy and separation, desire and repugnance, help shape people's bodily postures and self-understanding.

The perceptual salience of others within people's everyday environments is enhanced by mutual responsiveness. Most entities are unaffected by people's gaze, even most of the animals otherwise sensitive to people's presence, and they are also indifferent to whether people listen to, and in some cases touch or move them physically. By contrast, people normally recognize one another's gaze and meet it or look away. Perceptual explorations not only actively set and move the body; they are also perceptually accessible to and salient for others, and that salience is practical as well as perceptual. There is no way to avoid responding to others' perceptual attention, since doing nothing in response to another's gaze or touch—ignoring it—will often be noticed and understood as a significant response. Approaching others, moving away, or remaining evidently unaware of or indifferent to their presence or behavior similarly stands out as perceptually salient and practically significant. Human bodily postures and the expressive and receptive capacities they enable develop in significant part through mutual responsiveness. That mutual responsiveness is not merely a predictive or anticipatory orientation toward how others will likely behave. People expect others to respond to them in a dual sense that conjoins prediction and assessment, such that they often respond to misplaced expectations not only as failed predictions and mistaken responses but also as misbehavior by others.[16]

Attending to and learning from the behavior and attentional direction of other organisms and their outcomes is widespread across animal taxa (Laland

---

15. Tomasello (2019) informatively differentiates the social, cooperative character of early human development from the developmental patterns in other great apes.

16. Lance (2015), Millikan (1995), and chapter 5 below discuss this ambiguity or conjunction of predictive and normative expectations more extensively.

2017, part 1), but active teaching of others is mostly confined to the hominin lineage (Sterelny 2012a, 2021; Laland 2017, part 2; Tomasello 2019). I argue in later chapters that the many ways people teach one another are not just distinctive forms of human behavior, but belong to more complex patterns of normative assessment. Both active teaching of others and these more complex forms of normative response contribute to the "ratcheting" (Tomasello, Kruger, and Ratner 1993) of practical and expressive behaviors and innovations that allows their retention and refinement over generations. Iterated cycles of niche construction have thereby rapidly and repeatedly transformed human ways of life and dramatically enhanced their complexity, diversity, and environmental range.

A second, closely related feature of human biological environments is their combination of practical differentiation, spatial dispersion, and interconnection. Humans typically migrate locally among practically differentiated settings. Those settings are mostly distinguished by the activities and interactions that characteristically occur there, the equipment already available or brought for the occasion, their physical configuration and relations to other spaces, and variations in who else is present, in which spatial patterns and temporal sequences and durations. Different people not only inhabit different environments; the same people move among many different settings in varied and overlapping combinations that make up vertically differentiated lives. People nowadays commonly inhabit spaces designed, built, and used for specific purposes, and usually practically subdivided in turn. We regularly move among those places and move equipment and materials to them. Those movement patterns are facilitated in turn by spaces and equipment devoted to those transitions: roads, rails, sidewalks, doorways, crosswalks, and trails, but also boats, trains, elevators, wheelchairs, and bicycles. The bodily postures and skills that enable people's diverse activities and the requisite capacities for movement, recognition, and response are developed, deployed, and refined in sustained interaction with the environments that they make both accessible and interdependent.

A third characteristic feature of human environments is their revised and often drastically loosened geographic and ecological constraints. Humans cannot live just anywhere, even on earth. People occasionally move through many spaces which they do not and cannot inhabit: under water, in the air, on steep slopes, inside volcanoes, in forest canopies, or in microhabitats too small or poorly configured for human bodies, along with others that require more substantial reconstruction or augmentation to become sustained human habitats. Humans nevertheless regularly do live amid an extraordinary range of terrains, climatic conditions, and seasonal variations.[17] This environmental diversity might suggest that human ways of life and bodily skills and capacities are relatively independent of immediate circumstances and overall habitat, but that inference does not follow. To understand the significance of human geographic and ecological diversity, we need to consider how those diverse locations and habitats have been configured and materially reconstructed for and by human habitation. Those reconstructions also include the developmental patterns and bodily postures through which human inhabitants adjust their own postures, skills, performances, and mutual responsiveness to accommodate varied habitats.

The geographic and ecological flexibility and adaptiveness of human ways of life should not be mistaken for a loosening of the biological constraints imposed by environmental intra-action to sustain bodily life. The basic biological needs of human bodies are instead pervasive foci of niche construction across these diverse human developmental environments. People can attend to the practically diverse interactions in multiple environmental settings because those settings have been reconstructed to provide for those inexorable biological needs. How people understand and meet those needs has been adjusted in turn to accommodate how they are provided for. The locations, organization, and functional orientation of people's diverse environmental settings are shaped by the niche-reconstructive maintenance and transformation of their ongoing physiological functionality. Among the many familiar provisions for extended human habitation are supplying, distributing, preparing, and making available food and potable water; enabling and sequestering bodily excretion and removing, concealing, and disposing of bodily wastes and equipmental trash; sheltering bodies and functional activities from meteorological, climatic, or tectonic disruption; maintenance of perceptually felicitous conditions of lighting, sound modulation, ambient temperature, and olfactory sensitivity; protection from bodily vulnerability to violence, predation, disease, or environmental "disasters"; facilitation and regulation of sexual interactions and relationships; partial separation and coordination of child care and development with other interactions among adults; and the timing, location, and facilitation of sleep and other preventive or ameliorative care for bodily health and aging. The smooth and mostly unobtrusive provision of such environmental facilitation of human physiology

17. This range looks more constrained from a geological perspective. Human ways of life arose within an unstable global and local climatic regime (Potts 1996), but subsequent practice-differentiated evolution happened in a more stable and congenial climatic period. Climate change may soon empirically test the robustness of human environmental flexibility.

and development through pervasive infrastructure and supportive work has been mostly overlooked in conceptions of social life as distinct from human biology. Social theorists do often rightly attend to cultural *differences* in how physiological and developmental needs are provided for but less often to the latter's centrality to and pervasiveness amid this cultural diversity. Serious, sustained disruption of that provisioning, whether through warfare, environmental disaster, epidemic disease, equipment failure, or bodily injury provides potent reminders of why that distinction is illusory.[18]

A fourth pervasive feature of niche-reconstructed human environments is that they are replete with conceptually articulated expressions, from articulated speech or writing to images, musical sounds, mimetic, ostensive or demonstrative performances, and physical markers or other signs.[19] While these performances and artifacts are often themselves situated within designated spaces such as theaters or lecture halls, they also belong together in patterns that cross the boundaries among many differentiated spaces and associated patterns of behavior. More will need to be said about the significance of describing these performances and circumstances as "conceptually articulated," but it is important to note from the outset their pervasive material presence, perceptual salience, and interpretive and expressive uptake. People often encounter and are engaged by such performances, which are integral to the practically configured environments they inhabit. These expressive performances also typically call for and depend on distinctive bodily postures to recognize and respond to them appropriately. As Kukla points out, "Linguistic transactions are just one integrated layer within our micronegotiations and place ballets. . . . Language is an activity, and our conversations are micronegotiations. It is one form that our niche-building and spatial negotiation takes. . . . Ritualized forms of conversation, including my examples just above of shouting in a New York subway car and so forth, can function as place ballets, which involve whole bodies and not just abstract exchanges of words" (2021a, 43). These characteristic features of human biological environments—

---

18. Crosby (2016) compellingly explicates how bodily injury makes visible the pervasive but often invisible interdependence of social interaction, physical structuring of everyday surroundings, and biological neediness and vulnerability. The COVID-19 pandemic also highlights how people's normal social interactions are intimately interdependent with biological vulnerability to the life cycles of other organismic ways of life. COVID-19 is discussed further in chapter 9.

19. Other organisms offer partial analogues to these aspects of human developmental environments, including birdsongs and other vocal performances, bodily expression in mating patterns or territorial defense, and olfactory marking of territories or paths. These aspects of human developmental environments nevertheless differ not only in their scale and pervasiveness but also in *conceptual* articulation. Discussion of this difference is deferred to chapters 6–9.

the material reconstruction of practically diverse settings and ways to move among them, the socially and culturally differentiated patterns of mutual responsiveness within those settings, the niche-constructive provision for human biological needs, and the discursive articulation of these differentiations, performances, and their significance—are closely intertwined, even though I had to characterize them sequentially. Moreover, these recognitive and responsive capacities must be developmentally reconstructed in each generation. The forms of teaching, learning or imitating, and discursive coordination and sense making that enable that reconstruction are themselves distributed throughout the environments people inhabit.

Ecological-developmental biology and niche construction theory emphasize the developmental reconstruction of bodies and their dynamic postures. This ongoing reconstruction takes place in responsive interaction with organisms' environments, which are themselves thereby maintained and transformed in turn. The importance of this ongoing reconstruction of bodies and environments has been obscured by some influential but problematic conceptual separations that help sustain conceptualizations of social life as distinct from biology. First and foremost has been separation of *evolved capacities* of organisms from the diverse *developmental realizations* of those capacities in specific circumstances.[20] This distinction takes two characteristic forms in accounts of human capacities and ways of life. The first is the distinction between human genotypes and their phenotypic outcomes, with the presumption that evolution only concerns the genotype as a species-characteristic that emerged early in its evolution. The second is the related distinction between the anatomically modern human bodies that people share with all more recent ancestors and the historically and culturally variant skills and performances that this common anatomy made possible. The previous chapter showed why it was a mistake to identify a genotype from commonalities in gene sequences; these sequences only acquire genetic significance through ecologically mediated developmental processes.

Discussion of body postures and skills lets us similarly reject any sharp distinction between evolved anatomy and acculturated skills and behavior. The point is not to deny anatomical continuities across the hominin lineage, any more than the rejection of genotypes denies continuities and changes in DNA sequences. We can identify anatomical components and relations by examination of the fossil record but mostly cannot thereby distinguish between anatomically specified capacities, such as for upright posture, bipedal

20. Tomasello (2019) shows why we should understand evolution of human cognitive capacities as evolved developmental trajectories, and how to do so.

locomotion, or articulated speech, and their ecological-developmental realization in specific environments. What capacities human anatomy provides
and what affordances it enables are indeterminate without specification of
the developmental environments in which those capacities are realized and
deployed. Some specification of those environments and developmental patterns can be inferred from marks on anatomical remains, but that is indirect
and partial. Tim Ingold made this point with great clarity, pointing out that
even human traits and capacities that develop with near-universality, such
as upright bipedal locomotion and articulated speech, only do so through
reconstruction of the right developmental environments:

> Walking is a skill that emerges for every individual in the course of a pro-
> cess of development, through the active involvement of an agent—the child—
> within an environment that includes skilled caregivers, along with a variety
> of supporting objects and a certain terrain. . . . The vast majority of human
> infants do learn to walk, moreover they do so within a fairly narrowly defined
> period . . . provided, however that certain conditions are present in its envi-
> ronment. This last proviso is absolutely critical. Infants deprived of contact
> with older caregivers will not learn to walk—indeed, they would not even sur-
> vive, which is why all surviving children *do* walk, unless crippled by accident
> or disease. (1995, 191)

Moreover, how they walk—their gait, stride, balance, endurance, or facility
with different terrains—varies with their developmental environments.

Environmentally mediated development is even more clearly manifest
in speech. Normal human infants develop amid verbally articulated speech
by adult caregivers responsive to the infants' own emergent vocalizations.
Nearly all do learn to speak, but their speech and comprehension develop
differently in different linguistic environments. They diverge not only in the
languages they speak and understand, but also in fluency, articulateness, vocabularies, accents, and verbal spontaneity. To be sure, human children removed from their linguistic communities early in life and raised elsewhere
learn speech patterns shaped by subsequent interaction with the new environment. Older children and adults also typically manage such transitions,
with varying degrees of fluency and residual influence by and competence
with previous developmental patterns. The wrong lesson can easily be drawn
from these phenomena, however. Individual people do not have a general
or universal capacity to acquire language as a property or capacity intrinsic
to human bodies or to anatomically or genetically "modern" human bodies.
Language abilities develop not as a general process of biological maturation,
but as genetically accommodated developmental patterns of interaction with

linguistic environments. Immersion within an extant linguistic practice is an indispensable condition for the development of linguistic capacities.

Several liminal cases of development of linguistic abilities illustrate this claim and its significance. The efforts to initiate nonhuman primates into suitably accessible linguistic practices show the indispensable significance of development in an already-extant and practically-perceptually accessible linguistic environment (Lloyd 2004). This point is reinforced by the emergence of pidgin languages among forcibly assembled human communities whose members shared no common language, and then of creole descendants in subsequent generations who developed in the linguistic environment of a pidgin (Bickerton 2008). These include the oft-discussed case of the sign languages developed over several generations by deaf children in Nicaraguan orphanages (Dor 2016, 173–82).[21] Once again, the point is not to deny roles for genes or their evolutionary history in the emergence and reproduction of linguistic practices and capacities but to insist that these only become genes "for" vocal and linguistic abilities through developmental expression in specific social-material, discursive environments. Similar points can be made about genes whose expression is relevant to development of upright posture and bipedal locomotion, and other widely distributed capacities and orientations.

This ecological-developmental conception recognizes that human lives are processes of environmentally mediated development from birth to death. Human evolution consists in continuity and change in those developmental patterns over generations through the reliable availability or recurring reconstruction of relevant developmental resources.[22] Children acquire stochastically recombined gene sequences from their parents, along with the cellular structures and processes that regulate and express those sequences and their energetics. They also encounter developmental environments and help reproduce and transform those environments in turn. Our initial discussions of bodily postures have focused on niche-constructive and responsive interaction with diverse, multifaceted environments, and how the postures and responses of other human bodies are integral to those environments. These

---

21. I have discussed these cases at greater length (Rouse 2015, chap. 3), especially Kanzi, the bonobo whose capacities developed in the research laboratories of Sue Savage-Rumbaugh (Savage-Rumbaugh, Shanker, and Taylor 1998; Savage-Rumbaugh and Lewin 1994).

22. Tim Ingold (2022) argues that even proponents of an extended synthesis are still overly reliant on a notion of inheritance that is problematic even when extended to include the inheritance of life cycle components other than genes. I have not specifically thematized a rejection of that way of characterizing relations among generations as Ingold does, but my exposition in this chapter converges with Ingold in eschewing any constructive use of the term.

concerns transform the significance of previous conceptions of social practices. The recognition that patterns of situated social practice are integral to
human biological development and evolution lets us refine and further develop a conception of what practices are, how they change over time, and
why they matter for understanding human ways of life. That conception of
practices as forms of biological niche construction in turn allows for a better understanding of human evolution and development, and the distinctive
capacities and evolutionary challenges that thereby arose.

Practices as Human Developmental and Selective Environments

The environments to which human bodily postures, skills, and performances
are responsive extend well beyond people's current material and social circumstances. In one respect, that point is familiar. Language provides a capacity for symbolic displacement, enabling people to talk about and act with
regard to situations distant in space, time, or modality. We need to bracket
these discursive capacities for the moment, however, both to recognize
other ways in which human environments are displaced beyond immediate
physical surroundings and to provide more adequate understanding of those
discursive capacities themselves. People's circumstances are themselves the
outcome of extended and iterated niche construction, in ways continually
maintained and transformed by people who inhabit or move through those
spaces, and how people engage one another within them. People mostly live
in spaces extensively modified, built, and often rebuilt to facilitate human
habitation, which they continue to maintain and reconstruct. The presence
of some people and not others and their ongoing patterns of postured responsiveness to one another are also integral to those environments. Those spaces
and their characteristic patterns of intra-activity are not self-contained but
are organized in significant part by their relations to other spaces and performances. The task for this concluding section is to call attention to the spatial
dispersion and temporal extension of the environments toward which human bodily postures are directed and responsive, and the ways those postures
form correspondingly interlinked bodily repertoires. Subsequent chapters
work out how such extended, interactive patterns of situated performances
are developed, sustained, and transformed, and why we should understand
them biologically as ecologically mediated development and evolution.

I have already noted that humans inhabit and move through multiple
functionally differentiated spaces, reconstructed and equipped over time to
facilitate associated activities. People typically work, sleep, eat, play, excrete,
travel, socialize, and acquire the resources needed for these and other activities

in places reconstructed or adapted for those purposes. Our bodily postures and performances are responsive to how those spaces are configured, including the postured performances of others as integral components. None of these activities stand alone, however. They belong together as components of multifaceted, vertically differentiated human lives. The organization of these local environments and the postures people take up within and toward them are interdependent with other spaces and postured performances and their roles within and among their participants' lives.

In inhabiting or moving through diverse settings, people understand themselves and one another in more or less defined, overlapping roles, with more or less distinct styles. These roles range across occupations and offices, kinship relations and life stages, genders, sexualities, ethnic and racial identifications, and political, religious, and other social affiliations among others. These role-enactments provide predictive insight into what people do, conjoined with normative standings (obligations, rights, proprieties, relevant virtues and vices, and more) to which people can be and sometimes are held accountable.[23] They also carry affective significance for themselves and others with varying degrees of first-personal normative grip on people's postures and performances. Interpersonal roles and styles are constituted relationally with one another and their settings. The expectations, accountability, and affectivity they carry with them interact with other codefined and cross-classifying roles and styles of enactment: parents, children, siblings and other kinship relations; employees, managers, customers, owners, and coworkers; members, officers, and so-called outsiders; citizens, immigrants, visitors, and aliens; neighbors, friends, enemies or opponents, acquaintances, partners, and exes; adults, adolescents, senior citizens, and children; social classes separated by wealth, income, social status and identity, and domicile; believers, heretics, and unbelievers; drivers, pedestrians, and passengers; insiders and outsiders, clergy and laypeople, and comparable dividing lines according to knowledge, training, or credentials; players, spectators, coaches, and referees; and a host of other interconnected identifications, along with many recognizable styles of inhabiting or enacting those positioned interrelations. These roles and styles do not just embody relations and standings vis-à-vis other persons, however, but also bestow role-related significance on and receive it from clothing, equipment, buildings, spaces, and boundaries, among others. Moreover, humans occupy multiple roles and standings at a time, with

---

23. I talk about standings rather than statuses because these normative considerations are not established, determinate norms but are instead at issue in the practices where they are articulated and deployed. This difference is discussed in chapter 5 below.

overlapping and conflicting as well as complementary obligations, opportunities, and allegiances. Such parallels, conjunctions, and conflicts in roles and standings do not merely condition the postures and performances through which people enact those roles; their public manifestation helps shape the environments to which those postures and performances respond.

These differentiated spaces, roles, and styles, and the associated postures, skills, standings, and their recognizable manifestations are not just contemporaneous structures of human environments. They are temporally extended patterns whose temporal relations and divisions are integral to their configuration as and within human environments. Humans respond to one another, their settings, the things around them, and the events taking place, in part through their temporal sequencings. Among the many temporally extended patterns that people now understand, situate themselves within, and respond to are the spans and temporal divisions of lives; trajectories in and of careers; personal relationships; histories of institutions, communities, nations, and so-called civilizations; sequential and cyclical patterns of hours, days, months, and years; figurative lives of buildings, neighborhoods, or artifacts through stages of planning and design, construction and manufacture, aging, decay, breakdown, repair and preservation, and demolition, discard, or disintegration; economic cycles; the emergence, migration, proliferation or diminution, and extinction of species and other populations; times of war and peace; and patterns and variations in local and global climate.

The situations in which people currently live and act are also not merely outcomes of prior development and niche construction that structure people's postures and the circumstances to which they are responsive. What people do is situated amid temporal patterns that encompass their current postures and performances and project partially open futures. In acting, people have proximate aims in current circumstances, but also point toward more extended future projections. Many performances aim to refine or enhance the capacities and skills they exercise, to advance further or improve the life course, career trajectory, or other temporal pattern to which they belong, or to change the circumstances or the pattern of practice to which they are responding. Human actions do not merely have niche-constructive effects through cumulative outcomes over time. They are also often performed and understood as critical or transformative responses to *possibilities* for reconstructing or reconfiguring their actual social-material circumstances and the larger patterns of social-material practice that encompass them.

These temporally extended patterns, prospects, and outcomes could only be part of the environment to which people's postures and performances are responsive if people can track and project them. To some extent, of course, all processes of organismic development track and respond to the larger patterns of evolution and development that constitute lineages and lives. Organisms develop through a sequence of environmental interactions, and the patterns of that development and the bodily postures organisms take toward current circumstances *embody* a tracking and projection of that interactive pattern. In some cases, this evolutionary and developmental history is manifest anatomically or contextually to subsequent scientific reconstruction. Similarly, the sequence of evolved forms and processes that constitute a lineage also implicitly tracks the complex sequence of environmental encounters that formed them. Human bodies and ways of life not only implicitly track developmental and evolutionary histories, however. People also mark and articulate many of the relevant differences and possible responses to them in conceptually articulated performances. Temporal and modal vocabulary and grammar both respond to and articulate these extended orientations. These conceptual performances and the articulative/communicative repertoires that enable them are also part of people's developmental environments. Those repertoires are taken up in the bodily postures and performances with which people respond to and transform their lives as environmentally situated organisms. Spoken and written language; images, diagrams, and models; musical, dramatic, and danced performances; ostensive gestures, and systematically interconnected equipmental complexes all exhibit expressive capacities through which people articulate, mark, and assess circumstances, performances, and ways of life, and thereby make sense of ourselves and our situations.

These capacities for conceptually articulated performances and their uptake are part of the interdependent, temporally extended patterns of niche-constructed settings, roles, and performances that make up human ways of life. Conceptual capacities and the normative concerns with meaning, justification, and performance that condition their uses are *part of* these larger patterns of an environmentally intra-active way of life.[24] If this claim is correct, then sense making itself is a biological capacity and concern that can only be made sense of in turn by situating it within evolving and developing ways

24. Earlier, I challenged distinctions between genotypic or anatomical capacities and their developmental realization in specific environments. I now similarly criticize distinctions between conceptual capacities and their realization in specific discursive environments. Brandom (2002, 50) distinguished two ways one thing might depend on another. X is sense-dependent on Y if understanding X requires understanding Y and X's relation to it. X is reference-dependent on Y if its being X depends on relations to an actually existing Y (see also Dennett 1991a, 324). Conceptual capacities are reference-dependent on the discursive environments in which they can develop, although the expressive capacities enabled in those environments extend well beyond what is currently expressed.

of life. There is no possibility of representing human ways of life and their expressive articulation from a "god's-eye view" from "sideways on."[25] The practice idiom was developed in part to understand the spatiotemporally extended, socially differentiated, meaningfully articulated, and normatively accountable patterns of human performances and ways of life delineated in this chapter. Chapter 1 showed that conceptions of social practices have mostly failed in this aspiration. Their failures occurred in part because they were unable to reconcile the embodied, situated patterns of human life with their conceptually articulated ways of making sense. They also failed in part by not providing an adequate grasp of how these *patterns* of situated performance held together as practices in ways that explicate the normative accountability of their constitutive performances and the spatiotemporal extension and continuity of the patterns themselves. A third aspect of their failure was their inability to reconcile these social relations with the biological patterns of life and death, development and evolution, that make up the lives of human organisms and their lineage within which those social relations are realized. The remaining chapters aim to show how to understand human ways of life as both embodying and responsive to the worldly, temporally extended, socially differentiated, conceptually articulated, and normatively accountable configurations of their environments, as forms of ecological-developmental, evolutionary niche construction.

25. I borrow "sideways on" from McDowell (1994). Both expressions make sense of the limits of human sense making from within. Both the spatial relations implicated by McDowell's phrase and the social-hierarchical and bodily-perspectival connotations of "god's-eye view" indicate unattainable ways of making sense.

4

Practices

Practices: The Very Idea, Revisited

We can now reconsider the concept of a practice, informed by developments in evolutionary biology and the discussion of bodily postures and environments in the preceding chapter. On most accounts of practices, what people do is conditioned by a background of interrelated performances. What they are doing, why those performances are done, how and why they do them in this way, and the significance of their outcomes are intelligible through their place in larger patterns of performance. Particular performances both respond to those patterns and contribute to their ongoing reproduction and transformation.

The previous chapter argued that people's bodily postures are oriented by circumstances extending well beyond their immediate surroundings. Humans have reconstructed those surroundings to accommodate and support diverse activities and social roles, and also to facilitate and track relations among the different settings, activities, and roles taken up in individual lives. Those roles and activities depend on supportive or conflicting roles and activities taken up by others living their own lives, and those relations among people and what they do make up the environments we each inhabit. The biological environments for people's postured performances are spatially dispersed, temporally extended in both directions, and intimately entangled with what others are doing in their own overlapping developmental environments. The key task for a niche-constructive conception of practices is to understand how those spatially dispersed and temporally extended interrelations are constituted and sustained as biologically significant developmental environments that condition their component performances.

Practice-based social theories now mostly recognize that the material circumstances of those performances are inseparable from the practices they enable and condition. A naturecultural conception of practices as biological niche construction construes the materiality of practices more specifically. The bodily postures of people's movements and actions are grounded on and directed toward those circumstances. A niche-constructive conception identifies the relevant circumstances as the developmental and selective environments of humans as living organisms, but so far that identification simply indicates the task ahead. The central claim of a niche-constructive conception of practices is that practices are the characteristic shape of the organism-environment intra-actions within which humans develop and the human lineage evolves. The work of spelling out what that claim means concretely still remains to be done.

Earlier, so-called classical conceptions of practices posited common elements or features of their constituent performances—performative regularities, shared presuppositions, or normative governance—to account for how they belong to a practice and are conditioned by their interrelations.[1] "Post-classical" conceptions of practices (Turner 2007) avoid postulating performative regularities or implicitly shared presuppositions or normative authority by starting with locally constituted "arrays" or "nexuses" of performances or practices, or interconnections within such constellations (Blue and Spurling 2017). A niche-constructive reconception of how performances belong to and are conditioned by practices offers an alternative to both strategies. This construal does not rely on postulated performative regularities, shared presuppositions, or governing norms. Many performances of a practice do resemble one another in some respects, or are similarly intelligible to participants, observers, or social theorists. Objects and events have manifold resemblances or differences, but understanding how practices are sustained and transformed does not derive from any constitutive similarities or commonalities. Understanding how practices are composed and sustained over time and how their continuation conditions subsequent performances nevertheless requires *some* further specification for the practice concept to provide insight. Indicating diverse local groupings of performances does not account for how the indicated patterns or interconnections exemplify social life more generally.

This book's central claim is that understanding human ways of life requires an expansive understanding of human biology that incorporates a naturecultural reconception of social practices. That biological background helps us understand how practices composed of spatially distributed and temporally extended groups of performances are held together by the coupled

---

1. Bourdieu (1977, 1990), Giddens (1984), MacIntyre (1981), Brandom (1979), and Taylor (1985b) exemplify classical conceptions of practices.

interdependence of those performances rather than by supposed performative regularities or constitutive presuppositions or norms. This section
begins with further reflections on organisms and their ways of life, setting
the stage for understanding the interdependence that unifies practices and
how that unity conditions the character and intelligibility of its constituent
performances.

A living organism is a material process whose components interact to
maintain and reproduce that process amid changing circumstances. Organisms' morphology, physiology, and behavior develop throughout their lifetimes, and their cyclical lineages evolve. An organism nevertheless remains
the same process across its developmental changes, and lineages continue
through evolutionary change and diversification. That unity is normative
in the sense that individual organisms can succeed or fail to remain alive
to produce offspring, and entire lineages can continue or become extinct.
Continuity of descent is in most cases the only measure of biological success
or failure. Life patterns replaced in developmental transformations within a
lifetime and ways of life reproduced in different forms are no less and no
more biologically successful than ones whose morphology and physiology
are stable throughout individual lives or across many generations.

The diverse and overlapping patterns of practices in the human lineage
introduce a new dimension of biological normativity interwoven with this
familiar assessment of biological success or failure along the single dimension of maintaining and reproducing the life process. The next chapter takes
up the normative complexity of human ways of life. Here, however, the one-
dimensional normativity of other biological life processes has a different significance. The processual unity of individual organisms and their lineages
provides a more adequate basis for comprehending how various situated performances belong together as continuations of a single practice or an interconnected nexus of practices.

That processual unity incorporates organisms' developmental and selective environments. Aspects of an organism's physical surroundings are only
configured as an environment by how they afford material resources, enabling, challenging, or threatening conditions and life-relevant information
for the organism's ongoing way of life. Environments are not goal directed
as such, but they are integral components of an organism's goal-directed life
process and are materially transformed by that involvement. Organisms are
usually treated as relatively self-contained and autonomous entities, defined
phenotypically or genetically. They supposedly live *in* an external environment, *interact* with it in influential ways, and evolve due to that environment's
differential reproductive significance for their lineages. Integrating evolution

with development and recognizing the ecological mediation of organismic development revises that conception. The *intra-active* interdependence of organisms with their environments,[2] and thereby with one another, becomes biologically central.[3]

Conceptual separation of organisms from environments has long encouraged another distinction that ecological-developmental reconceptions of biology undermine. If organisms interact with an external environment, with a selective significance that drives the evolution of populations inhabiting the same environment, then interactions among organisms in that population would be distinct from how those organisms individually or collectively interact with their environment. The organism's way of life would be *distinguished* from its interaction with developmental and selective environments rather than as *encompassing* them intra-actively. So conceived, the organism's way of life would be confined to its internal physiology and development, its individual behavioral patterns, and social interactions among conspecifics. Natural selection would then change that way of life over time as combined effects of genetic mutation, differential effects of its external environment, and random genetic drift, but also thereby omitting frequency- and density-dependent selection from the evolutionary process (Brandon and Antonovics 1996).

Conjoining niche construction theory with ecological-developmental biology conceives evolution more comprehensively. Evolution is not limited to changes in the lineage's genetic composition or its individual and social way of life (phenotype) in response to an external environment. Organisms' developmental and selective environments coevolve with the lineage, rather than externally causing evolutionary change. Evolution consists of changes in patterns of organism/environment intra-action over multiple life cycles, through conjoined effects of phenotypic plasticity, mutation, niche

2. As noted in chapter 2, Barad (2007) distinguishes *intra-action* from *interaction* according to whether the entities involved can be identified and specified apart from that involvement. All relations between organisms and their developmental and selective environments are intra-active, however, so I usually retain the familiar term "interaction" for organism-environment intra-actions. I use "intra-action" here to highlight that Barad's distinction is again at issue concerning where intra-action among conspecifics belongs with respect to organism-environment boundaries.

3. This extensive interdependence among organisms with overlapping environments has more strongly unifying interpretations, as itself an organism (named Gaia by Lovelock [1979] and Margulis [1998]), a "hypersea" encompassing the bodies of terrestrial organisms as extending the original ocean-environment of living organisms onto continental land masses (McMenamin and McMenamin 1994), or a "biosphere," a geologically distinct region defined by the mutually reinforcing effects of living organisms. My argument is agnostic about these interpretations.

construction, natural selection, and drift.[4] The population density, behavior, and environmental intra-actions of conspecifics are integral to organisms' developmental environments and the selection pressures their lineage encounters. Conspecifics thus belong to one another's environments, as do the ways of life and developmental patterns of other organisms affecting their development and life prospects.

This recognition complicates the mathematical form of differential equations modeling the evolutionary dynamics of biological populations (Lewontin 1983, 283; Odling-Smee, Laland, and Feldman 2003, 17–18). The modern evolutionary synthesis long recognized effects of environments on organismic lineages, but not the reverse: $dO/dt = f(O,E)$, but $dE/dt = g(E)$, that is, changes in organisms over time are a function of both environmental conditions and prior states of the organisms, but environmental changes are only a function of its own prior states. Recognizing the importance of niche construction turns this model system into *coupled* equations with feedback in both directions: $dO/dt = f(O,E)$, and $dE/dt = g(O,E)$, indicating that changes in organisms and in their environments are each functions of prior states of both. The nonlinearity of these coupled equations thus loosely models the coupled interdependence I ascribe to components of niche-constructive practices. Evolutionary change in populations is not a linear outcome of adaptation to exogenously changing external environments but rather a nonlinear interaction characterizing how organisms individually and collectively reconstruct the environments that exert selection pressure on the distribution of life cycles within that population. Social relations among conspecifics are among many contributing elements to those patterns of change.

The account developed here takes practices as a more complex form of biological interdependence, constituted by the coupled interdependence of their constituent performances with one another, their circumstances, and other practices.[5] This extended naturecultural conception of practices supplants conceptions of practices in philosophy and social theory that either posit common features among performances of a practice—regularities, constitutive presuppositions, or governing norms—to account for how they belong to and are conditioned by the practice or that start from particular groups, bundles, or nexuses of performances picked out pretheoretically.

4. Griffiths and Gray (2001) reconceive the evolutionary process similarly.
5. Tomasello et al. (2012) and Tomasello (2019) understand the evolution of human cooperation as a collaborative interdependence among individual *persons* that emerges in early human ontogeny. I instead emphasize interdependence among performances of practices and practices themselves; the cooperative relations Tomasello et al. analyze as interdependence among agents are partial aspects of a more encompassing interdependence of the *practices* they take up.

Practices are temporally extended patterns of interdependent performance. How an action is performed is not only shaped by current and prior performances to which it responds. Actions contribute to ongoing patterns of performances, and would only accomplish their intended effects in concert with what others go on to do in sufficiently conducive circumstances. People take up and contribute to a practice in open-ended anticipation of others' subsequent performances. They implicitly project appropriate alignment of subsequent performances with how their own performance builds on or responds to its predecessors. Practices often do not continue as anticipated, but the interdependence of present and future performances is still evident in how subsequent performances introduce or respond to those changes. Mark Risjord takes jazz improvisation to exemplify a related conception of practices: "Since [jazz performers'] instruments have different capacities and play different roles, they must recognize different affordances and have a practical understanding of how they interlock. More importantly, their actions not only influence each other, they partly constitute each other. When the bass plays one note and the piano plays some others, they each provide part of the environment which changes the way the harmony sounds. Each contributes something that makes the other's action possible, and the harmonic and rhythmic structure arises as a consequence of this joint action" (2014, 233–34). That mutual responsiveness does not entail that these performances always work together successfully, but failed couplings are among the circumstances to which subsequent performances respond. These relations of coupled interdependence account for how performances belong together within an ongoing practice. They are also a resource for understanding similarities or differences among their constituent performances.

Risjord's discussion emphasizes how interrelated performances belong to performers' current practical-perceptual circumstances, but the considerations ending the previous chapter show how the relevant environment is also spatially and temporally extended. Risjord does not emphasize this point, but jazz improvisation also engages spatially and temporally extended environments. Jazz performers do not merely respond to other instruments playing together with them and to the listening audience. Their responses to one another belong to a practice of jazz improvisation whose history and prospects inform its performances, including how they differentiate themselves from past patterns and open new possibilities for intelligible performance and its assessment. Practices are constituted as articulated and differentiated components of human ways of life which together make up ways of making a living within and from biological environments. This alternative conception of practices may initially seem abstract and hard to envision. Two concrete examples provide illustrations before proceeding further.

Practical Interdependence

Recent work on early hominin evolution emphasizes teaching as an influential form of social cooperation nearly unique to humans. Many organisms actively observe and learn from others, but instructing or guiding others is critical to enable "high-fidelity" social learning and the ratcheted niche construction through which humans developed differentiated practical interactions (Laland 2017; Sterelny 2012a, 2021; Tomasello 2008, 2014, 2019). People now teach one another in many ways, some formal, some not, including explicit verbal instruction, correcting hands-on performances, or providing exemplary models, and these activities take place in many institutional or other interactive settings. These diverse forms of teaching are also now situated within different practices. To develop an initial sense of how practices can involve constitutive interdependence without performative regularities or shared presuppositions or norms, consider college classes as one practice of teaching and learning. Interactions among students and teachers in those classes illustrate the intimate interdependence of postures and setting which identifies those situated interactions as belonging to a practice, and as components of more encompassing practices.

How one comports oneself as instructor or student in a college classroom depends closely on others' comportment and the material setting. Students and teachers come into class with varied goals, backgrounds, specific preparation, and affective orientation, and these considerations affect how their interactions take place. Instructors may lecture, pose questions and work from students' responses, lead general discussions, break the class into smaller groups, or lead them on field trips, but how these proceedings unfold partly depends on students' response. How closely do they listen, with what willingness to interrupt with questions, elaborations, or criticisms? How do they respond to the instructor's questions, other students' comments, or in small groups? College classes are aimed at students with extensive educational background and some life experience, but how instructors proceed typically depends on the extent, character, and diversity of students' background and orientation, the relevance of that background to the course, and how well the instructor understands students' backgrounds and interests. Are students already motivated to engage with the material, or does the instructor need to convince them of its interest and importance? Can they respond constructively to readings on first encounter, or do they need further guidance to formulate relevant or perspicuous responses? Students similarly adjust to instructors and fellow students: the pace, audibility, clarity, or complexity of lectures and accompanying visual materials; the authority, confidence, and enthusiasm conveyed by the instructor's posture, expression, dress, and markers of social standing, and the comparable orientation of other participants; the kinds of questions posed and how answers are utilized; how discussions are organized and guided; how fellow students respond to different interventions; the assignment of essays, examinations, projects, or presentations, or other work to be evaluated; how those assignments relate to what is done in the classroom; and so forth. Both students and instructors respond to one another more subtly with changing postures and orientation, attention and enthusiasm, and by deference, recognition, or challenge to one another.

Everyone's responses are also affected by the room's physical arrangement: its size and seating configuration; the lighting, acoustics, and temperature control; the availability, placement, and functionality of blackboards, podiums, computing consoles and projectors, laboratory facilities, or relevant teaching aids from maps to periodic tables. College classes normally extend through multiple sessions, and these considerations may change in response to previous sessions and participants' satisfaction and engagement. Patterns that do not change despite evident mismatches with what other participants are doing typically have consequences for subsequent interactions. What holds these behaviors together as a practice is not common goals or behavior—participants in college classes often diverge in those respects— but the dynamic, ongoing interactions among participants and settings in response to how their postures, performances, and circumstances work together constructively or are misaligned.

I have focused so far on participants' bodily postured responsiveness to one another as salient within their respective practical-perceptual environments. That is one aspect of interdependence among performances of a practice. These proximate practical-perceptual interactions work together with other considerations: how those performances are affected by other practices, how they are talked about and verbally coordinated, how participants assess and respond to them in light of different normative concerns, and how those interactions enable or encourage some performances and achievements, constrain or block others, and allow resistance or creative redirection.

Classes are somewhat unusual in their relatively fixed scheduling of time, term, and place, stable group of participants, conventionally established and advertised content and structure, and institutionally fixed contribution to a larger program of study or work for all concerned. Classes nevertheless vary in these respects as well. When does formal class activity actually start or end, with what breaks? How regular is attendance, by whom? How closely does what happens fit the course description or syllabus? What level of commitment, engagement, intellectual depth, disciplinary orientation, or individual style is displayed? How well does it align with institutional commitments or expectations and participants' aspirations and goals? Moreover, answers to these questions are typically coupled with other variation in what students, teachers, visitors, or observers do or occasional intervention by others such as parents or supervisors.

Emphasizing internal variation within the practice might prompt responses that its characterization as "college education" already introduces some regularities, norms, or presuppositions common to its instances. That response is misleading in at least three ways, however. First, this common characterization conceals considerable diversity in what is done under that designation, for whom, how, and why. Second, what the designation "college education" means or should mean, where and how it applies, and how and why it matters, is contested within the practice. Designation as "college education" does not mark a determinate common feature or concern shared by its instances, but is instead the site of conflicts among participants concerning what they are doing, to what ends. The goals of higher education have been construed in diverse ways: technical expertise, critical reflection, or developing moral character; vocational preparation, flexible skills, or meaningful life-orientation; preservation of class privilege or opportunities for social mobility; continuation of tradition or support for innovation; individual advancement, maintaining an educated class as a social, economic, or political resource, or achieving a preferred distribution of its outcomes among different groups. These issues play out differently across historical or cultural contexts despite links and influences among them. Moreover, educating undergraduates is sometimes regarded as instrumental to faculty research rather than a primary aim, with significant effects on how teaching is undertaken.

Finally, these issues are shared across institutions and among classes taught in any one institution only in the sense that these instances of the practice do not stand on their own but affect one another and are influenced by other practices with more limited links to classroom teaching. Students do not normally take only one course, and instructors also typically teach others; the activity would be very different otherwise. What is accomplished in one course is instead conjoined with others, with overlapping groups of participants. When courses fail to fit together constructively to further the divergent aims of students or other concerned parties, those misalignments are at issue in the ongoing reproduction of the practice. That issue is addressed

or ignored in different cases, and how these practices fit together is also partially interdependent with other practices that prepare for it or for which it is
preparatory—parental guidance, high school and graduate education, career
trajectories, differentiations of social class and ways of life, and more. College teaching and learning also engage with other practices that provide its
resources and institutional standing; organize, support, and administer it; or
interact with it economically, politically, and culturally in other ways. Much
more could be said to develop this example. The point to emphasize is that
what happens in college classes exemplifies how what holds these interactions together as a practice is not determined by performative regularities,
shared presuppositions, or governing norms but is shaped by how their constitutive performances, prepared settings, and relations to other practices and
performances are interactive and interdependent through ongoing mutual
adjustment.

A second initial example addresses a prominent conceptual resource for
thinking of practices as constituted by governing norms or rules. Many
practice-based conceptions of social life draw on the notion of a "constitutive
rule" (Rawls 1955; Searle 1969, 1995, 2010, 2018; Haugeland 1998, chap. 13), exemplified by the rules of a game or sport or a political or legal system. Games
are canonical models for relatively autonomous domains of practice organized by governing rules.[6] The institution of a game and its governing rules
constitutes entities and statuses that only exist within the bounds of the game
under the authority of its instituted rules. Strikes, outs, baserunners, and
RBIs only exist amid the playing of baseball or softball as constituted by rules.
Rooks, ranks and files, knight forks, and capturing en passant are likewise introduced by the rule-governed play of chess. Games are an especially attrac-
tive philosophical model because they seem (almost) entirely self-contained
and autonomous. Players have a life outside the game, and game-playing has
a place within their lives, but these considerations are mostly "out of play"
within games, with limited significance for players' game roles, actions, or
status. Rule-governed performances of other partially autonomous practices,
such as political institutions or legal systems, are less insulated from external
concerns.

---

6. Practice theories taking Wittgensteinian "language-games" as models for practices and
forms of life mostly ignore his denial that games have defining features, or implicitly limit its
implications to linguistic meaning rather than a greater diversity and variability among social
practices and "constitutive rules." Searle's (1969) influential distinction between constitutive
and regulative rules foregrounded game examples, whereas Rawls made them one example
among many.

Understood as practices in a sense relevant to understanding human ways of life, however, even games are more complex phenomena. There are occasions when a more confined conception of games as rule-governed practices applies—when two teams meet in an English Football Association match, or two players contest a world chess championship, the rules of football or chess are taken as settled and authoritative even if their proper application to cases is contestable. The play of the game also takes place entirely within the stadium or around the chessboard. As practices sustained over time, however, the game of football or of chess includes more than what happens on the pitch or the board, whether in accord with or violating the rules. These considerations include where games are played and how they come to be played there; who plays the game or has other roles as referee, spectator, substitute, coach, commentator, bettor, or security guard; how the game fits into an overall way of life for all concerned; and how the rules, scheduling, and their governing concerns are determined. Even within the confines of play, rules take into account but cannot determine the materiality of game play. Philosophers' favored example of chess is misleading, because that game is (almost) formally specifiable, apart from the requisite "compliance and inertness" of the pieces and board (Haugeland 1998, chap. 13), and its often wrongly discounted physical demands. Understood in this broader sense of what a game is, the rules are not constitutive of the practice, but instead answer to many issues arising within the practice.

Consider volleyball, a game invented in 1895 for one purpose—lunchtime recreation for businessmen whose lives and bodies were too sedentary for basketball—which has subsequently evolved into different versions. In its primary competitive descendant, volleyball places extraordinary physical demands on highly skilled athletes. It is perhaps the most reciprocal of team sports and among the more strategically complex. An alternate competitive version is played outdoors on sand or grass with a smaller number of players. The outdoor versions drastically simplify team play, are played with a heavier and softer ball, and emphasize versatile skill, stamina, and variable conditions of wind, sun, heat, and playing surface. Volleyball is also played recreationally, in versions that minimize the skills required, lack formal rules, vary the equipment and court dimensions, and often emphasize some cooperation between teams to sustain play. It is sometimes taught to children in schools in another version, Newcomb, in which the ball is caught rather than hit to emphasize strategy and reciprocal team play in the absence of technical skill.

As a practice, volleyball and its varied and oft-changing rules answer to many other concerns. Who can or would want to play? Where and how can they play or learn the requisite skills? What skills are minimally required of participants and how are they developed in advanced forms of play? Where are equipment and play spaces found or acquired? How is play institutionally located and its skills taught, with what ancillary commitments to schools or other team-sponsors, sport-specific clubs or leagues, pickup games at public venues, or national or professional teams? Many considerations affect how rules are set and interpreted. These include whether the rules should reward higher levels of skill or athleticism or reduce competitive advantages; facilitate play at younger or older ages, different or mixed genders, or shorter heights; encourage or discourage strategic complexity; favor the offense or the defense; enable intelligibility, visual spectacles, or predictable durations for spectators, at live venues or on television, and target only aficionados or seek broader appeal; limit dangers of injury to players; align better with how the game is actually played, or how it is played elsewhere; reduce costs of facilities or equipment; simplify interpretation for referees; support a lifestyle or image preferred by athletes or advertisers; and many more. Games also develop over time due to changing skill levels; size, power, or speed of players; tactical innovations; coaching expertise; spectatorial or commercial interest or lack thereof; amid other factors. These changes often lead to perceived needs for rule changes to maintain continuity with previous forms of play, initiate favored changes in play, or satisfy proposed rationales for or virtues of play.

Understanding games as autonomously and constitutively rule-governed requires imaginatively abstracting from other practices and ways of life connected to the play of the game and determination of who plays where and when. As only-partially autonomous activities, games are not rule-determined practices. They are instead complex, interconnected practices whose many constitutive concerns and interrelations shape the development, interpretation, and application of rules. The effective authority of the rules in practice often diverges from their authoritative expression in rulebooks. Which rules are put in place and how they are interpreted and enforced are closely interconnected and sometimes misaligned with how the game answers to other things people do and what is at stake in those interconnections. The game is always much more than just rules and rule-following or violation. The next chapter has more to say about the normative accountability of practices and their constitutive performances. With these examples in hand, however, we can think more carefully about how belonging to a practice bears on its performances in the absence of appeals to performative regularities, shared presuppositions or commitments, or governance by determinate rules or norms.

Practices as Nearly Decomposable Systems

This section develops a more comprehensive conception of how linkages among performances constitute practices and interdependent networks of practices. The following section then takes up how such constitutive links among performances and practices are now the basic makeup of human biological ways of life. Both sections build on Haugeland's (1998, chap. 9) adaptation of Herbert Simon's classic analysis of nearly decomposable complex systems. Haugeland aimed to understand how cognition is embodied and embedded in the world. My aim is to understand how practices are embodied and embedded as organism/environment relationships in the human lineage.

Complex systems in Simon's sense are "made up of a large number of parts that interact in a nonsimple way. In such systems the whole is more than the sum of the parts, not in an ultimate metaphysical sense but in the important pragmatic sense that, given the properties of the parts and the laws of their interaction, it is not a trivial matter to infer the properties of the whole" (Simon 1981, 195). Simon addressed "hierarchically organized" systems, meaning only that the systems are composed of parts—subsystems—that are themselves complex systems. The defining feature of those systems is that interactions *within* a subsystem are different in kind from interactions *among* subsystems. This difference enables a principled analysis of the larger system into component subsystems and the components into their component parts. Haugeland summarizes relevant distinctions that together characterize the organization of nearly decomposable systems: "A *component* is a relatively independent and self-contained portion of a system in the sense that it relevantly interacts with other components only through interfaces between them (and contains no internal interfaces at the same level). An *interface* is a point of interactive "contact" between components such that the relevant interactions are well-defined, reliable, and relatively simple. A *system* is a relatively independent and self-contained composite of components interacting at interfaces" (1998, 213). The italicized terms are not intelligible apart from the others, and their referents would not *be* components, interfaces, or a system in others' absence.[7] Decomposable systems need not be static, synchronic structures; systems such as living animals or operating machinery move through space with internal degrees of freedom, but the boundaries among their components move accordingly (Haugeland 1998, 214). Haugeland's point extends to encompass

---

7. Systems, components, and interfaces are thus both mutually sense-dependent and mutually reference-dependent (Brandom 2002, 50).

systems of practices as *processes* changing over time. Those changes become intelligible through their systemic interrelations, even though the system's operation not only relocates the boundaries among component practices but also changes their configuration within the system.

Simon and Haugeland recognize that "relevant interaction" and "boundaries" or "contact points" among them are also only specifiable within this interconnected pattern, although neither highlights the point. Interfaces, after all, are only a special case of boundaries (those whose relevant interactions are well-defined, reliable, and simple), and relevant interactions are those specified by the system in question. Sometimes those interactions are demarcated spatially and physically, as in complex molecular structures formed by covalent, ionic, or hydrogen bonds among atomic or other submolecular components. In an economic system, by contrast, the interactions might be movements of money, goods, and services—which can "move" economically while remaining physically in place, or vice versa—across boundaries of ownership, with some owners legally rather than physically "incorporated" within others. Haugeland (1998, 214) notes that spatial-corporeal relations may be poor guides to functional subdivisions of organizations into divisions, departments, and smaller units, but relevant flows of information, authority, and their uses also might cut across these formal organizational structures.

Simon characterizes the key distinction of interactions *within* a component of a larger system from interactions *between* components in terms of relative strength and weakness (1981, 210) or "intensity" (1981, 199). Haugeland adds two further glosses. First, "'intensity' of interaction here means something like how 'tightly' things are coupled, or even how 'close-knit' they are—that is, the degree to which the behavior of each affects or constrains the other" (1998, 215), even though the relevant forms of coupling and constraint vary greatly among mechanical, biological, inferential, electrical, athletic, economic, or kinship systems. Second, Haugeland also differentiates these "well-defined, reliable, and relatively simple" interfaces among components as relatively "low-bandwidth" interactions in contrast to the "high-bandwidth" interactions within a component.

The initial relevance of these considerations for understanding practices is that commonalities or similarities among situated performances are not needed to account for their belonging together within a practice. What matters is how various performances are linked by temporally extended and spatially dispersed patterns of multifaceted, coupled interdependence. Whether and how performances of a practice occur depends on whether and how other performances have occurred, other relevant circumstances, and anticipation of subsequent performance. Relevant forms of interdependence include how some performances respond to or solicit others; how they depend on and respond to their material settings; how they contribute to larger projects along with others; or how they otherwise only make sense for participants as enabling and continuing interrelated performances.[8] Moreover, the relevant forms and patterns of interdependence can change over time, partly as effects of how performances anticipate or respond to one another.

Alongside the close interdependence among performances as constituents of a practice are more localized and limited interactions between practices at interfaces. Differences between close, multifaceted interdependence among performances of a practice and its more limited but still consequential interfaces with other practices can be analyzed at different levels of scale and complexity. Many practices are composed of other practices; the closely interdependent performances making up a practice at higher levels are themselves composed of other patterns of more intimately interdependent performance.[9] Relatively simple, low-bandwidth interfaces among component practices at one level can be understood at higher levels as *relatively* intimate, high-bandwidth interdependences of a complex practice interfacing with other practices. In such hierarchically arrayed patterns of practice, the simpler forms of interdependence arising at interfaces among complex patterns of practice nevertheless affect performances of the lower-level practice in ways mediated by their own more intimate interdependence. In such cases, how one practice continues may depend in more limited or focused ways on whether and how another practice complements, supports, or challenges it in specific respects. In other cases, different practices can partially overlap in ways that affect how some of their performances take place, with indirect effects on others in turn. Practices often change over time due to changes in practices and performances with which they interface. These changes occur both through effects of other practices on those performances and how practitioners change what they do to build on, mitigate, or circumvent those effects. Moreover, characterizations of boundaries among interrelated practices are themselves often at issue within the practices, complicating any sharp or rigid demarcation of one practice from another.[10] A naturecultural conception of niche-constructive

8. These performances are reference-dependent due to *participants'* understanding of their sense-dependence (Brandom 2002, 50). Sense-dependence concerns whether one thing could make sense to someone without also *understanding* another thing, but here participants are the sense-makers. They would not undertake some performances in the recognized absence of others, and that unwillingness to act senselessly makes their actions reference-dependent on others.

9. Recall Haslanger's description of different levels of practice that come together in a public high school cited above in chapter 1.

10. Chapters 5–7 below show that discursive articulation, coordination, and critical assessment of practices belong to the practices themselves, but also to discursive practices with their own constitutive interdependence.

practices is not an overview of human ways of life from sideways on but is instead part of the discursive articulation of those ways of life from within.

Another example further illustrates these levels of interdependence within and among practices and their interfaces. Consider the practices that maintain a standing military force whose capacities for organized violence serve broader political purposes, such as the eighteenth century Prussian Army, the nineteenth century British Navy, or the late-twentieth century Israeli Defense Force. Many closely interdependent practices must be established and maintained in alignment. A military force is organized around specific forms of fighting capabilities, even though such fighting may only occur intermittently or not at all and military forces are often used in unexpected ways. The fighting practices around which they are organized involve specific weapons, fighting skills, unit size and organization, transport and supply logistics, forms of discipline, and communication among members. These elements are typically arranged in anticipation of actual or conceivable antagonists, fighting terrain, weather conditions, and durations of warfare or other deployments. These factors are nevertheless often not controlled by military organizations or their political governors. The misalignment of military preparation with its actual deployment is familiar and often deliberately invited or imposed by antagonists. Deployment of large military forces requires practices of command and obedience that must be instituted and maintained to be effective when needed. These practices also include forms of political control—military forces are established and maintained for political ends, but their capabilities for organized violence are existential threats to their governance. Maintaining a force over time requires recruiting and training new participants to replace those exiting due to age, service completion, injuries, or death. Troops must be housed and supplied even when not deployed, but they must also be readily mobilizable with appropriate notice for possible deployment.

Each component practice—recruitment, training, command and control, housing, supply, mobilization, fighting, and exit transitions—involves intimate interdependence among constitutive postures, environments, and performances that differ significantly from one another. Recruiting, impressing, or drafting members of a military organization, for example, differs from training, commanding, feeding, or sending them into battle (the Prussian, British, and Israeli cases exemplify different practices of enlisting, impressing, or drafting members of their military units). These activities are nevertheless more closely interdependent with one another than with other practices with which they have simpler interfaces. Despite the many differences between how soldiers are recruited and how they are trained for their new roles, these practices affect one another. How one trains new recruits or draftees, with what effects, depends on their demographics, their experience and abilities or limitations, their motivation for and understanding of their roles and activities, and how long they serve. The kinds of training members undergo may in turn significantly affect who is recruited and how recruitment is undertaken. By contrast, although a military force depends on reliable supply of weaponry and other equipment, how their provision is contracted, manufactured, transported, inspected, purchased, and eventually disposed of is not usually closely interdependent with military practices themselves. Significant changes in logistic arrangements may have minimal effects on recruitment, training, and deployment of troops if provisioning is not interrupted. What matters logistically is only the timing, quantity, quality, and reliability of supplies, not how they are achieved. Logistic arrangements mark interfaces between military practices and the manufacturing, purchasing, inspecting, distributing, and disposing of military supplies.

Analyses of nested interdependence among practices and their constituents can move in the other direction, however. Recruitment and training interface with one another with limited intensity of interdependence compared to the localized day-to-day interactions among the postures and performances that implement each aspect of military practice. How recruiters or draft boards entice or enforce civilian enlistment in a military organization and how enlistees respond, or the daily interactions among military trainees and officers guiding their developing skills and affects, are the constituent performances of recruitment or training practices. Changes in recruitment can profoundly affect military training, but those effects are channeled by interfaces between civilian and military life. Given different recruits, military training may need different, high-bandwidth interactions to transform *these* civilians into a more or less cohesive, disciplined, and effective fighting force.

These boundaries are not given or fixed, however, whether differentiating less intense interdependence among diverse military practices from the more tightly connected performances that implement them, or more intensive interdependence among specifically military practices from more limited interfaces with other practices. Military organizations might outsource their recruitment or the assessment and remedial training of less qualified recruits to commercial or civilian agencies much as they purchase equipment. Alternatively, they might bring manufacture and supply in-house to tailor technical development more closely to military aims, and they might place manufacturing quality control under close supervision rather than just end-product inspection. How various performances interact with and depend on one another in practice, and how practices interact with and depend on one another, can change.

Rearrangements of the interdependence among performances of a practice and among practices with more limited interactions can also take more
complex forms. The fighting techniques, equipment needs, and mission-
orientation of a military force might become subordinated to the interests
of a "defense industry" due to growth in the size and political influence of
those industries. Those shifts might be encouraged or sustained by career
crossovers between military officers and industry leaders. The more intimate
interdependence linking together some practices within a larger practice
might thereby extend beyond military practices of recruitment, training, and
deployment to comprise a more extensive military-industrial complex. That
development in turn might engender simpler, more limited interaction of a
military force with the political system that ostensibly controls it.[11] In a different direction, a military coup or gradual erosion of civilian control might
militarize state institutions and practices, so that boundaries between military
practices and political life are no longer well-defined interfaces. These shifts
in boundaries within and among practices illustrate more concretely the difference between an interface between practices and the "high-bandwidth"
interdependence within a practice. That illustration in turn should clarify
how what unifies performances into a single practice or multiple practices
that interact in more limited ways is not common elements or shared presuppositions. Those differentiated interrelations are nevertheless more than just
amorphous or localized "bundles" or "arrays" of performances. The identification of practices and their constituent practices or performances instead
turns on the kinds and depth of interaction and interdependence among performances or practices and their relevance to how practices continue and
change over time.

Practices as Human Organism/Environment Intra-actions

Simon's and Haugeland's conceptions of complex systems has a second important implication for understanding practices as structuring human ways
of life. Simon characterizes complex systems at a high level of generality and
abstraction which blocks clearer or more precise differentiation of interactions within system components from what occurs at their interfaces. How
interactions in different kinds of system can be stronger or weaker or more
or less intense resists general specification. Different systems involve different

---

11. Readers may decide for themselves whether this hypothetical transformation of a military force into a "military-industrial complex" resembles actual examples.

kinds of interaction within components and at interfaces. Haugeland provides an instructive example.

> An electronic component, like a resistor, is a relatively independent and self-contained portion of a larger electronic circuit. This means several things. In the first place, it means that the resistor does not interact with the rest of the system except through its circuit connections—namely, those two wires. That is, nothing that happens outside of it affects anything that happens inside, or vice versa, except by affecting the currents in those connections. (To be more precise, all effects other than these are negligible, either because they are so slight or because they are irrelevant.) Second, it means that the relevant interactions through those connections are themselves well-defined, reliable, and relatively simple. For instance, it's only a flow of electrons, not of chemicals, contagion, or contraband. (1998, 212–13)

The previous examples further illustrate this point. Interactions among students and teachers in classrooms are more intensive and high-bandwidth than how classroom activity depends on university administrative practices or other practices affecting college teaching. Reciprocal interdependence among players and referees on a volleyball court or among military trainers and recruits is more multifaceted and intensive than with other considerations that shape volleyball as a sport or those forming new military recruits' initial skills, dispositions, affects, and interrelations. Formulating a more detailed *general* specification of these differences in intensity or bandwidth would be difficult. Haugeland nevertheless uses that distinction effectively to understand relevant boundaries in intentional or cognitive systems. A comparable use enhances understanding the different forms of goal-directed interdependence that constitute human organism/environment intra-action as a biological system.

Haugeland argued that relations between brain and body or body and worldly objects are not interfaces in a cognitive system. About Rodney Brooks's robot Herbert, which roams a moderately complicated laboratory environment, finds empty soda cans, picks them up, and puts them in a bin, Haugeland writes: "Each of Herbert's highest-level subsystems is somewhat mental, somewhat bodily, and somewhat worldly. That is, according to Simon's principles of intensity of interaction, the primary division is not into mind, body, and world, but rather into 'layers' that cut across these in various ways. And in particular, the outer surface of the robot is not a primary interface" (Haugeland 1998, 218–19). Haugeland argues that human cognitive systems also lack simple, well-defined interfaces through which light or sound affects perceptual receptors. Nor are there interfaces through which sensory receptors send inputs to the central nervous system and receive instructions for how to move the body. The relevant cognitive systems closely couple neural circuits in the brain with perceptual and motor nerves, movements to track and focus on aspects of a situation, and shifting patterns of light, sound, and odor reflected from and dispersed through that situation throughout those exploratory movements. The relevant close coupling or high-bandwidth interactions produce multichannel, real-time, cognitively-relevant interdependence among neural connections, bodily movements, ambient light, sound or odor, and reflecting surfaces. Changes in one pattern affects others in complex, ongoing patterns of coupled interdependence.

The significant difference is that these closely coupled, high-bandwidth interactions within an integrated component are more intensively interdependent than is assumed in most accounts of perception and action, which treat them as interfaces between mind and world. Haugeland argues that these "interrelationist" accounts of perception and action misconstrue how they contribute to cognitive systems.

> Interrelationist accounts [of the dependence of mind on body and world] retain a principled distinction between the mental and the corporeal—a distinction that is reflected in contrasts like semantics versus syntax, the space of reasons versus the space of causes, or the intentional versus the physical vocabulary. (Notice that each of these contrasts can be heard either as higher versus lower "level" or as inner versus outer "sphere"). The contrary of *this* separation—or battery of separations—is not interrelationist holism, but something I like to call the *intimacy* of the mind's embodiment and embeddedness in the world. The term 'intimacy' is meant to suggest more than just necessary interrelation or interdependence but a kind of *commingling* or *integralness* of mind, body, and world—that is, to undermine their very distinctness. (1998, 208)

Haugeland is arguing for the *cognitive* intimacy of mind, body, and world; I argue for the *practical* intimacy of people's relations to other people as integrally interdependent with their biological environments. The practices and constituent performances now making up human ways of life similarly resist merely interrelationist interdependence among social and biological components. Understanding practices as biologically niche constructive recognizes that the manifold cultural relationships and interactions within human societies are more complex and varied forms of the ecological interdependence that constitutes an organism's and its lineage's way of life. Differentiated practices now structure the environments where people sustain themselves physiologically through environmental intra-action, develop throughout their life span, and reproduce others in developmental patterns that are responsive to selection pressures on our lineage. Bodily micronegotiations and extended practical engagements with other people are intimately entangled with these larger patterns of practice.

My argument that practices are biologically niche constructive rather than nearly decomposable social components of a biological way of life begins by acknowledging the familiarity and apparent plausibility of interrelationist conceptions. Our practice-differentiated ways of life share with other organisms a basic biological normativity in which success sustains lives and ways of life, and failure is death or extinction.[12] In this limited respect, the hominin lineage has flourished despite having diminished to a single species, with massive growth in numbers and geographic range since its recent evolutionary origins as a small population of east African primates. Human biological flourishing and the complexity and differentiation of how people interact with and depend on one another are often thought to insulate us from environmental vulnerability. These evolutionary shifts thereby seem to support the relative autonomy of social practices from human biology. Many aspects of people's lives, including activities and skills; the goods and services we produce, exchange, and consume; our institutions, behavior, and interpersonal relations; and linguistic and artistic expression above all seem remote from immediate biological significance or constraints. Human biology may instead seem mediated by social institutions and divisions of labor according to social roles. Individual humans now typically do not meet basic biological needs through their own efforts. Social institutions and practices instead provide for and regulate supply, distribution, and consumption of food and water, availability of and access to shelter from weather or predation, sexual reproduction and child development, disposal of waste products, and management of disease and death.

Human sociality now even seems to insulate individual human lives from biological vulnerability. People die, but now often only at the end of a lengthening natural life span.[13] "Early" death often suggests distinctively social failings:

12. The next chapter characterizes this basic biological normativity as one-dimensional. The diverse normative concerns arising within practice-differentiated ways of life introduce a second dimension of normative assessment, enriching and complicating how practices change and how the lineage is open to nonarbitrary assessment.

13. Relative declines in mortality from infectious disease are quite recent. Developmental economist Charles Kenny, using the Our World in Data website, concludes that, "looking at overall life expectancy, in 1900, the average person born on planet Earth could expect to live to

deliberate or accidental violence; severe economic and political inequality; unhealthy behaviors, lifestyles, or environmental exposures that were socially enabled and encouraged; or restricted access to effective medical intervention. Retrospective consideration of earlier statistical distributions of human mortality thereby suggest that the earlier patterns manifest social failures rather than biological limits. These aspects of people's recent biological history have encouraged anthropocentric conceptions of people as social agents in a social world disconnected from our physiology, biological development, and evolution as a population. The complexity and centrality of people's social relations and practices thereby seem increasingly discontinuous from other organisms' ways of life despite their manifest biological continuity.

The preceding cautionary paragraphs were written before the SARS-CoV-2 pandemic, which revised and strengthened my response. During this pandemic, familiar forms of sociality and social inequality often intensified people's biological vulnerability. Adjusting these practices to accommodate a novel virus has led to social distancing; use of biologically protective equipment in social interactions; newly consequential classifications of people; renewed or revised efforts to disinfect human environments; and shifting relations among political, medical, economic, and scientific practices. Commentary on the pandemic nevertheless mostly frames these events as social responses to a biological threat to social, economic, and political life. These social, economic, and political practices have nevertheless always been constitutively intertwined with human biology. Ongoing relations among human bodies, their microbiomes, diverse animals and plants, and many other organisms are integral to those supposedly social practices. An emergent virus with different contagion patterns, manifest symptoms, systemic effects on bodies, and varied immune responses highlights the historical particularity of previous patterns of proximity and interaction among human and other bodies. The shape of subsequent accommodations "in practice(s)" among people, vaccines, and evolving populations of SARS-CoV-2 variants are far from settled, but this family of coronaviruses has joined other microbes circulating amid people's ongoing, situated interactions with one another. That

---

the age of thirty-three. By 2000, that expectation had doubled. Today, life expectancy is above seventy years" (2021, 145–46). Social practices, notably agriculture and densely populated urban ways of life, were biologically consequential both in facilitating life supports and enhancing vulnerability to infectious agents.

the initial viral variants likely evolved and proliferated in the social practices of live animal markets or microbiology laboratories only reinforces the point. I return to this telling example in chapter 9 below.

This book, planned and begun before this instructively deadly exemplification of its thesis, turns familiar conceptions of the relative autonomy of social life inside out. Social or cultural practices and institutions are not relatively autonomous components of human life that merely resulted from earlier evolutionary adaptations. Those adaptations were indeed important. Physical anthropologists and others studying early hominin evolution rightly call attention to many evolved anatomical, physiological, and neurological traits of human bodies that remain integral to contemporary ways of life.[14] Recent work on hominin evolution (e.g., Sterelny 2012a, 2021; Laland 2017) emphasizes that these traits were not simply inherited, but coevolved with cooperative social foraging as a distinctive ecological way of life, complementing the book's argument. My reasoning nevertheless turns on other considerations. These early evolutionary adaptations are common ground between my account and many interrelationist accounts of social practices and institutions as relatively autonomous components of human ways of life. I argue instead that subsequent practices are integral aspects of human biology that must develop anew in each generation and continue to evolve in complex patterns of ecological interdependence.

The current chapter and its predecessor lay out the book's basic conception of horizontally and vertically differentiated practices as the evolved shape of organism/environment interdependence in human ways of life. The remaining chapters examine other central and interconnected aspects of that conception, beginning with the two-dimensional normative accountability effected by the evolved, differentiated interdependence of niche-constructive practices. This account of the diverse normative concerns arising within a practice-differentiated way of life then frames the book's remaining topics: how language and other discursive practices fit within that way of life and conceptually articulate people's developmental environments; how people's

14. Familiar examples include upright posture, opposable thumbs, brain size and organization, diminution of the digestive tract and its energetic and behavioral constraints, neoteny and dependent care, vocal articulation and auditory discrimination, and more (Stringer and Andrews 2012). Comparative primatology emphasizes cooperative dispositions among human infants and children markedly different from young great ape behavior (Tomasello 2019). As noted in chapters 2 and 3, these anatomical continuities are only realized in specific developmental environments that affect how those features are manifest.

different situations within and among those practices powerfully enable and constrain their lives and prospects; and how to understand the character and limitations of human capacities for critical assessment and transformation of those practices amid our interdependence with many other organismic ways of life.

5

Normativity

Practice-Constitutive Normativity

Normativity, broadly construed, concerns what is open to nonarbitrary assessment, whether as good or bad, successful or failing, correct or incorrect, meaningful or senseless, right or wrong, just or unjust, appropriate or inappropriate, justified or not, and so forth. As Brandom (1979) once noted, which phenomena are normative is itself a normative matter of how to understand and respond to them correctly or appropriately. The complex normativity of human ways of life has seemed to provide compelling reasons to distinguish a social world from human biology as a natural phenomenon. Scientific understanding of physics or chemistry provides no defensible basis for understanding natural phenomena as normative. Evolutionary biology does provide a limited basis for nonarbitrary evaluation of organisms and their lineages but only concerning successful continuation of those lives or lineages. The human lineage has been extraordinarily successful in this respect so far, but human lives, actions, and institutions are open to assessment in many other ways: moral, political, instrumental, epistemic, aesthetic, social, and many more. These diverse normative concerns seem grounded in how people respond or ought to respond to one another's activities and achievements, and hence seem integral to human social life.

Research programs in human sociobiology or evolutionary psychology attempted to account directly for many aspects of human behavior and social relations as successful evolutionary adaptations (Wilson 1975; Barkow, Cosmides, and Tooby 1992). A naturecultural conception of practices as biological niche construction proceeds differently by showing that the normative diversity and complexity of human ways of life are an evolved and extended form of biological normativity. Organisms and their lineages are complex, interdependent processes of environmental intra-action, with a constitutive

goal of sustaining those processes in changing conditions. Continuing fulfillment of that goal is threatened by predation, resource competition or depletion, disease, or other deleterious environmental changes extending to the
possibility of mass extinctions. How an organism develops and how its way of
life evolves in response to those threats affect whether the organisms and lineage sustain themselves successfully. Within a lineage, individual organisms
vary in their responses, and that variation leads to evolutionary changes in
both the population of organisms and the environments where their survival
is at issue. The physiology, development, behavior, and evolution of most organisms is nevertheless only normative along a single dimension of success
or failure in sustaining life and producing living descendants. It does not matter biologically whether both remain relatively stable or change extensively
through the combined effects of phenotypic plasticity, mutation, environmentally driven natural selection, organismically driven niche construction,
and random genetic and developmental drift. The "goal" toward which that
process is directed, successfully or not, is internal to the process, with no
determinate character apart from homological continuity with its own prior
incarnations.

The evolution of a practice-differentiated way of life introduces a second dimension of nonarbitrary normative assessment into life cycles of the
hominin lineage. We saw earlier that practice-differentiated ways of life in
our lineage are a complex form of conjoined material and behavioral niche
construction. That differentiation now involves building and maintaining
needed material settings and equipment along with developing and sustaining the requisite skills and behavioral patterns. In some cases, notably languages, human neural, anatomical, and practical-perceptual development
also coevolved with those practices to take their current forms. Like the (human) organisms who participate in them, however, and like other organisms
in this respect, practices are patterns of environmental intra-action which
only continue to exist through their ongoing reproduction. That reproduction is internally goal directed in much the same way as organisms and lineages are, and practices can succeed or fail in maintaining themselves within
human ways of life.

This evolved two-dimensional normativity encompasses the full diversity
of normative concerns usually understood as socially or rationally grounded
norms. These concerns comprise two dimensions of normativity rather than
distinctly biological and social-normative registers, because they conjoin to
comprise a more complex "space" of normative assessment. Understanding
this biologically grounded conception of normativity begins with recognition
of two key aspects of the one-dimensional normativity of other lineages. The

first consideration is that biological normativity arises from organisms and their lineages as Aristotelian energeia, processes whose goal is to continue that process. Sustaining an organismic lineage serves or satisfies no external or encompassing normative concern; nature is not better or worse for the continuation or extinction of any lineage. To that extent, biological normativity is consistent with modern scientific conceptions of nature as anormative. A second aspect of biological normativity is that the goal of a biological process is only determinable anaphorically by reference to a process already underway. In biological terms, those processes are only goal directed as homological patterns of descent, not analogical patterns of morphology or function. Their goals cannot be correctly characterized by general type-specifications of a process currently instantiated by an organism or its lineage. The goal of a life process is *its* continuation in whatever form its descendant processes take. Both considerations are relevant to understanding the two-dimensional normativity of a practice-differentiated way of life.

A second dimension of biological normativity first emerges with the evolution of cooperative activities that are behaviorally niche constructive. Behavioral niche construction is behavior whose presence in the developmental environments of other organisms helps produce "descendants" of that behavior. Many organisms engage in social learning or socially coordinated behavior. Those behavioral patterns are only open to assessment one-dimensionally for their instrumental contribution to the organism's deictic goal-directedness toward the sustenance and reproduction of life and lineage. Cooperatively interdependent human behaviors can also be instrumentally assessed as contributions to sustaining the lineage, but they also establish new forms of partially autonomous goal-directedness. As an example, Brandom highlighted the expressive freedom enabled by discursive practices:

> When one has mastered the social practices comprising the use of a language sufficiently, one becomes able to do something one could not do before, to produce and comprehend novel utterances. One becomes capable not only of framing new descriptions of situations and making an indefinite number of novel claims about the world, but also becomes capable of forming new intentions, and hence of performing an indefinite number of novel *actions*, directed at ends one could not have without the expressive capacity of a language. (1979, 194)

Languages do have instrumental significance for human ways of life, and the evolution of language was undoubtedly sustained in significant part by its adaptive significance in coordinating early hominins' social foraging and collective defense against predators and competitors. With discursive practices underway, however, new normative considerations arose concerning the expressive capacities those practices enable.

The novel goal-directedness that Brandom describes as "expressive freedom" exemplifies how situated, interdependent performances that make up a niche-constructive practice can institute new normative concerns as constitutive goals or "ends." Rudimentary forms of "protolanguage" likely originated and were sustained by instrumental contributions to an emerging way of life as cooperative foragers (Sterelny 2012a, 2012b, 2021; Bickerton 2009, 2014; Dor and Jablonka 2000; Tomasello 2008). With this initial expressive repertoire established, however, it also provided new capacities recognizable and appreciated by other speakers. Language enables new expressive and ostensive capacities, new forms of personal interaction, new ways of coordinating action, and more complex imaginative constructions of stories, poems, or theories with their own constitutive forms of virtuosity or excellence. Realizing these new capacities not only depends on encountering other speakers and listeners familiar with and responsive to prior uses of words in patterns mostly intelligible to one another. That mutual interdependence would not matter unless other speakers were also appreciative of these new capacities and responsive to the normative concerns constituted by participation in the practice. Those latter concerns are the ends of the practice, not as something it enables that is external to its continuation but instead as an aspect of the normative significance of the practice itself within one's life. Like the organismic way of life to which they belong, practices are Aristotelian energeia whose constitutive goal is to continue the practice. What confers significance on those interdependent performances is one another's ability to recognize, appreciate, and build on what they achieve together by sustaining the practice and its constitutive capacities and ends.

The dependence of language use on alignment with others' linguistic skills and performances is highlighted by the extinction of many languages through the dissolution of the groups of people who spoke them or in the face of competition from other languages. Those developmental environments no longer exist, and consequently those languages can no longer be spoken. Even if someone were to make similar sequences of sounds or marks, these would no longer be *words* without other speakers capable of responding to them and prepared to do so. Languages are sometimes described as living or dead, and this locution is appropriate. Scholars may reconstruct a language that no one speaks, but without other speakers to respond to those utterances by picking up on their expressive possibilities those reconstructed sounds or marks would be analogous to the corpse of a once living language rather than continuing its expressive life. The cognitive capacities or goods enabled and sustained by participation in ongoing linguistic practices are vividly illuminated by the plight of colonial plantation workers forcibly brought together without a common language among them or spoken around them. The expressive limitations of the pidgin communicative repertoires they initially developed in its absence sharply contrast to the expressive capacities of the creole languages that evolved as subsequent generations grew up in the linguistic environments of pidgin communication (Bickerton 2008).

Brandom highlighted *expressive* freedom as a capacity conferred by participation in a linguistic practice, but as patterns of mutually interdependent ongoing performance, language use exemplifies a more general and pervasive human phenomenon. The economic practices of selling and buying goods and services, producing those goods and developing the needed skills, and the more arcane financial and marketing practices now surrounding them are also closely coupled activities that depend on one another's continuing contributions. Their component activities would not be undertaken in the absence of ongoing continuation of those practices and their constitutive ends. We often think of economic practices as merely instrumental, responsive only to concerns satisfied by the outcome of production and exchange of goods and services. Particular exchanges are instrumental, but the *practices* of production and exchange to which they belong also have constitutive ends. Jürgen Habermas's (1968, chap. 6) reconstruction of familiar narratives of emergent "modernity" called attention to this internal purposiveness as a "legitimation from below" in emerging mercantile economies as "subsystems of purposive-rational action" that challenged the "legitimation from above" of traditional cultures and political practices. Recognizing that legitimating role shows that these new modes of economic life were not merely instrumentally rational (*Zweckrational*); these practices established normative concerns for fairness, mutuality, and equal standing. The economic cases nevertheless also highlight that when a practice does achieve separable outcomes, *continuation* of those achievements and contributions to a more encompassing way of life belong to the practice's own constitutive ends.

Many other human activities similarly sustain ongoing patterns of situated, interactive, and interdependent performance that constitute new relationships, activities, or capacities as constitutive ends. These include personal interrelations from greeting practices to sustained friendships or sexual relationships; teaching and learning; political governance and citizenship; measuring and tracking time or space; health and body care; games and sports; artistic expression and its display and appreciation; scientific investigation and its deployment; ritual forms of celebration, mourning, worship, or protest; armed conflict; and so on. This list could be extended indefinitely without exhaustion because new practices with their own constitutive concerns often diverge from or interconnect with other practices already underway. The many patterns of interdependent performance making up a practice-differentiated way of life enable people to recognize, fulfill, and sustain the normative concerns constituted in and made intelligible by those mutually supportive activities.

The constitutive interdependence of the situated performances making up a practice and the nearly decomposable system of practices making up a practice-differentiated way of life amalgamates different forms of coupling. Some activities can only be performed at all if done interactively: buying and selling, conversation, competitive sports or games, teaching and learning, greeting and acknowledging, or meetings for many purposes or occasions. Unless other people's performances sufficiently accord with one's own in relevant circumstances, one is at best attempting, rehearsing, or simulating those activities. Some activities materially depend on other performances or practices. They can only be done or done well with suitable equipment or materials, in conducive settings, or under the right conditions. Sometimes a practice and its constituent performances depend indirectly on other performances or practices: transporting people or things to or from a site, satisfying participants' material needs for food, water, waste disposal, shelter or protection, or providing them with requisite skills or background knowledge. Often participants could not *intelligibly* contribute to a practice without presuming other performances have or will take place: marking a piece of paper can only cast a vote or sign a contract if others recognize and authorize it as such and enable, enforce, and regulate its anticipated consequences. The connections between what one does physically and how it matters to do so are sustained by relations to other practices and their performances. These continuing interconnections let a practice make sense as having an intelligible end to which it is recognizably connected, but that end is embedded in mutual support among various performances and their niche-constructive descendants. As subsequent chapters emphasize, making sense is also a constitutive normative concern enabled by and requisite for participation in language and other conceptual practices.

I amalgamate these diverse forms of interdependence among practices and their performances because all contribute to shaping the normative concerns and responses to them that constitute practices as energeia, resulting in the two-dimensional normativity of a practice-differentiated way of life. If one's own performances and the practices and way of life to which they belong constitutively depend on what others do, how circumstances are arranged, or whether and how other practices continue, then assessing and responding to the presence and continuation of those prerequisites are integral to what one does. Many normative concerns arise from how situated performances of an ongoing practice depend on one another, and how performances of one practice depend on or support others. Their dependence on other performances and practices would not matter unless people were responsive to normative capacities and concerns enabled or fulfilled by maintaining or developing those practices within an interconnected system.

To understand the constitutive normativity of practices and a practice-differentiated way of life, consider the concrete example of nursing care for patients in a medical setting such as a hospital or long-term care facility. Nurses at work regularly undertake many activities: visiting patients in their rooms and observing them; taking and recording vital signs such as temperature, blood pressure, pulse, and others displayed on monitors; engaging in conversation, both to establish rapport and elicit reports of symptoms; administering medication or other therapy; maintaining appropriate prophylaxis; assisting in other medical procedures; communicating with supervisors, attending physicians, nursing colleagues, administrative staff, and patients' family or visitors; and more. To undertake these activities *as a nurse* is in part to be responsive to normative concerns such as patients' health and well-being, responsibilities as professionals in a hierarchical organization, and ethical issues of privacy, patient autonomy, social justice, and avoidance of harm. Why these particular concerns arise and what responses they call for emerge within the practice as issues that shape its development. Nurses need not and often do not understand or apply these concerns in the same ways or with equal diligence or skill, but they are responsible to and for those concerns as at stake not just in their own performances but in those of others around them and their equipment and facilities. Nurses do not just regulate their own performance, but take responsibility for how equipment, facilities, and the efforts of others answer to what is at issue in caring for patients as a nurse. Even if one performs some of nursing's characteristic activities, not to be oriented by and responsive to those concerns, however imperfectly or divergently, is not to participate in the practice, but at best to undertake a simulacrum of nursing.

These constitutive concerns shape the postures from which nurses are perceptually and practically responsive to circumstances. Some features of their surroundings (ought to) be salient in light of orientation toward those concerns while others recede from awareness or attention. The parenthetical "ought to" indicates that the salient reconfiguration of nurses' postured orientation in taking up the practice's constitutive concerns is neither a de facto regularity in their bodily postures nor a norm that perhaps no nurse actually satisfies. It is instead a situated normative concern that arises both from how nurses generally do comport themselves and what concerns they or others hold themselves accountable for.[1] What nurses encounter and recognize solicits some responses and not others. A nurse taking patients' vital signs in the morning may attend to those measurements sequentially and efficiently but will stop and respond immediately to a patient's very low or high blood pressure reading for its more urgent significance for patient care. To a first approximation, the features of practicing nurses' situations to which they attend and respond divide among those directly involved in patient care and those responsive to nursing's interdependence with other practices. The former include how nurses visit, observe, and interact with assigned patients, use equipment, perform procedures, or communicate findings to others involved in a patient's care, with what attitude and focus. The latter start with recognition that they should take up these activities only in the right settings—not just any medical facility but where one is authorized and assigned to work, at the right times, in accord with its procedures.

These concerns extend to encompass other practices and concerns through which nurses live and sustain a vertically differentiated *life* incorporating nursing. Nurses must have somewhere to go—to live—between nursing shifts. They need to eat, sleep, and engage in other sustaining and fulfilling life activities, including taking up other practices responding to those needs and concerns, from parenting and maintaining friendships and familial relations to shopping, commuting, or recreational activities. Other organisms' ways of life also include multiple life-sustaining activities they take up at different times, often in different places, and sometimes in different social groups. The lives of other organisms are nevertheless not vertically differentiated by the place of those activities amid horizontally differentiated practices. In people's practice-differentiated ways of life, whether and how we take up those activities are constitutively and normatively interdependent with the postures and behavior of other people whose lives are vertically differentiated in alternative ways.

Nursing is characteristically an occupation or profession that in modern economies involves monetary remuneration that enables participants to take up or depend on other practices. The constitutive normativity of occupational practices does not differ fundamentally from others, however, from playing a sport or practicing a religion to raising children or speaking a language. These and other practices require participants to take up postures that

---

1. This dual character of the normativity of practices is taken up more extensively in the next section.

are practically and perceptually responsive to circumstances that matter for sustaining the practice. The characteristic ends enabled by participation, the normative concerns whose satisfaction helps enable that mutual support, and those bearing on how the practice fits among or depends on others, together shape participants' normative orientation. Some of those practices play central roles in the lives of their participants while others are more peripheral, although participants in the same practice often vary in that respect. Those practices are also interdependent with participants' vertically differentiated involvement in other practices along with the many other practices making up people's evolved, horizontally differentiated ways of life. The interplay among the diverse normative concerns enabled by a practice-differentiated way of life is what establishes its characteristically two-dimensional normativity.

Taking up the practical-perceptual postures characteristic of nursing or other practices, exercising their skills, and making judgments responsive to their animating normative concerns presuppose *already-extant* practices in which one participates together with others. Appropriate settings and equipment, and other people participating in related roles as patients, administrative staff, and other medical practitioners, must already be in place for people to *practice* nursing. The practice, after all, involves caring *for* patients, *in* medical settings, *with* others; absent those persons and places, those postures cannot be taken up, and the relevant practical-perceptual discriminations and responses cannot be made. One can rehearse, simulate, or merely attempt to take up those postures in insufficiently supportive environments, but those are derivative postures whose character depends on the postures that actually engage circumstances that are mostly appropriately aligned with the practice's performances and ends.

People can and do initiate novel practices but they can do so only by establishing a place for them amid other practices already underway. Participants must be recruited, and that normally involves situating the new practice among others in which prospective participants are already involved. It also involves responding to normative concerns that prospective participants already have or that make sense in relation to those concerns. Any specific practice such as nursing, along with other practices on which it is interdependent, could have been arranged differently or not initiated or sustained at all. Alternative arrangements cannot actually arise, however, except by making a place for them amid an extant array of continuing practices complementary to or competitive with those already in place. In this respect, the normative concerns at issue in a practice and what is at stake in the ongoing resolution of those issues resemble the normativity of biological lineages more generally. The prospects for successful continuation and reproduction of biological lineages, their "fitness," cannot be assessed at any taxonomic level, including the performative lineages that make up practices, apart from the actual environments with which they interact.[2]

The Constitutive Ambiguity of Normative Concerns

The two-dimensional normativity of a practice-based way of life involves a complex evolutionary relationship between practices and the (human) lineage to which they belong. Practices can only be maintained if that lineage successfully reproduces, and if people can participate in those practices while maintaining or replacing themselves as living participants. Some practices—agriculture, food provision and preparation, sanitation and waste disposal, or those protecting from adverse weather or disease—contribute directly but partially to sustaining human lives and continuing the human lineage or its subpopulations. All practices, however, must fit into a more encompassing way of life, both to meet their participants' biological needs and to accommodate other practices they take up. Human ways of life have in turn become more dependent on maintaining that diversity. Our practice-differentiated lineage has sustained itself and can continue to do so even as many of its constituent practices change significantly or disappear and are replaced or circumvented. Sometimes that continuation requires adjusting, revising, or replacing some practices to fit into a changing nexus of practices as part of the global diversification of human ways of life; sometimes it requires adjustments elsewhere to enable a particular practice or nexus of practices to persist and satisfy their constitutive goal-directedness.

At first glance, the basic biological normativity of the hominin lineage as a whole, its success or failure in maintaining itself over time, might seem to remain alongside but distinct from the novel concerns constituted within practices making up our way of life. There would then be two distinct normative registers, a biological concern for survival and reproduction of the lineage, and the diverse normative concerns raised by the continuation or extinction of particular practices and their constitutive ends. A practice-differentiated way of life does not merely add the normative concerns constitutive of practices to the one-dimensional normativity of the biological lineage, however. The

2. Biologists often ascribe fitness to individual traits or genetic alleles, as stand-ins for the reproductive prospects of populations of environmentally situated organisms exhibiting those traits or possessing those alleles. "Fitness" does not denote the actual likelihood of reproducing organisms, lineages, or practices, which might be idiosyncratic, but instead an idealized propensity to reproduce indicated by counterfactually robust models of factors that *would* affect reproduction.

interdependence of those practices within a biological way of life transforms its normative accountability as a whole. People whose postured orientation develops through taking up a practice must then adjust that orientation not only to align sufficiently with the postured orientations and performances of other participants. They must also be responsive to the practice's more limited dependence on other practices and to the vertically differentiated demands of any other practices they take up in their own lives. How nurses treat patients, for example, needs to accommodate the place of nursing within a network of other activities, including its place within their lives and those of other nurses and patients. These patterns of mutual adjustment among practices and their performances then generate further normative concerns for how to accommodate and sustain the lineage's component practices as ends together with one another and with the biological needs of their participants as living organisms. The normativity of a practice-differentiated lineage *as a whole* is no longer one-dimensional, because the continuation of the lineage is shaped by a wider range of normative concerns for how its component practices and vertically differentiated lives are sustained together. Integrative concerns— examples include social justice, personal security, interpersonal and political community, balance among vertically differentiated practices, or how these concerns make sense together—affect how a practice-differentiated way of life evolves. The remainder of this chapter addresses how this more complex, two-dimensional normativity arises in a practice-differentiated way of life, and why it is a complex, evolved form of biological normativity.

Recent work on early hominin evolution and accounts of cultural evolution rightly emphasize the importance of evolved cooperative dispositions and skills, together with practices of teaching and learning that enable stable continuation of skills and achievements and ratcheted refinements or innovations that are also subsequently retained (Sterelny 2012a, 2021; Laland 2017; Tomasello 2019). These evolved capacities, dispositions, and practices are nevertheless not sufficient to understand the practice-based diversification of human ways of life. Achieving and sustaining such diversity is complicated. The increasing geographic, climatic, and trophic diversity of human populations and the increased complexity of interdependence among multiple practices with overlapping participants require a combination of sufficient stability and adaptability. Each practice encompasses a differentiation of roles, skills, tasks, and degrees of involvement. Practices also need to accommodate the recruitment, training, and assessment of novice participants alongside or amid their normal proceedings. When those practices are situated amid many other practices, they must utilize, revise, or circumvent many differences among participants: in expertise, commitment, familiarity with prior performances, complementary roles, and changes in other practices or circumstances, each of which can also vary in different locations. Familiar conceptions of practices that emphasize performative regularities, shared presuppositions, or governing norms ascribe too much stability and uniformity to the practice-based differentiation of human ways of life.[3] In this respect, these practice-based accounts of social life parallel the familiar but misguided efforts in biology to locate stable patterns or structures that either persist through development and evolution or causally determine them. In both cases, we instead need to understand human ways of life and the practices that partially compose them as irreducibly processual due to the coupled interdependence among those processes' changing components.

A thoroughly processual conception of a practice-differentiated way of life recognizes practices as temporally extended patterns of situated, interdependent performance. What each participant does depends not only on past performances but also on the anticipated future continuation of the practice and others on which it depends. Practices do need sufficient prior stability to establish a projectible pattern, but that initial stability only enables further continuation of a practice. It cannot determine whether or how the practice continues. The normative force binding participants to a practice, however, is that internal, normative goal-directedness: people participate in practices for the sake of what the practice is doing, what it can or will do, and how their participation affects or shapes their own developing lives. Participants' subsequent performances, even in aiming to continue the practice and fulfill its animating concerns, depend on others' future participation and changes or continuities in the associated circumstances.

The need to sustain practices through multiple generations of participants calls attention to a constitutive ambiguity in the normativity of practices.[4] Consider how new participants are recruited into extant practices so that their postured orientation toward their surroundings is partially shaped by those practices' constitutive normative concerns. Current participants initiate prospective new ones into the practice's activities, skills, and normative concerns, whether by explicit instruction, ostensive indication, or just inadvertently modeling its skills, performances, and concerns. To recognize a practice as responsive to a normative concern, in one sense, is for that con-

3. Similar difficulties afflict conceptions of human social life that emphasize establishment and maintenance of coordinated equilibria among incentives shaping people's behavior (e.g., Guala 2016; Cronk and Leech 2013), but this book does not address these and other conceptions of social life or cultural evolution.

4. Lance and Blitzer (2012) rightly argue that this ambiguity characterizes all invocations of normative concerns.

cern to play a predictive role in how the practice continues. Anyone who understands a practice as partially constituted by normative concerns could recognize and predict what participants in the practice will *normally* do under various circumstances, by predicting that they mostly will do whatever is called for by that normative concern. In another sense, however, normative concerns always outrun or exceed this predictive sense of what is normal. The question of how that concern applies to subsequent cases, and what it calls for in *this* case, is at issue in the ongoing evolution of a practice and its animating normative concerns. That is the sense in which answering to normative concerns always requires exercise of authoritative judgment.

Understanding the normative binding of people and their situated performances to ongoing practices requires the predictive and the authoritative senses of normativity to work together. Kant (2000) classically separated these two aspects of normative responsiveness by distinguishing reflective from determinative judgment, which map onto what I call the authoritative and predictive senses of normative concerns. Determinative judgments start from a norm or rule and apply it to cases. Reflective judgments start with a case and work out what normative concern or rule ought to apply. The difficulty is that these are not separable processes, because they do important normative work together. Any application of a normative concern to cases is reflective, by working out in practice what that normative concern means in or for those circumstances. It also implicitly requires reflective judgment about the relevant reference class of circumstances and performances. Normative concerns must nevertheless also have sufficient determinacy to provide an intelligible basis for their reflective application. The relevant sense of intelligibility is itself a normative concern arising from and sustained by the interdependence of various practices with discursive practices of social learning, training, ostension, and explicit teaching through which participants recognize, respond to, and take up practices. Discursively articulable intelligibility—making sense— is a multifaceted normative concern that emerged as central to the differentiation and continuation of a practice-based way of life.[5]

The predictive sense of normativity plays an important role in sustaining practices in both dimensions. Consider first its role in recruiting and training new participants in a practice. Acquiring the requisite postures and skills to participate in a practice involves learning to recognize practice-relevant differences in circumstances and determine what participants then (predictively) ought to do in those circumstances. Others can initially guide that

5. Chapters 6 and 7 address the character and importance of discursive practices and sense making within a practice-differentiated way of life.

recognition, but new participants must eventually develop abilities to recognize such matters on their own. The process commonly begins with novices imitating what authoritative participants do and gradually developing in practice the situationally appropriate postures, perceptual saliences, and practical orientations. Unless the normative concerns constitutive of a practice have sufficient predictive reliability to enable (some) new participants to pick up on those concerns and develop the requisite postured orientations, perhaps with guidance and correction from others, those practices would not be sustainable.

The predictive normativity of practices also helps sustain the interdependence of different performances and roles within a practice and their more limited dependence on other practices. What one does in a practice normally depends on what other participants do, and participants must therefore rely on others to act in sufficiently predictable ways. Teachers can only teach, or even know how to teach, if students mostly respond as students normally do, in the predictive sense. Students likewise rely on teachers acting in sufficiently predictable ways so that they can know what to do in response. Similarly, if performances of one practice depend in some respects on what is accomplished or determined by other practices, then the latter must have sufficient stability and predictive reliability for participants in the former to anticipate and rely on mostly proper performance of the latter. The requisite predictive understanding nevertheless also incorporates and must take account of predictable variations in performance. Predictive understanding of a normative concern is not limited to how that concern might be addressed ideally or even correctly but must also be able to take into account predictable failings and variations in responses.

Although the predictive sense of normativity thus necessarily has a role in sustaining practices, it is not sufficient to understand how they are held together normatively, even when amended by predictable variations in uptake. First, even the predictive normativity of a practice cannot be identified with a constitutive regularity *in* its past performances, even a noisy regularity. The reason is that the predictive sense of a practice's constitutive normative concerns as ends incorporates the predictive capacities of those who are *learning* to recognize the relevant normative concerns. Those recognitive skills are normative in a different way. As Haugeland once commented about the normativity of perceptual discernment:

> The relevant point can be brought out through a pair of complementary analogies. . . . The "chicken-sexers" of epistemological lore [are] semi-skilled laborers who, through long training, can learn to tell the sex of a newly hatched

chick. But they don't know *how* they can tell; it's as if they acquire a new perceptual primitive. . . . There are [also] various "hidden picture" puzzles and novelties . . . in which, through diligence and practice, one can come to be able to see, often with startling clarity, an image that was previously quite imperceptible. (1998, 327)

In the former case, the predictive capacities involved might be partly constituted by primitive skills in discernment acquired by some learners and not susceptible to further analysis.[6] In the latter case, the relevant predictability might be partly constituted by their ability to recognize mostly the same patterns already recognized by current participants, even if that mutual recognition has no common basis.

A deeper reason for the insufficiency of predictive normativity is that, although what people do as participants in a practice partly depends on a predictive sense of how other participants will respond to a situation and how other practices will be carried out, their subsequent performances are responsive to what then happens, with their predictive sense of its normative concerns changing accordingly. Similarly, people participate in practices with an anticipatory orientation toward a predictive sense of how other practices will affect what they do, but they nevertheless then *normally* defer to the judgment or performance of that practice's participants if it conflicts with their predictive anticipation.[7] The requisite predictive character of a practice may therefore change over time while still maintaining its continuity as a practice sustained by the interdependence of its situated performances. The normative concerns it embodies might also be dependent on the order in which new cases occur as models for subsequent assessment. Predictive uptake is also complicated by the possibility that success may be indicated by performance of another practice with its own constitutive concerns.[8]

The inadequacy of the predictive sense of normativity, as a predictably "normal" response to a normative concern, is well illustrated by difficulties

---

6. Further experimentation showed that chicken-sexers respond to olfactory rather than visual cues, though they claim to recognize the sex of chicks visually, but that does not affect Haugeland's point. The requisite predictability of past performances might result from learners' more holistic capacity to pick out combinations of considerations that collectively enable them to respond appropriately. These considerations need not be the same for all so long as each picks up on enough relevant considerations with an independent criterion for judging their success.

7. The term "normally" here implies deference to the judgment of participants in another practice in the predictive sense of normativity.

8. Brandom (1979) lucidly characterizes these and other reasons why the predictive normativity of a practice is not sufficient to determine how it ought to continue.

confronting a prominent conception of normativity that does take the predictive sense as sufficient. Dennett's (1987, 1991b) conception of the intentional, design, or physical stances as uncovering "real patterns" in the world relies on the predictive sense of normativity of the phenomena picked out by a stance, even though the resulting patterns may be noisy and imperfectly predictive. On Dennett's account, for example, one ought to regard a system as intentional if one gains nontrivial predictive ability by attributing to it beliefs and desires and inclination to do what one rationally ought to do in their light. One difficulty with Dennett's conception is that a purely predictive orientation is systematically blind to any distinction between a system that answers to a normative concern and one that merely seems to do so under a limited range of circumstances. That distinction bears on the projectability (Goodman 1954) of stance-based interpretations, which is crucial to attributing concepts of adaptive design or intentional directedness. The predictive sense of a normative concern would then discern a real pattern in the world but one merely simulating the normative concern in question.[9]

A deeper difficulty is that Dennett's account only concerns an observer's interpretation of a normative concern. In the most relevant cases, however, taking a stance toward a system as responsive to a normative concern is not just a synchronic predictive orientation but an interactive response to it. People do not merely revise their predictions of the behavior of those who seem to behave less than rationally; they typically respond to that behavior, whether by objecting to it, requesting clarification or explanation, intervening to constrain or block its untoward consequences, or otherwise responding to its normative significance within practices in which they also participate. Intentional attributions are normally not just observations and predictions of others but belong to the attributor's engagement with others in a mutually transformative practice.[10] The reciprocal failing of Dennett's predictive conception is that it cannot adequately accommodate the interpreted system's interpretive response to its interpreters. For the design stance, an organism well-designed to respond to stable environmental features may not be well-designed to respond to an environment in which other systems predict and subvert that response (Sterelny 2003, chap. 2). Similarly, Dennett's account of the intentional stance cannot

9. On the importance of projectability for empirical assessment of concepts, see Goodman (1954). Sellars (1997) argues more specifically that a de facto correlation cannot account for normative assessment of intentional classifications. Haugeland (1998, chaps. 11–12) then argues that Dennett's account of intentionality discerns a real pattern in the world, but only an "ersatz" imitation of intentional directedness toward objects.

10. Brandom's account of discursive normativity thus requires that "scorekeepers" in the game of giving and asking for reasons have their own score as fellow participants.

adequately account for an interpreted system that *takes* the intentional stance toward other systems, including toward the interpreter.[11]

For these and other reasons, the predictive, normalizing sense of normativity is not the primary sense in which normative concerns or ends are constitutive for practices, despite its important roles for initiates and outsiders.[12] Someone who takes up a practice as a novice does so successfully only on making the transition from merely imitating experienced participants to recognizing for oneself which considerations are relevant and what to do in response and mostly responding accordingly. Completion of that transition, from novice who merely imitates others to participant competent to judge appropriate responses to situations in a practice, is itself a matter for normative assessment rather than a determinate matter of fact, even though such assessments answer to evidence (Kukla 2000, 2002). Kukla's and my point here is clearly evident for learning to speak a language. There is no determinate point at which a child or other language learner makes the transition from merely imitating others' utterances to becoming a competent speaker performing speech acts on her own (defeasible) authority, even though that transition is eventually undeniable. Caregivers or teachers often respond to language learners *as if* they were competent speakers before that transition has happened, however, to scaffold their acquisition of the relevant skills by modeling appropriate responses.

To participate in a practice is thus to acknowledge its normative concerns *in practice* as authoritative for one's own performances. To be a nurse, to participate in the practice rather than merely imitate its performances, is not to attend to and care for the well-being of one's patients because that is what nurses do. Rather, nurses do what is needed to attend to and care for the well-being of patients because that concern is authoritative for them. That is what it means to recognize and respond to a practice's normative concern on one's own, as a more or less capable, authoritative participant. Practices can shift their constitutive concerns, however, so that what participants do now might look like they are just going through the motions from the perspective of its earlier incarnations, or vice versa. Such changes are nevertheless *shifts* in a practice's constitutive concerns rather than their disappearance, even if

11. Ironically, Dennett's proposal that systems are intentional if the intentional stance yields predictive benefits suffers from that reciprocal failing. Dennett's view is controversial, so predicting that people mostly *will* equate intentionality with predictability from a stance is predictively unreliable, even if Dennett were right.

12. The apparent initial plausibility of Dennett's conception of interpretive stances results from methodological restriction to circumstances resembling these constructive roles for predictive normativity.

the revised concerns are centrifugal, driven more by other practices in which its participants are engaged. In that case, the practice's constitutive concerns become primarily instrumental.[13]

We can now see why both the predictive and the authoritative senses of normativity must work together.[14] Initiation into a practice at first seems to give established participants authority over initiates, both individually and collectively. They provide ostensive orientation to relevant differences in circumstances, provide expressive articulation of those differences and how they matter, and exemplify and guide response to those circumstances according to the practice's constitutive concerns. Any challenge to an instructor's or guide's authority can be overridden by the aligned responses of other established participants: "this is simply what we do," in the predictive sense (Wittgenstein 1953, 1:217). Initiates' failures to accord with those authoritative recognitions, performances, and assessments would then be a failure to get it rather than challenges to the practice, its constitutive concerns, or other participants' understanding of those concerns.

Here we encounter the most important reason why the predictive sense of normativity is never sufficient, however. It is not enough for new initiates to learn to recognize what participants in the practice normally do in various circumstances in response to its constitutive concerns. Initiates must also be gripped by this recognition, to let that recognition also be authoritative over their own subsequent postures and the recognitions and responses they enable.[15] And here is where initiates to a practice, both individually and in mutual alignment, have countervailing authority that makes initiation into practices and the consequent reproduction of those practices over time a mutually transformative relationship. Novices might only recognize the norma-

13. MacIntyre (1981, chap. 14) limited practices to activities whose constitutive normative concerns are not primarily instrumental for other activities. A naturecultural conception of practices is more inclusive than MacIntyre's.

14. Davidson (1986) argues that predictive and authoritative senses of normativity are jointly necessary for linguistic meaning. He models interpretation of speakers as involving both a prior theory that models the predictive sense of the normativity of linguistic meaning and a passing theory that models its authoritative sense.

15. In Wittgenstein's remark that when interpretation and justification of a rule run out, teachers can only say, "This is simply what we do," that utterance expresses neither a performative regularity nor a regulative stipulation, but an interplay between the two in practice. The descriptive sense of "this is what we do" works together with the sense in which parents authoritatively say, "We don't hit other children, do we?" (Wheeler 2000, chap. 6), typically on occasions when children do hit others. Neither of Taylor's two practice-theoretical readings of Wittgenstein's remark is adequate (1995, 167–68, cited above in chapter 1), because both readings need to work together.

tive concerns of a practice, in the stronger sense of letting them be authoritative over their own performances, if the predictive sense of those concerns were to be revised in some respects. Other practitioners may dismiss their revisionist responses as mistakes or as evidence of incompetence, negligence, or failure, but the constitutive ambiguity of normativity shows that this assessment cuts both ways. An inability to bring new participants into accord with this assessment instead becomes a failure on the part of the practice to sustain its animating concerns, so construed.

The resulting ambiguity between mistakes, innovations, or reforms can arise in different ways. If established participants take up an educative role to correct initiates' revisionist responses to an ongoing practice, they may try to understand how and why those initiates see or do things differently in order to clarify how and why that response is misguided and correct it. Sometimes that strategy succeeds, and new participants can then take up the practice in closer accord with the previously predictive sense of its constitutive concerns. At least three other outcomes are possible, however. First, established participants might instead come to recognize for themselves (authoritatively rather than merely predictively) that what the initiate sees or does better realizes what the practice was already aspiring to in its prior performances. Even for established participants, the predictive sense of normativity does not exhaust a practice's normative concerns, which always potentially outrun their own predictive determination. Second, however, since a practice can only be sustained if it gains new participants, previous participants may need to take some initiates' alternative conceptions or performances into account predictively. They might acknowledge that those alternatives now express normal or permissible variations in how the practice's constitutive concerns are recognized and realized, despite their own disagreement with those responses. Third, they might instead take a stand against a supposed decline or abandonment of the practice and its constitutive concerns that they take to be embodied in those revisions. They then implicitly accept the possible disappearance of the practice, so construed, if the normative concerns they take as authoritative are not taken up and sustained by others. These latter alternatives become increasingly significant to the extent that new participants' revised uptake of a practice aligns with one another in conflict with prior performances.

Recognizing that the authoritative sense of normativity has priority over the predictive sense explains not only why the predictive sense is not fully reliable but also why it cannot and should not be fully reliable. People's acknowledgment or endorsement of a normative concern is not fully predictive of what they do for some familiar reasons. Normative concerns are notoriously not causally or lawfully determinative: people make mistakes, backslide, or override one concern in favor of others, including those arising in other practices they take up. The temporally extended relations among performances nevertheless provide deeper reasons for a gap between the predictive sense of normativity and the sense in which those concerns are authoritative for participants. Patterns of prior practice, even when they include discursive expressions of what the practice aims to accomplish, do not uniquely determine how that pattern should continue.[16] People may confront and answer to the same concerns that already animate an ongoing practice, despite different conceptions of those concerns, how to respond to them, and how that response matters. We just saw how such differences can arise as new participants take up a practice, but established participants may also have divergent responses to earlier performances and their outcomes. Participants in practices also differ in their experience and skill, sometimes leading to different performances, with different explications or rationales. Participants may respond in novel ways to relevantly novel circumstances or disagree about whether those circumstances are relevantly novel. Moreover, people can be misidentified as participants who would thereby be responsive to a practice's normative concerns; who is or is not a participant in a practice is sometimes at issue in the practice. In these and other ways, the authority that normative concerns have for participants in a practice can diverge from the predictive sense of normativity, even though that predictability provides a baseline that enables its own authoritative reinterpretation. Despite the many possibilities for performative divergence and interpretive conflict within practices, however, the constitutive interdependence of their performances constrains those differences and brings the predictive and authoritative senses of normativity closer together in their implications. The distinctive, two-dimensional normativity of a practice-based way of life arises from the resulting temporally extended interplay between these opposing tendencies, which we can now address.

The Temporality of Normativity

We have seen that what holds a practice together is not any common element or feature of its performances but instead their coupled interdependence. Participating in a practice and sustaining it over time does not involve

16. Practice-based accounts of normativity take the partial indeterminacy of regularities or rules as a principal lesson from Wittgenstein (1953) on rule-following, although not surprisingly, they respond to that lesson differently.

doing what other participants do or continuing the same patterns of situated performance as before. It only requires that participants' performances and the relevant circumstances be sufficiently aligned to sustain their coupled interdependence over time. "Alignment" in this context is not equivalent to a performative regularity or shared presuppositions.[17] Performances may be only imperfectly aligned, with considerable "friction" (Tsing 2005). They may be mutually supportive or enabling even though the performers do different things or do not share assumptions, intentions, or determinate norms. Consider the example of college classes from the preceding chapter. Teachers and students might understand and assess what goes on in a class differently, behave in partial conflict with what each would ideally prefer from the others, and yet each continue to participate because their interactions still contribute to one another's roles and goals. Competitive sports provide an especially clear example, since opposing players' actions are mutually aligned as performances that depend on one another to constitute a game—without the efforts of opponents, one is at best practicing or simulating the game—even though they also directly oppose one another. Moreover, participants' understanding of a practice and their own involvement in it often changes in response to how the practice develops.

Misalignments among performances of a practice and their circumstances continually arise for multiple reasons despite their need for mutually supportive alignment. Human ways of life are fraught with complications, confusions, conflicts, miscommunication, disruption, and incompetence. People's lives are also vertically differentiated by participation in multiple practices, and what they do in one context may need revision to align with their involvement in others, or to respond to misalignments with other aspects of a practice. Participants in practices typically encountered different histories of past performance and different training and skill-development in the practice, which affects their own performances. People may differ in how they understand or desire for how the practice will continue, and thus in their expectations and demands for what others will do or what its circumstances will be. Their participation in a practice may be contingent on their particular

17. Wartenberg (1990) describes an agent's ability to exercise power over another as mediated by "social alignments" among others whose actions are oriented by and coordinated with what the dominant agent says or does. I use the term more comprehensively. First, it encompasses how performances align or misalign in any practice, not just those sustaining power over others. Second, it encompasses how actions align with apparatuses and material circumstances. Chapter 8 addresses power as a normative, biologically grounded phenomenon, partly by critically comparing a naturecultural conception of power to Wartenberg's and other sophisticated accounts of social power.

understanding of its constitutive concerns, which may conflict with those of others. Circumstances are also not always conducive to their intended performances. As a result, people often set themselves to recognize and respond to their situation in ways misaligned with what others do or with other circumstances affecting their performance. Encountering such misalignments, participants may adjust what they do, rearrange the circumstances, point out difficulties, call for others to adjust, persevere in the face of incongruity or failure, or reconfigure their lives to abandon participation. Issues arise within a practice wherever performers encounter misalignments due to the resistance or crossed purposes of others or the recalcitrance of equipment or other circumstances. Their responses to these issues can be misaligned in turn, whether due to differing recognition or conception of the issue, conflicting efforts at adjustment, or ineffective modifications of recalcitrant circumstances.

What does account for whether and how practices hold together over time and how the performances of those practices are mutually accountable is participants' *need* for effective alignment of their own performances with those of others in conducive circumstances. Most of what people do depends on such alignment—that is, on successful continuation of a practice. Absent appropriate support, those performances would not be effective. Indeed the expectation of such support—the predictive sense of the normativity of practices—is crucial to the intelligibility of those performances to the agents themselves or to others, even when that expectation is thwarted by misalignments that threaten to render their performances unintelligible. Participating in an ongoing practice is not just doing what has been done before; it involves adjusting continuation of that pattern to sustain the practice's constitutive goal-directedness in response to changing circumstances, including changes in what other participants do. Their performative strategies thus answer to one another, and as we shall see, participants often try to hold one another accountable for such accommodations. Each depends on supportive alignment with one another in an ongoing practice, and the shape of that alignment depends on how a prior pattern of interdependent performance is sustained and revised. Misalignments with others' performances or other circumstances thus motivate adjustments of one's posture and orientation to reestablish supportive alignment, including efforts to induce others to adjust.

Practices are consequently shaped by a dynamic internal tension. Out of an evolved need to participate in some of the practices making up their developmental environments, people adjust what they do in response to misalignments they encounter, including attempts to influence others or change the circumstances. They do so to sustain the practice and its constitutive goal-directedness. Other participants adjust what they do in turn. Each does so from only partial and partially overlapping acquaintance with how the practice has developed so far amid differences in their own aspirations and needs. As a result, although everyone to some extent aims to adjust their performances and involvement to sustain or restore sufficient alignment with what others do and satisfy the practice's constitutive ends, that aspirational alignment is a moving target. Authoritative responses to misalignment may even directly counter some predictive aspects of past performance, taking the latter as telling indications of what *not* to do in light of the misalignments that resulted. Even in cases of such abrupt shifts in a practice, however, the force of that "not" arises from how it sustains other continuities with past patterns of situated performance.

The ongoing need to align what one does with the actual and anticipated performances of other practice participants has led to many iterations of ratcheted niche *re*construction. Through such extended, multigenerational material and behavioral niche construction, humans have evolved more extensive and articulated forms of interdependence. One consequence is that individual persons now have much greater difficulty providing for their own basic biological needs—food, water, air, shelter, sexual and other interpersonal relations, dependent care, and avoidance of disease, predation, disaster, or violence—without effective participation in many practices. These practices are in turn partially supported by other practices in which many of their own participants are not proximately involved, and those basic needs have evolved in turn. They have also been complemented and complicated by the normative concerns that emerge within practices as integral to their own goal-directedness: the goods, achievements, skills, virtues, obligations, and other considerations recognized or proposed as conducive to the practice, or as otherwise worthy of recognition and encouragement by those who take part. Someone imagined as an isolated individual whose life was not vertically involved in multiple practices would have even greater difficulty fulfilling the many other concerns now central to human ways of life. Whether these concerns involve specific forms of recognition and standing, enhanced instrumental capacities, or the extraordinary variety of goods and achievements only specifiable and appreciable through understanding the relevant practices (MacIntyre 1981, chap. 14; Brandom 1979, sect. 3), their fulfillment constitutively depends on sufficient alignment with what others do in interrelated practices.

What initially seems at stake in resolving issues arising within an ongoing practice is whether that practice continues, including whether its characteristic normative concerns are recognized and either satisfied, revised, or abandoned. Some responses to misalignments in practices may lead other participants to modify or discontinue subsequent participation. That outcome can then affect what is at issue for others, including possibly diminishing their own motivation to participate. Practices only exist through current and subsequent generations of participants continuing to act in ways that intra-actively depend on one another. If enough people stop acting in these ways, or act in sufficiently discordant ways, the practice falls apart or disappears. Confronting that prospect, which would remove conditions needed for their own performances to be intelligible or successful, some participants may modify what they do or say to encourage others to remain involved, accommodate their divergent needs or interests, discourage them from opting out, or change inhibiting circumstances. Tradeoffs thereby arise in the course of ongoing shifts in a practice's patterns, between whether a practice continues and what it is becoming—that is, its constitutive goal-directedness. Only some of the requisite modifications would be acceptable to various participants, or effective in sustaining its constitutive concerns, and some would require further adjustments elsewhere.

Articulating the World (Rouse 2015, chaps. 3–5) argued that the emergence of such tradeoffs is a consequential evolutionary novelty in the hominin lineage.[18] Other organisms develop and evolve, but their constitutive life patterns are one-dimensionally normative. What is at stake in their evolving patterns of environmental interaction is only *whether* their lives and lineage are maintained and reproduced. Human evolution has become two-dimensionally normative in that what is at stake in how people respond to issues arising within their lives and life patterns—the practices in which they participate and others on which that participation depends—is both *whether* their lives and lineage continue and *how* they are lived. The latter concern encompasses which practices are sustained, toward what ends. What normative concerns are at issue in how those practices continue depends on many considerations: how the practices change over time, how those changes affect other practices, which practices are taken up by whom, and how they reshape their participation in response to one another's involvement in changing situations. This two-dimensionally normative way of life results from the niche-constructive evolution of multiple partially autonomous patterns of practice. What people do in what circumstances is proximately accountable to their involvement in ongoing practices shaped by partially autonomous normative concerns. Those developing responses also answer to other concerns arising from the

---

18. Differentiating evolutionary novelties from homologues of ancestral traits is central to the biological subfield of developmental evolution (Wagner 2000; Müller 2011; Love 2006).

place of those practices within vertically differentiated individual lives and from interdependence among the many practices that horizontally differentiate human ways of life.

In discussing how goal-directed participation in practices is accountable to other practices and to what other participants do amid changing circumstances, I have spoken of normative concerns, standings, issues, or stakes rather than norms, goods, virtues, rules, commitments and entitlements, obligations, laws, or other determinate normative statuses. This approach to understanding normativity supplants two familiar strategies for understanding the content and authority of normative concerns. Normativity is often identified with accountability to determinate, authoritative norms whose authority is established rationally. Such strategies are familiar from Kantian accounts of moral norms such as Rawls (1980), Korsgaard (2009), or Herman (2022), or the consequentialist accounts of ethical norms surveyed in Driver (2012). An alternative, expressivist strategy distinguishes normative attitudes (the feelings, judgments, valuings, sanctions, and other responses to performances or circumstances which *take* them as subject to normative assessment, often by so assessing them) from the normative statuses (goods, rights, obligations, virtues, successes, justifications, meanings, etc.) attributed by those attitudes. Expressivists—Gibbard (1990) or Brandom (1994) are familiar examples—characteristically develop accounts of how normative statuses are instituted as authoritative by interrelations among normative attitudes, which then are the basis for assessing the appropriateness or adequacy of the attitudes or takings that originally projected or instituted them.

Understanding normativity as biologically grounded changes this order of explication. Biological normativity is not accountability to (already-)determinate norms, but instead concerns how various events and processes are dependent on and hence accountable *to one another* for continuation of that mutual interdependence. For the biological normativity of organisms and their lineages, each component of a life cycle "ought" to adjust to how others develop and evolve in ways that would sustain the process to which they each belong and on which they all depend. The force of that "ought" comes from the prospect of death or extinction as biological failures. Dennett (2014) argues that other organisms thereby have *reasons* for their behavior, anatomy, and physiology, even though they cannot understand, express, or act *on* those reasons. I am transformatively generalizing his point to encompass the normative concerns to which those attributed reasons would be responsive. Which changes in organisms' ways of life those concerns would provide reasons for, if the organisms in question could reason with one another and act on those reasons, depends on which other changes take place and what other concerns arise.[19] Reasons and reasoning are not the *ground* of normative accountability (or would not be so on Dennett's attribution of reasons to nonreasoning organisms), but are instead integral to a temporally extended *process* through which performances and circumstances answer to one another in conjoined interactions over time. Organismic adaptations and niche-constructive changes to their environments have no predetermined direction of fit. Sometimes a lineage undergoing environmental change would be sustained by the organisms developing and behaving differently to adapt to those changes. Sometimes the lineage would be more readily sustained instead by niche-constructive changes to the environment to make it more conducive to those organisms' development and reproduction. Organisms and lineages often fail to do what they ought in either case. Moreover, adaptive responses and niche-constructive accommodations to environmental change are linked and often work in tandem. An organism's phenotypic plasticity and a population's evolutionary adaptation to environmental shifts introduce further environmental effects. Similarly, any niche-reconstructive response to environmental change or to internal changes to organismic development or populational variation also usually changes organismic functioning, development, or behavior in some respects.

The resulting conception of normative accountability bears some resemblance to expressivist approaches but also differs in several important respects. Expressivist conceptions of normativity typically start with evaluative *responses* to various performances: feelings, reactive attitudes, interpretations, judgments, challenges, justifications, and the like. Talking about biological development and evolution in normative terms instead encompasses the whole pattern of organism-environment intra-action. Biological normativity does not simply concern how organisms and lineages ought to develop and evolve to maintain themselves in response to changing environments, for a biological environment is part of this ongoing, interdependent process. Everything an organism does and everything that happens in its environment can significantly affect what "needs" or "ought" to be done to sustain the internal goal-directedness of its constitutive processes. "Needs" and "ought" are in scare quotes to parallel Dennett's (2014) claim about organisms having "reasons" for their behavior despite being unable to articulate or express those reasons or respond to them *as* articulated reasons. I

---

19. The available variability in response to environmental change constrains the normativity of biological processes. What an organismic lineage "ought" to do in the face of changing environments depends on its genetic variation, phenotypic plasticity in those environments, and capacities for migration or perturbative niche construction.

am similarly arguing that the physiology, behavior, and niche-constructive effects that make up an organism's environmentally situated way of life are open to *nonarbitrary* assessment of their significance for the organism's and the lineage's internal goal-directedness, even though those organisms do not explicitly engage in that assessment themselves.[20]

The emergence of a practice-differentiated way of life then extends these considerations into more complex forms of mutual accountability. Like the (human) way of life to which they belong, practices are internally goal-directed patterns of interdependent performance. The end of a practice is not something external to it, but is instead the considerations or concerns for the sake of which participants take up and sustain participation in the practice. We have seen that the ongoing fulfillment of those concerns—the patterns of situated, interdependent performance which enable one's performances to align with others to let them be intelligible and appropriate in that context—can then be threatened by various misalignments among those performances and circumstances. Such threats to the continuation of a practice parallel how other organisms' lives and ways of life can be threatened by changes to their environmental circumstances or to their bodily processes responsive to those environments. Responses to those threats to a practice and its constitutive concerns similarly have no predetermined direction of normative fit. Sometimes participants need to change what they do to accommodate changing circumstances, including changes in what other participants do; sometimes what is called for are material changes to the circumstances in which the practice takes place, including how other practices support or obstruct it; sometimes sustaining a practice would require a shift in or reconception of its constitutive normative concerns, the end(s) of the practice; and at the limit, a practice may disappear because its constitutive ends can no longer be fulfilled or are no longer authoritative for its participants.

Discursive practices do enable explicit articulation of normative issues or concerns, along with determinations of how they bear on a practice or nexus of practices, but these expressions are themselves part of those practices.[21] The normative assessment and governance of a practice is not simply a matter of discerning a definitive standard, virtue, or good that already governs or defines that practice. The articulation and application of those concerns

---

20. Dennett would not use scare quotes for "needs," "ought," or "reasons," because he does not acknowledge a difference between one- and two-dimensional normativity. The significance of that difference for whether to attribute articulated reasons to other organisms emerges in chapters 6 and 7 below in considering how discursive practices articulate reasons.

21. This point is developed in greater detail in chapters 6 and 7 below.

is an aspect of the practices for which norms, goods, or other specifications of their issues and stakes are postulated. Practices are partly held together by efforts to express their constitutive goal-directedness and the normative concerns at issue in sustaining them. Such expressions play an important role in coordinating and guiding what participants do, how they understand their own participation, and how the practice is aligned with other practices with which it intersects. These efforts to articulate the issues and stakes in practices can nevertheless also be misaligned with one another or with what participants do. What people do may not fit how they make sense of what they do, whether to themselves or others. Some participants also make different sense of practices they take up together. Such misalignments may arise through mutual misunderstanding, misdirection, or disagreement, and their direction of fit likewise cannot be settled in advance. Sometimes people adjust what they do to accord better with the norms, goods, or virtues proposed to assess and resolve misalignments among performances and circumstances; sometimes these normative articulations are revised to accord better to how participants actually respond to those issues; on yet other occasions, revisions to relevant metanormative concerns reshape what would count as accord.[22]

Recognizing that talking about the normative interdependence of performances of a practice is part of the practice brings out the anaphoric character of such talk. The terms "issues" and "stakes" provide an explicitly anaphoric normative vocabulary. Participants in a practice—a temporally extended and spatially distributed pattern of interdependent performances—can encounter the same issues and answer to the same stakes without sharing a conception of what those issues and stakes are. Linguistic practices provide familiar examples of participants' divergent conceptions of their mutual accountability to the same issues and stakes. People can refer to an object by anaphorically citing someone else's uses of referential expressions (e.g., "the agent Donald Trump called the FBI's spy on his campaign") without endorsing all of the commitments undertaken in using those expressions (Brandom 1994, chap. 8). What is at issue in inferential relations among the multiple expressions used in this anaphoric way is what, if anything, they are talking about, how *it* should be identified and characterized, and how those characterizations bear on one another. Such disagreements may persist but are only disagreements

---

22. Springer (2013) develops a related conception of normative concerns as temporally constituted by people's accountability to one another *for* those concerns, emphasizing their discursive communication and adjudication.

at all if those uses are understood as mutually accountable, so that it matters to resolve them in ways that apply to all.[23]

The character and bounds of an object at issue in uses and mentions of referential expressions are not just analogous to the normative interdependence of performances of a practice; they are prominent instances of this phenomenon. Determinations of objects and their properties are normative issues in discursive practices rather than being somehow metaphysically predetermined: the causal or lawful entanglements of those aspects of the world may shape how those issues are or should be resolved, but they are caught up in larger patterns of practical engagement rather than settled once and for all causally or nomologically.[24] As a telling domain of examples, I have shown (2015, chaps. 8–10) how the alethic-modal lawfulness of scientific concepts interacts with their normative accountability within scientific practices and other aspects of human ways of life. Scientific practices, in this sense, are not self-contained "social" practices, but incorporate the phenomena that those practices articulate conceptually. They also encompass how that articulation enables holding the practice and its conceptualizations accountable to those phenomena and to other practices and phenomena.

The normative interdependence of performances within an extended pattern of practical interaction works similarly.[25] Efforts to say what normative concerns should govern the practice and guide assessment of its constitutive performances belong to the practice. Participants' concern for whether and how that practice continues constitutes the anaphoric accountability of its situated performances to one another for what is issue and at stake in the practice. They confront the same issues and stakes despite different "conceptions" of what the issues are and how they matter (their stakes), because what each of them does and can do depends on how the practice as a whole evolves from previous patterns of interdependent performance. "Conceptions" is in scare quotes to highlight an expansive use of the term, for which I hereafter

23. Sometimes the reference of anaphorically interrelated expressions and their characterization is resolved by divided reference, because they anaphorically conflated two different referents or modes of reference. That resolution then still applies to all of them, however.

24. Chapters 6–7 discuss the normativity of object boundaries and identity.

25. Temporal externalism (Rouse 2014; Tanesini 2014; Ebbs 2000; Jackman 1999) in philosophy of language extends earlier semantic externalisms of Putnam (1975), Burge (1979), and Kripke (1980) who claim meaning is accountable both to relevant material circumstances of word use and to social practices encompassing those uses. Temporal externalists extend that accountability to past and future developments in those discursive practices. On the view of practices developed here, normative semantic temporal externalism is subsumed by the temporality of normativity more generally.

eschew scare quotes. In this expanded sense, people's conceptions of practices in which they participate are not limited to what they could articulate discursively. Explicit characterizations of a practice are aspects of a more encompassing conception partially embedded in the bodily postures they take up, including the practical abilities, affective orientations, and perceptual saliences those postures incorporate. Discursive practices are not separable from that postured responsiveness, as the next two chapters argue in greater detail. How people can describe and explicitly assess a practice and their participation in it are part of that postured involvement in ongoing practices.

Participants in a practice contribute to an *already-ongoing* pattern of situated, interdependent performances. What they do both responds to its prior and current patterns and is oriented toward its continuation in partially modified forms. They typically have encountered only a partial, selective subset of the practice's prior performances and current patterns, and typically vary in the extent and depth of their grasp of the practice's development over time and in different settings. They do, however, normally understand others as engaged in activities that support, enable, and sometimes conflict with their own performances, and in undertaking those performances they rely on continuation of sufficiently supportive roles by others in sufficiently conducive circumstances. Each understands and copes with their situation and its challenges and conflicts as it encompasses not only one another's performances within that situation, but also their conceptions of the situation they share and the issues it raises for all of them. To a significant extent, their understandings of the practice and one another's place in it are embedded in the body postures they take up and what they thereby recognize as perceptually salient and practically significant. In responding to the situation by modifying what they do and inducing others to realign their own performances, each participant aims to redirect future continuation of the practice toward accord with how each understands its constitutive issues and stakes. What the issues "really" are, and what is at stake in how those issues are resolved and thus in whether and how the practice continues, are determined by interactions among those subsequent performances. The determination of those issues and stakes thus always lies ahead of how they are conceived at any time.

Emphasizing the constitutively open-ended accountability of performances within and to a temporally extended practice does not deny that the resulting patterns and institutions can be relatively stable or that expressions of what is at issue and at stake in the practice can be widely shared or endorsed. Practices are often sustained by achieving a more or less settled resolution of issues they confront. Such resolutions are nevertheless also temporally extended projections that remain open to new issues. Issues may arise from within an ongoing practice, from changed horizontal alignments with other practices, or from new participants for whom the practice is differently situated within their vertically differentiated lives. Recognizing that people contribute to normatively interdependent practices despite partial misalignments among their performances and assessments highlights a central problem for "classical" theories of practices, however. Achieving and sustaining a pattern of regular performance or of accepted, practice-constitutive presuppositions or normative concerns cannot explain the continuity, intelligibility, and normative accountability of practices and their performances. Any such regularities or other commonalities are instead consequential and sometimes fragile achievements that are themselves in need of explanation.

Two-Dimensional Normativity

The two-dimensional normativity of people's practice-differentiated way of life originated with the constitutive interdependence of their participation in multiple practices. One dimension of normative concern initially arises from within a practice. The closely interdependent mutual alignment of people's situated performances which sustains a practice enables them to undertake activities, develop skills, and achieve outcomes not possible without that mutual support in conducive circumstances. The goods thereby achieved (i.e., those practices' constitutive ends) are often only recognizable as such by those whose postures developed in contexts that make those goods salient and intelligible. Languages, economic exchange, food acquisition and preparation, religious worship, childrearing, manufacturing, scientific research, games, transportation systems, and the many other practices that now overlap in human ways of life each raise characteristic normative concerns. These concerns may include the skills and background understanding required to participate; the goods toward which it is directed, the hazards to be avoided, and the ability to recognize and appreciate both; the virtues and vices that contribute to or detract from effective participation; the suitability or unsuitability of materials, equipment, and settings; appropriate and inappropriate ways of timing, locating, coordinating, and rewarding or correcting one another's performances; and many other issues that motivate, sustain, or challenge supportive alignments among a practice's ongoing performances and circumstances. Sometimes these normative concerns are explicitly articulated or contested in terms whose significance and use are part of the practice. Even then, however, these concerns and their discursive expression are also embedded in people's developed bodily postures and postured responses to one another's performances in various circumstances. Participants' active, affective responsiveness to such practice-constitutive, proximate normative concerns plays a critical role in sustaining practices as ongoing patterns of interdependent situated performances.

These proximate forms of practice-constitutive normative concern are only partially autonomous, however. One reason for their limited autonomy is the material circumstances they incorporate, which both enable and constrain how the practice continues. Those circumstances may have been transformed by iterated cycles of niche construction, but those transformations also depended on the causal capacities of their circumstances. The emergence of multiple partially overlapping, interdependent practices also matters to their evolutionary significance in the human lineage, however. Although what participants do is proximately responsive to normative concerns internal to the particular practices in which they are engaged, their participation must also respond to their involvements in other practices and to how other practices bear on those proximate concerns and involvements. These overlapping involvements often lead to adjustments in participants' performances that partially misalign with what others do or with surrounding circumstances. These misalignments can reverberate through multiple practices due to how one practice depends in more limited ways on supportive alignment with others. Shifts in one practice that remove or change its supporting contribution to others may raise issues in those practices as well, and responses to those issues may then affect whether that originating change can be sustained, or whether it has further implications for other practices not originally affected.

The two-dimensional normativity of practices is often manifest in distinct but interdependent forms of normative assessment. A good example is the relation between intelligibility and truth in linguistic practice. What can be said and understood as meaningful—whether at the level of words, sentences, inferences, narratives, or other discursive performances—is primarily dependent on its relation to patterns of situated language use with which speakers and listeners are familiar. Meaningfulness arises from how people can pick up on those uses (predictively) and respond to or extend them (authoritatively) in ways recognizable (predictably) by others and taken up by them (authoritatively) in turn. Which utterances should then be taken as *truthful*—not just in the paradigmatic case of assertions, but in the broader sense of truthfulness that includes authoritative commands, successful ostensions, compelling narratives (Roth 2020), or other felicitous speech acts—then depends further on how those utterances are situated within more encompassing patterns of practice.[26]

---

26. Heidegger's (1962, 1975) more encompassing sense of truth situates correctness or incorrectness of sentences or judgments amid a wider range of comportments answerable to how

These two forms of assessment are nevertheless interdependent: truth values can change due to shifts in how utterances are understood and meaningfully related to other situated uses of those words, while recognition and acceptance of new truths can change how other utterances should be understood.

Other practices have their own forms of two-dimensional assessment. We can distinguish proper procedure, skillful use of equipment, or social propriety from successful outcomes, even though these assessments similarly affect one another as practices evolve. The apocryphal surgeon's report that an operation succeeded even though the patient died illustrates both the distinction between these dimensions of assessment and the need to adjust their relation to one another. Whether the appropriate adjustment to that misalignment is to change how or whether that surgical procedure is performed or assessed, or to make supportive changes elsewhere to enable the attributed surgical success to produce better patient outcomes, is then at issue in the development of those surgical practices. Those changes are responsive to the place of surgery among other practices and their concerns, including the ethical issues and professional tunnel vision that the apochryphal story was presumably intended to highlight. Moral assessment also has dual dimensions. It can be directed toward agency—the intentions or character-traits expressed in action—and toward action itself as consequential. Deontological, virtue-theoretic, and consequentialist moral theories often argue for the moral priority or exclusivity of one of these aspects of action, but they are genuinely competing accounts of the moral domain only because assessment of action normally involves both considerations.[27]

These aspects of the two-dimensional normativity of a practice-differentiated way of life are usefully clarified by comparison to Alasdair MacIntyre's (1981, chap. 14) conception of practices as constitutively normative. MacIntyre also understood practices as sustained by participants' orientation toward the practices' "internally" constituted goods or ends—the kinds of performative excellence or achievement that may only be adequately recognized or appreciated via the postures, abilities, and saliences developed within the practice. He then distinguished such goods from the more widely applicable

---

things (in the broadest sense of the term) are. Sellars (2007, chap. 14) is comparably sensitive to the normative significance of "things in the broadest sense of that term" as they "hang together in the broadest sense of that term" within the "space of reasons" opened by human ways of life.

27. Hegel's moral theory most insistently takes morality to encompass both dimensions of agency and its consequences. Pippin (2008) thoroughly examines Hegel's moral theory; Brandom (2019, chaps. 11–12 and 14 section 7) provocatively discerns this aspect of Hegel's view in *Phenomenology of Spirit*.

normative concerns arising from how those practices belong to more encompassing ways of life. As MacIntyre noted:

> What is distinctive of a practice is in part the way in which conceptions of the relevant goods and ends which [its] technical skills serve—and every practice does require the exercise of technical skills—are transformed and enriched by these extensions of human powers and by that regard for its own internal goods which are partially definitive of each particular practice or type of practice. Practices never have a goal or goals fixed for all time . . . but the goals themselves are transmuted by the history of the activity. (1981, 180–81)

MacIntyre thereby understood both practices' constitutive ends and other goods "external" to them more narrowly than in the naturecultural conception of practices developed here. He defined practices by particular internal goods and the virtues required to achieve them. MacIntyre then contrasted those goods to "external" goods as purely instrumental concerns exemplified by monetary reward, social status, popular acclaim, or political influence. Such external goods are only contingently related to particular practices and their internal goods.[28] MacIntyre's (1980, 1988) aim was to assess various historical traditions according to whether they could sustain goods internal to practices or could only recognize them as subordinated to "merely" external, instrumental goods.

Part of the constitutive history of practices within a practice-differentiated way of life, however, is how they contribute to or challenge and are supported or undercut by their place within that more encompassing way of life. I therefore understand practices more broadly than MacIntyre does as any patterns of interdependent performance held together in part by how those performances are assessed and revised. I also construe external goods more inclusively and relationally, as normative concerns at issue or at stake in how various practices are interdependent and mutually responsive. These more encompassing normative concerns include any basis for assessing the (possibly contested) internal goods or ends of a practice for their normative significance within or contributions to a larger nexus of practices. Mutual critical adjustment of internal and external assessment is partly at issue in the niche-constructive evolution of practices, rather than serving as a substantive

---

28. MacIntyre also distinguished internal and external goods from the empirical accountability of "technical" skills, whose normative governance comes from the internal or external goods they serve. I understand technical skills as also exemplifying practice-constitutive normative concerns, but this disagreement does not bear on why comparison to MacIntyre clarifies the philosophical significance of two-dimensional normativity.

ground for assessing historical traditions as MacIntyre proposes.[29] What is important about this difference in our views, however, is its implications for the wider range of integrative normative concerns that arise within our practice-differentiated way of life with its distinctively two-dimensional normativity. Attending to these implications brings this chapter to its conclusion.

One consequence of the anaphoric temporality and two-dimensional normativity of practices is the constitutive tension we discerned in how practices develop over time. In one direction, the dependence of what people do on supportive alignment with others' performances leads to a constant pull toward adjusting their performances, skills, and concerns to accommodate or overcome misalignments with what others do. People need to adjust what they do to achieve and sustain support from other participants in practices and from other practices. That is not quite a pull toward social conformity or a conception of practices as performative regularities or stable equilibria among incentives, however. Practices are often sustained with different contributions from participants, partially different conceptions of the practice and its goals, partially conflicting expectations of one another, and something more like satisficing rather than maximal fit among what participants do. In this respect, the evolution of practice-based ways of life in our lineage is comparable to evolutionary processes more generally, in producing jury-rigged compromises that reflect contingent histories and material limitations rather than forms and ways of life that might have been conceived by a rational designer. So one side of this constitutive tension among performances of practices is a pull toward reducing or accommodating misalignments among performances and other circumstances.

That need to sustain patterns of mutually aligned performances is nevertheless countered by tendencies toward fragmentation and dissolution. Practices are spatially dispersed and temporally extended, allowing only partial overlap among participants' familiarity with prior performances and the skills and capacities developed in those divergent settings.[30] As a result, people often conceive differently the practices in which they each participate

29. MacIntyre's conception of internal goods and their supportive virtues thus expresses a historically particular, substantive conception of the issues and stakes in contemporary practices and "modern" ways of life, which I do not address here.

30. The spatial dispersal of performances of an ongoing practice may cause splits among practices analogous to allopatric, peripatric, and parapatric speciation in which different species form due to geographic isolation and geographically localized variations within populations. Sympatric speciation due to divergent niche construction within an overlapping geographic range is more controversial in biology, but analogous forms of differentiation among practices are likely more common. Each of these analogues is well illustrated by divergent linguistic

and how their performances ought to fit together. Their willingness and ability to adjust what they do to sustain or restore a supportive alignment with others in an ongoing practice is often constrained by accommodation of their participation in other practices. Practices are also sometimes challenged or undercut by loss of supporting alignment with other practices, destructive incursions by practices that block their continuation as before, or recalcitrant circumstances. Such challenges and incursions can lead to further divergences, as participants respond differently to shifts in the practice's constitutive alignments. The differentiation of human ways of life into multiple partially interdependent practices thus yields shifting patterns of dynamic interaction.

The pervasive role of discursive practices only partially mitigates this dynamic of instability and stabilization among practices with opposing pulls toward fragmentation and accommodation. The ability to call out others and try to direct attention, coordinate and correct performances, express normative concerns, and hold one another accountable to those concerns does often play important roles in sustaining practices. The ability to describe practices and roles within them and to say what is at issue and at stake in their further continuation also aids such coordination and the initiation and training of new participants. It can guide assessments of how well ongoing performances are supportively aligned and how best to respond to misalignments. The ability to express normative concerns that (should) govern a practice nevertheless can also bring out more clearly or even initiate differences in participants' conceptions of the practice and what sustaining that practice requires. Expressing what is at issue and at stake in whether and how a practice continues allows other participants to challenge such claims, reason about them, and evoke affective response to the relevant differences. It is not only the performances and situations of an ongoing practice that can be misaligned, but also their discursive characterization and how people try to guide or influence one another's participation. The ability to talk about a practice and what is at issue in whether and how it continues is itself part of the practice, as the next two chapters discuss more extensively. The discursive expression of normative concerns is thus also a potential locus for further misalignment and instability.

The temporality of normativity as a biological phenomenon and its two-dimensional articulation in human ways of life then have important consequences for how to think about critical assessment of practices and ways of

---

dialects and descendant languages, whether from geographic or social isolation of participants or their linguistic utterances.

life. Over the course of evolution, countless organisms have lived and died, and myriad species and larger taxa have emerged and then become extinct. Some emerged under propitious conditions and flourished before passing away. Other individuals or taxa encountered straightened circumstances, more successful competitors, or were caught up in occasional mass extinctions. We can meaningfully assess them as having succeeded or failed to maintain life or lineage, including whether they developed or evolved to accommodate changing conditions, but there are no further plausible, nonarbitrary grounds for assessing their ways of life or their evolutionary adaptations as better or worse. In one respect, the hominin lineage is no different. Our lineage emerged from common ancestry with other primates amid stressful and often shifting climatic changes in a range of east African habitats (Potts 1996). Within a geologically short span that lineage dwindled to a single surviving species that nevertheless successfully migrated globally and achieved massive demographic growth. Sooner or later our species and its lineage will pass on in turn, outlived or supplanted by others, or encountering environmental challenges we cannot accommodate.

New possibilities for assessment and self-transformation nevertheless emerged *within* human ways of life. The niche-constructive diversification of human ways of life into multiple interdependent practices enables tradeoffs among which practices are sustained and how those practices change to resolve internal misalignments or accommodate other practices. Because what people can do depends on the present and future performances of others and how different practices interact, human ways of life are more readily open both to significant and rapid transformation and to stabilizing responses. People may take up different practices and those practices may be adjusted in ways that significantly affect others. New degrees of freedom and new possibilities for assessing and changing people's situations thereby arise from the ongoing reproduction of those ways of life. The emergence of language and other discursive practices dramatically expands those degrees of freedom by enabling participants in practices to talk about the practice and the issues it confronts, reason about them, and hold others to account for how performances fail to align with others.

What is at stake in resolving the new kinds of issues that human individuals and populations thereby confront is not only whether human lives and ways of life continue but what they are thereby becoming. That turns out to concern which practices are maintained and revised, how they change, and how other practices change to accommodate them, including the discursive practices that articulate normative concerns at issue in those developments. The next two chapters develop in greater detail what discursive practices are and how they are part of the niche-constructive adjudication of what is at issue and at stake within particular practices and the practice-differentiated way of life to which they belong. The most striking consequence of a two-dimensionally normative way of life, however, is that its constitutive goal-directedness is no longer limited to survival of the lineage. We saw earlier that practices depend on one another in ways that open a further range of normative concerns for how various practices fit together within individual lives and our encompassing, practice-differentiated way of life. Familiar examples of such integrative normative concerns include social justice, participatory governance, individual freedom, epistemic insight, personal ties of kinship and affiliation, religious commitments, and various collective capacities or achievements. In this more complex field of goal-directedness and critical assessment, people do not just act either to advance normative concerns internal to particular practices or to maintain life and lineage. They are also responsive to issues in how the practices they take up depend on, support, and answer to one another or fit together as normatively significant. It becomes possible and intelligible to stake one's life, the practices one takes up, or even the continuation of the lineage on how that way of life develops and evolves, whether for the sake of social justice, political equality, individual freedom, or the many other issues arising from the interdependence of those practices. The normativity of a practice-differentiated way of life thus does not just have two distinct components in the biological normativity of our organismic lineage and the "social" normativity of the many practices that horizontally differentiate our ways of life. The normative concerns at issue and at stake in human ways of life constitute a complex, two-dimensional field of concerns internal to particular practices and integrative concerns for how those practices fit together and which practices are thereby sustained. The biological continuation of the lineage and its practice-differentiated way of life then recedes into one among those many integrative concerns, even though its failure would also extinguish other practices and their constitutive concerns.

6

Language

Language as Niche Construction

The previous chapters worked out an initial reconception of practices as the developmental and selective environments for human animals. Practices are evolved forms of differentiated, cooperative interdependence among persons and their reconstructed material circumstances. People develop situated bodily postures, active capacities, and normative orientations by taking up these collaborative activities. This differentiated way of life enabled a newly complex, two-dimensional biologically grounded normative accountability. Language plays a central role in this way of life as ongoing behavioral and material niche construction coevolving with the hominin lineage. Languages only exist as lineal patterns of interactive expression and recognition. Their prominence in people's developmental environments brings about the differential reproduction of descendant expressive and perceptual patterns and capacities, with relevant forms of neural development and genetic assimilation in human bodies.[1] Language is a biological phenomenon that evolved amid and helped shape the evolution of a cooperatively interdependent and practice-differentiated way of life. Its differentiated forms became integral to the environments of subsequent human generations. The roles of linguistic performances in those environments generated selection pressures for their own reproduction and transformation, affecting people's bodies, cognitive capacities, and ways of life.

Discursive interactions are pervasive and salient features of human developmental environments. Humans normally develop in close proximity to and

1. Deacon (1997) and Bickerton (2014) address the coevolutionary role of neural development in language evolution. Dor and Jablonka (2000) emphasize the role of genetic assimilation of capacities for language acquisition.

involvement with others, and we are usually awash in spoken conversation from our earliest days of life. Children are not only audibly surrounded by conversations among proficient speakers whose bodily postures are visibly directed toward one another's gestural and vocal expression. Adult caretakers and older children address developing infants directly. From their first days of life, infants' tactile and emotional intimacy with others, visual focus on them, and utter dependence on their care and concern are difficult to separate from the vocal streams visibly and audibly directed toward them.[2] As children begin babbling and imitating sounds they hear around them, their caregivers typically respond verbally and gesturally to scaffold their conversational responsiveness. These vocal streams are also intertwined with practices and paraphernalia of caring for developing infants: feeding, clothing, comforting, cleaning, holding and carrying, entertaining and distracting them, and facilitating sleep. These practices, equipment, and linguistic expressions vary culturally, but human infants almost invariably grow up amid verbally expressive and equipmentally mediated bodily intimacy with caretakers.[3]

Human development is neotenous. Adults gradually initiate children into other practices, abilities, and expectations throughout their long developmental dependence. This involvement is facilitated by linguistic interaction with responsible caretakers and other children and often involves new linguistic forms and expanded vocabularies. Stories, song lyrics, rhymes, and other linguistic performances and practices extend and enhance linguistic abilities and enable the extension of human capacities for imaginative, projective, mnemonic, counterfactual, and subjunctive comprehension.[4] Linguistic expressions and their uptake are integral to diverse settings and performances and also commonly bridge those settings.[5] Linguistic perceptual and expressive abilities develop early and continue to develop through

2. The early development of congenitally deaf infants is more complex and varied in these respects. This topic is important and fascinating, but its complexity and requisite empirical detail would distract from the book's main arguments.

3. Di Paolo et al. (2018, chap. 9) more extensively describe these developmental initiations into discursive environments.

4. Dor (2015) argues that guiding imagination to extend people's shared circumstances beyond what is perceptually accessible provided significant selection pressures driving language evolution.

5. Young children do readily pick up on multilinguistic environments with different patterns of talk in different settings or with different interlocutors, but similar expressions reoccurring in different contexts usually aids understanding of both language and the situations in which those expressions occur.

mutually constructive involvement with novel social, material, and practical environments and abilities.

These pervasive and salient features of human developmental environments both illustrate and provide evidence for niche-constructive conceptions of language as coevolving with the hominin lineage, which have been extensively developed elsewhere.[6] Niche-constructive accounts emphasize the coupled coevolution of humans with languages in use. On these accounts, the initial emergence of rudimentary forms of communicative displacement allowed coordinated response to distant or hidden opportunities and hazards for cooperative foraging. Simpler expressive repertoires expanded as successful communicative cooperation generated selection pressures for ease-of-acquisition, reliability, and facility of protolinguistic abilities. Enhanced and more widespread discursive activity then led to neural reorganization and growth through people's development and evolution in verbal environments. Genetic and neurological changes played important roles in language evolution, but primarily as responsive accommodations to phenotypically plastic expressive innovations. Arguably, the subpersonal processes through which people construct sentences and follow others' constructions evolved through neural reorganization for efficient processing of these novel expressive and recognitive capacities.[7] The more complex structures of extant languages were eventual outcomes of these coevolutionary processes.[8] The hominin lineage evolved neurally and anatomically in discursive environments under selection pressure for ease of recognition and acquisition of an initially protolinguistic facility.

Languages only persist and evolve through their ongoing reproduction in successive generations. The developmental salience of verbal expressions in children's surroundings is enhanced by the prompting and guidance of

6. Niche constructive accounts of language evolution include Bickerton (2014, 2009), Dor and Jablonka (2000, 2001, 2004, 2010), Tomasello (2008, 2014), Sterelny (2012b), Rouse (2015, part 1), Dor (2015), Laland (2017). Deacon's (1997) emphasis on symbolic displacement rather than syntactic combination and recursion was an important precursor.

7. Bickerton (2014) persuasively argues that subpersonal neural processing to construct or interpret strings of words as intelligible sentences was a plastic neural reorganization to handle new kinds of "protolinguistic" input. Dor and Jablonka (2000) emphasize genetic accommodation of phenotypically plastic expression in a "stretch-and-assimilate" pattern. Selection for ease of acquisition and processing of verbal expressions also likely enabled greater expressive capacities, generating further selection pressures for enhanced abilities.

8. My account is agnostic between gradualist conceptions of the coevolution of hominins and languages and Bickerton's (2014) proposal of a multistage but comparatively rapid process in which displacement, symbolism, and enhanced neural processing of strings of verbal expressions evolved successively, although I incline towards Bickerton's reasoning.

caregivers and others around them. These pervasive aspects of human developmental environments let children integrate linguistic uptake and expression into their bodily postured openness and responsiveness to their surroundings. The most readily learnable and practically efficacious linguistic forms typically survive, and linguists have tracked some characteristic intergenerational transformations (Dor and Jablonka 2000; Tomasello 2008; Dor 2016). As rudimentary forms of linguistic expression, communication, and coordination became entrenched in and advantageous for human ways of life, the resulting selection pressures not only enabled more complex linguistic forms and uses to emerge. They also integrated discursive activity into other aspects of human lives, further changing the selection pressures affecting language acquisition and use. The emergence of structured linguistic expressions enabled a broader range of high-fidelity social learning and active teaching (Laland 2017; Sterelny 2012a, 2021). That high-fidelity uptake extended the ratcheted preservation and refinement of skills and practices, including linguistic practices themselves (Tomasello, Kruger, and Ratner 1993).

Niche-constructive accounts treat language use in both rudimentary and evolved forms as public activities integral to human developmental environments. Linguistic understanding is not first an internal mental capacity that people can then deploy to communicate and learn from others. Linguistic expressions in communicative use are instead integral to their surroundings. Linguistic understanding develops biologically as a perceptual/practical capacity for uptake of and participation in ongoing conversational engagements with others in partially shared circumstances. Capacities for internal monologue, mental rehearsal, conceptually articulated thinking, and critical reflection emerge alongside people's skillful uptake and participation in already-discursive human environments. Niche-constructive accounts of language thus encourage a broadly pragmatist strategy of explication and explanation. A pragmatist explanatory strategy begins with discursive practices and proceeds to understand syntactic and semantic structuring of those practices and intentional mental processes through their contribution to and dependence on the discursive uses of verbal expressions.

This broad explanatory strategy has been most extensively developed and argued for in philosophical traditions descended from Sellars (1997, 2007) and Davidson (1984, 2005b), among others. I do not here justify this order of explanation de novo. I instead conjoin this tradition with an ecological-developmental, niche-constructive conception of human ways of life to show that these influential research programs have not been *sufficiently* pragmatist. In different ways, their proponents develop what Haugeland (1998, chap. 9) called an interrelationist strategy.[9] They model language use as practices of reason-giving or rational interpretation that are nearly decomposable components of human ways of life. In these models, language use and linguistically expressible mental states or processes do depend on people's embodied involvement in the world but only via perceptual and practical interfaces between the causal or lawlike structure of physical events and a rational-normative ordering of language and social life. Biological understanding of people as living organisms plays no role in these accounts of language use as a social practice. Imaginative construction of interfaces between social-rationally normative discursive practices and the anormative inexorability of causal processes ignores the biological significance of people's involvement in the world as living organisms interdependent with their environments.

Brandom's model of discursive practices as "the game of giving and asking for reasons" is the proximate foil for my argument. Among those undertaking a pragmatist explanatory strategy, Brandom most explicitly examined language as a social practice. His account exemplifies one side of the split discussed in chapter 1 among practice-based conceptions of social life over the primacy and independence of bodily skills or discursive articulation. Understanding language use as a niche-constructive practice helps resolve this split. This reconciliation starts with recognition that understanding and speaking a language is itself a cultivated bodily skill requiring sustained interaction with verbal environments. Linguistic understanding develops in settings where a language is already spoken and other speakers typically respond constructively to children's or other learners' efforts to understand and participate in the practice. The acknowledgment of linguistic understanding and response as pervasive in human life and integral to how people live comes from the other side of the split, however. Speaking a language is not an isolated skill but informs and articulates other activities and capacities. Perhaps the most important reason to start with Brandom is that much of his richly articulated account of discursive practices can be appropriated within a niche-constructive conception to capture the integral role of discursive practices in people's practice-differentiated ways of life. In this respect, my critical appropriation of Brandom's model of discursive practices does not reject it for an alternative; it recontextualizes his systematic analysis with a different philosophical orientation. Turning Brandom's model inside out also exemplifies my broader approach to the practice-based tradition in social theory, which reconstructs practices as our niche-constructed developmental environments.

9. See also chapter 4, section 4 above.

Social-Rationalist Inferentialism and Its Discontents

Pragmatist accounts of language start amid already-ongoing discursive practices. They work out from within the practice an account of participants' prior, basic know-how which, if correct, could adequately explicate those abilities. Brandom models discursive practices as "the game of giving and asking for reasons," a broadly rationalist vision of humans as concept users who track and assess relations among one another's sayings and doings. Speakers play this game by making claims and reasoning from or to other claims. Speakers take responsibility for the commitments they thereby make or accept, through willingness and ability to give reasons for those claims if challenged. Making claims—asserting—is then the paradigmatic discursive performance, and other speech acts are meaningful only derivatively: the expressions they employ acquire conceptual content from their use in assertions and inferences. Sentences are the primary bearers of conceptual content as the minimal assertible expressions. Subsentential expressions are then contentful indirectly by contributing to the content of claims. The conceptual content of a claim is a function from what would serve as reasons for or other entitlements to the claim to its inferential consequences.

Reasons for or consequences of a claim do not determine its inferential significance in isolation but instead depend on other claims as collateral premises. Speakers differ over which claims should serve as premises and thus assign different semantic significance to the same linguistic expressions. The result is a strongly holistic account of conceptual content which incorporates ongoing adjudication of differences in speakers' conceptual commitments and entitlements. Understanding meaning and using language meaningfully is not grasping something shared or agreed on, but instead tracks and responds appropriately to conceptual differences among speakers. Brandom models this responsive tracking of one another's claims and inferences as "discursive scorekeeping": scorekeepers are also players with a score of their own. Subsequent performances change the scores everyone implicitly keeps on one another's discursive commitments and entitlements, which vary with differences among scorekeepers' own commitments and attributions.

Brandom thus construes linguistic meaning and understanding as a social phenomenon of mutual recognition among speakers. What people say and do is meaningful through how it responds to one another's prior and projected performances. Speakers are rational agents who track one another's performances and hold them accountable to accord with discursive practices and their projected continuation. Individual speakers have only limited, partial direct access to the spatially dispersed and temporally extended practices to which their own performances and assessments contribute. Their mutual interactions nevertheless provide wider indirect access. The content of what people say and do is not self-contained, but embedded in extended, partially overlapping interactions among many speakers' performances and mutual assessments. The resulting "space of reasons" is the indefinitely open-ended interconnection of those local discursive performances and assessments. That space is first and foremost conceived as social-normative standings of speakers and claims within a maximally inclusive community of rational scorekeepers. Brandom (1994, chap. 4; 2000, chap. 3) then argues that their scorekeeping practice is also world-involving through the discursive significance of reliable perception and action within the game model. Speakers' performances become accountable to a causally interactive world by answering to one another's reasoning.

Showing how intralinguistic performances and assessments within the game of giving and asking for reasons are accountable to causal interactions in the world is critical for Brandom's model. A central concern of practice-based social theories and modern intellectual life more generally has been understanding how normativity emerges from a supposedly anormative natural world. Brandom begins with the supposedly less problematic phenomenon of how participants in discursive practices *take* or *assess* one another's performances as normatively significant, and then shows how the normative status of those performances is instituted through interactions among those normative "attitudes."[10] Put differently, he starts with how performances *seem* meaningful (or not) and correct (or not) *to* discursive scorekeepers in order to generate a possible difference between seeming meaningful or correct and actually being so. This difference cannot be determined by speakers' attitudes alone if their utterances are to be conceptually contentful. Some claims would be wrongly made and some inferential connections should not be endorsed or drawn even if some or all participants in the game think otherwise. Utterances would not be meaningful claims and relations among them would not be inferences unless all were open to assessment and could be mistaken. Otherwise the game of giving and asking for reasons would be "merely" a game rather than a consequentially accountable aspect of human life. The scare quotes around "merely" highlight that games are also constitutively normative phenomena, and hence even chess or basketball is not "merely a game."[11]

10. This expressivist conception of normative statuses was critically discussed in chapter 5, which argued for an anaphoric conception of what is at issue and at stake in practices in place of Brandom's and other expressivist conceptions of norms.

11. Chapter 4, section 2 argued for this claim, using volleyball as an example.

Games are often taken as normatively self-contained, governed by rules that define them, and bracketed from normative significance outside of game play. The phrase "merely a game" is invoked in that sense: taking winning and losing, other successes or failures in a game, and whatever forms of excellence matter in play as internal to the game and unrelated to other concerns. The play of a game and its outcome may be *assigned* significance within other practices and assessed accordingly, but "the game itself" would be impervious to external considerations. Brandom needs to show that in modeling discursive practices "the game of giving and asking for reasons" is not "merely a game" in this sense.

Brandom's strategy for modeling human linguistic practices as the game of giving and asking for reasons is to show how claims and inferences within the game are about worldly objects and accountable to the objects themselves. Brandom's detailed explication of this strategy is a significant technical achievement. In modeling the objective accountability of what people say and do, he explicates linguistic representation *in* discursive practices. Starting from normative scorekeeping attitudes people take up within discursive practice, he generates a semantics of reference and truth without positing substantive referential relations between singular terms and objects or sentences and worldly "truth-makers." He also develops a pragmatist-expressivist account of logical and semantic vocabularies, including talk of propositional attitudes as expressive additions to *discursive* practice. Beliefs, desires, and other propositional attitudes are public normative statuses rather than internal mental states. People do not learn to speak and think by acquiring and using a logically and semantically structured language. Logical and semantic vocabularies instead have the expressive role of explicating aspects of what people already do in discursive practices. Brandom's own exposition would only succeed, however, if discursive practices so construed would answer to the world so as to vindicate his expressive strategy for explicating the normativity of discursive practice. The remainder of this section reviews why his account fails in this respect, and how that failure points toward an alternative, biologically based account of discursive practices.

The fundamental difficulty with Brandom's social-rationalist model of discursive practice is its disconnection from the materiality of language in the world. This disconnection is multiply manifest. The most prominent criticisms of Brandom's model target his accounts of the supposed interfaces between language use and perception or action. Brandom initially analyzes discursive practice as composed of intralinguistic moves—assertions and challenges, and inferences from and deference to them—which only then connect to perception and action as language entries and exits. Observation statements are language entries that provide default entitlements to judgments that can play a role in inference, while practical reasoning concludes with prescriptive judgments—language exits—about how to act. Moreover, Brandom embeds this strategy within a conception of discursive normativity that disconnects it more fundamentally from people's practical-perceptual immersion in the world. He treats perception and action as normatively inert causal processes. Perception or action do not figure into discursive practice directly, but only through speakers' *judgments* about one another's perceptual and practical reliability. Brandom's critics conclude that the space of reasons so construed would indeed be merely a game, composed of intricate, self-enclosed inferential moves without bearing on the world.[12] If discursive practices were like that, assertions and inferential norms would only answer to other assertions. In McDowell's (1994) picturesque formulation, reasoning in discursive practice would then be a "frictionless spinning in a void," constrained only by what participants in the practice are inclined to say or accept, and would thereby lack conceptual content.

This well-taken criticism of Brandom's model of discursive practice effectively undermines the account in its own terms. Unless he could successfully differentiate meaningful and justified claims from those that merely *seem* so, moves within the game of giving and asking for reasons would not be claims or inferences at all. They would only be "about" their relations to other moves in the game. Stopping with this global critique of his model, however, overlooks its potential to guide more adequate understanding of discursive practices. This global line of argument only shows, negatively, the debilitating consequences of treating language use as social-rational performances reducible to their normative score in a discursive game model. From that starting point, Brandom cannot reconnect those performances to the practical-perceptual settings in which and about which what people say matters. A more thoroughgoing critical response turns Brandom's account of discursive practices inside out by showing how language use is embedded in the world through people's postured engagement with practice-differentiated environments.[13] Such an inversion appropriates Brandom's impressive technical accomplishments within an understanding of language as integral to people's

12. McDowell (1994, 2010), Haugeland (1998, chap. 13), Rouse (2002, chaps. 6–7), Lance (1998), and Kukla and Lance (2009) develop versions of this criticism of Brandom.

13. Kukla and Lance (Kukla and Lance 2009; Kukla 2017; Lance 1998, 2000, 2015, 2017) pursue a similar strategic response to Brandom, albeit without exploring language as a biological phenomenon integral to human developmental and selective environments. For a more extensive engagement with Brandom that goes beyond the constructive appropriation here, see Rouse (2018; 2002, chaps. 6–7).

biological ways of life in massively niche-constructed, practice-differentiated, discursively articulated environments.

This inversion begins by recognizing that Brandom treats discursive practices as a nearly decomposable component of human ways of life by ascribing two kinds of interface between discursive practices and the world. Perception and action form the first kind of interface, as distinct causal interactions with a dual conceptual role. As language entries and exits, perception and action explicitly connect intralinguistic moves to the causal nexus to which those moves must be answerable. Perception and action also implicitly serve as interfaces between persons as keepers and targets of discursive scores and people as living, bodily beings. Brandom says little about this second role for perception and action, but careful reading of his work shows its central importance.[14] He then ascribes a second kind of interface between discursive practices and the world in representational relations between singular terms and objects. It is no accident that Brandom speaks of the normativity of discursive practices as objectivity; discrete objects are the locus of their accountability to the world. The remainder of this chapter indicates why these supposed interfaces between distinct components of human ways of life are instead more intimately commingled aspects of people's biological lives. The chapter concludes with a third interface implicit in Brandom's account even though unmentioned in his model of discursive practices and their social-rational normativity. By construing action as a causal phenomenon that only acquires normative standing through the discursive explication and application of governing norms, Brandom treats rational justification as the interface between discursive practices and the rest of human ways of life. Chapter 7 below shows how discursive articulation and reasoning are more intimately entangled in a practice-differentiated way of life.

Language Embodied

Brandom's strategy of first analyzing intralinguistic "moves" that then extend to incorporate perception and action fails at the level of the utterances themselves. Brandom argues that intralinguistic communication and justification are fundamental to discursive practice but that the noninferential authority accrued through reliable perceptual and practical abilities—empirical content—is not, since only the former is independently intelligible (1994, 221).

---

14. I have extensively analyzed (Rouse 2002, chap. 7) the dual roles Brandom ascribes to bodies and desires in differentiating conceptual perspectives, an ascription that implicitly relies on the Myth of the Given (Sellars 1997).

Even on his scorekeeping model of discursive practice, however, token performances in the game of giving and asking for reasons must be publicly accessible. Scorekeepers must recognize others' assertions and speakers must make
assertions recognizably. Language use is a thoroughly practical-perceptual
ability in which people's skills in vocal or graphic articulation mostly accord
with their abilities to notice and discriminate among those expressions. Intralinguistic inference and deference thus presuppose the perceptual and practical capacities that Brandom treats as philosophically secondary.

The vocal-articulative and auditory-discriminatory capacities to use and
comprehend spoken language are significant anatomical differences between
humans and other primates, and the match between those two capacities
is important evidence for the coevolution of humans and languages. One
important aspect of evidence for evolution by natural selection is the jury-
rigged and satisficing character of many adaptations. The matching of human capacities for vocal articulation and auditory discrimination of phonemic and morphemic differences exemplifies this feature; people's abilities
to produce sounds others can recognize and to recognize the sounds others
produce are fraught with error but not sufficiently to block mostly effective
communication. Moreover, such capacities develop in close, developmental
interaction with actual patterns of vocal articulation and uptake, rather than
as generally applicable, genetically assimilated structures of human perception and vocal expression. The phonemic and morphemic articulation of
languages are clear examples of patterns whose elements and differences are
intertwined with capacities to discern them.[15] These recognitive capacities
also conjoin the requisite auditory and vocal discriminations; the ability to
tell an auditory difference between two linguistic expressions is most clearly
manifest in reproducing that difference as recognizable by others. Moreover,
capacities for perceptual discrimination and semantic interpretation are not
separable. People do not first learn to discriminate the sounds, sound combinations, and word breaks of languages spoken around them and then assign
them semantic significance. They learn perceptual and semantic differences
in tandem, hearing patterns that make semantic sense (Ebbs 2009, chaps. 4–5;
O'Callaghan 2017, chap. 6).

Language is thus dually embedded in people's engagement with their biological environments. Spoken and written utterances are *perceptually* salient
features of normal human developmental environments. People's postures
are perceptually attuned to focus on words and speakers' expressive bodily

15. Dennett (1991) and Haugeland (1998, chap. 12) emphasize constitutive relations between
"real patterns" and capacities for pattern recognition.

posture as gestural, affectively expressive, and often ostensive.[16] This bodily orientation allows linguistic expressions to stand out perceptually from noisy auditory and visual backgrounds. Learning a language also prepares one to speak in that language in conversational or other discursive contexts. In that respect, speech falls between developed capacities for skillful movement as responsive to circumstances, and the effective incorporation of equipment within a bodily posture. Whether in skillfully wielding a hand tool or writing implement, maneuvering a vehicle, or controlling one's movements on skis or skateboards, people incorporate equipment into the active posture through which they encounter their environment, and they similarly take up languages as expressive "instruments." People do not take up a language quite as they grasp or handle a tool, but they take up postures that incorporate the expressive capacities enabled by languages spoken and understood around them.

A closer analogy to language as both perceptually salient and expressive, however, is the integration of bodily postures with the gravitational field within and against which people normally stand and move. Bodily postures, their grounding, and the earth's gravitational field together orient human environments around the differential practical significance of horizontal and vertical movement. People are expressively and perceptually immersed in language much as they move on the earth within its gravitational grip. We take up a language as we learn to stand, set ourselves, and move. People respond semantically to discursive surroundings, structuring utterances grammatically and rhetorically and forming phonemic patterns vocally, with appropriate modulation of tone, rhythm, volume, and gesture. People differ in overall linguistic facility and effective responsiveness to different discursive settings, but those differences parallel variation in bodily skills for balanced, situated movement. This dual character of understanding languages is especially evident to second language learners. Speakers acquiring a new language only gradually shift from haltingly perceiving and deciphering or constructing phonemically, semantically, and expressively significant utterances or written words to absorbing a linguistic practice within bodily capabilities that are smoothly and effectively oriented within a discursive environment.

16. Written texts are shorn of authors' tone, gestures, affect, and circumstances, instead standing out from a blank page and its boundaries, which seem to abstract words from extratextual relations. That appearance is misleading, however, since they only work linguistically when read, and readers' abilities, posture, and orientation replace dialogical bodily interactions among speakers.

The practical/perceptual significance of language is not exhausted by its perceptual salience and expressive deployment, however. Language users do not perceive vocal utterances as ambient features of their surroundings but as spoken *by* others and directed *toward* an audience.[17] Spoken language is addressed to someone, which often matters to what is said. Kukla and Lance (2009, chaps. 6–7) note that many utterances function primarily as vocative "hails" that single out others as called or as recognitive acknowledgments of those hails. All spoken utterances have a vocative dimension, however, even when only marked by tone of voice, facial expression, conversational context, or proper names. Vocatives can also work descriptively. Locutions such as, "If you are worried about the embarrassment of incontinence," "Would you like to earn an exciting vacation in only a few hours a day?" "Is there a doctor in the house?" or "Hey, you in the white sweater!" are vocative invitations to recognize oneself under those descriptions and respond to the call. The ineliminable vocative aspect of utterances embeds them in practical-perceptual settings that incorporate other people as expressive, responsive bodies.

Audience-constitutive vocative significance is also not just a perceptible feature of utterances, for it normatively reconfigures the encompassing situations of both speaker and audience. Utterances pick out the one(s) addressed, both as persons addressable by the speaker and as addressed in this way. They direct what is said toward the one(s) called, but also change the situation by calling for a response. That change takes place even if the addressee fails or refuses to recognize the call: going on as before is then a refusal or a lack of recognition. Vocatives can misfire in different ways. Addressees can *fail* to recognize themselves as called, *repudiate* the call by challenging the caller's authority to address them so (e.g., as a stranger, a social inferior, or one speaking "out of turn"), or *ignore* it by not even acknowledging entitlement to a response. In some contexts, speakers can "bait" others, for example with catcalls, racist or homophobic invectives, or other bullying that succeeds *by* misfiring as calls and thereby affirming the speaker's power to address others *in the face of* the target's repudiation (Kukla and Lance 2009, 147n13, citing Colleen Fulton).[18] The ability to call out others and call to them helps constitute an interpersonal relationship among people who might otherwise merely be in one another's proximity. Vocatives with a built-in, individuating call for uptake and acknowledgment have a further constitutive role in letting speakers

---

17. Kukla and Lance (2009, especially chaps. 6–7) discuss the vocative and recognitive aspects of language more extensively. My discussion is indebted to theirs.

18. Chapter 8 below considers how practice-differentiated environments are also thereby power-differentiated.

single out others and hold them accountable to the constitutive ends of practices and their normative interdependence.

Conversation is the proximate relationship structured by the interplay of vocatives and the uptake and acknowledgment they call for. Conversations are characteristically sustained until a participant closes them off and is acknowledged as having done so. They also establish and sustain or break off a broader range of interpersonal relationships, however. These situated interpersonal relationships range from the sustained intimacy of friendship or familial connections or the collegiality of working together to more transient micronegotiations. Saying, "Whew, it's hot" to someone sitting nearby at a bus stop on a sweltering summer day is not informing them or undertaking an empirical commitment but instead proposing a limited, situated interpersonal connection based in shared discomfort (Taylor 1985a, 259). Kukla (2021a) emphasizes such transient verbal micronegotiations as integral to urban life, where one regularly encounters and deals with diverse people, in varied settings, with limited shared background.

Brandom imagines declarative assertions primarily as moves in an abstract discursive game, understood and evaluated in relation to other similar discursive moves. Kukla and Lance (2009) instead show that the grammatical forms of utterances are often poor guides to their *situated* discursive significance. Everyday language use is instead mostly embedded in postured interrelations among people as they make their way through the world. A model of situated exchanges as shaped by a more concrete implicit "scorekeeping" would need to track interpersonal relationships and situational negotiations rather than just linguistic relations among assertions and associated discursive commitments and entitlements. Understanding how utterances affect people's places in one another's lives or reconfigure their current situation is a more basic consideration to which those performances are accountable.

Brandom's model might seem better suited to more impersonal verbal performances and exchanges in academic, political, or journalistic discourse. These verbal performances do often develop more extensive inferential relations, but their significance still remains conversationally situated. Academic discourse is not indifferent to its venues or their disciplinary or interdisciplinary contexts. The empirical commitments undertaken in academic publications are also practical efforts to extend or redirect ongoing conversations, and recognizing their proximal addressees and the conversational state of play among them is important both to what is said and how it is heard or read. As Bruno Latour once noted about scientific prose, "I do not say that because the literature is *too technical* it puts people off, but that, on the contrary, we feel it necessary to call technical or scientific a literature that is made to isolate the reader by bringing many more resources. . . . Reading a sentence after the other sentences, we have not suddenly moved from opinions and disputes to facts and technical details; we have reached a state where the discussion is so tense that each word fences off a possible fatal blow" (Latour 1987, 44, 46). Political discourse is comparably situated. Those who imagine it as extended reasoning about how best to conduct public policy apart from ongoing negotiation of situated relationships among political agents and constituencies are politically "out of the loop."

Articulating Objects

Picking out other persons or groups by hailing them is not the only articulative effect of discursive practices. These interactions enable ostensive articulation of people's partially shared circumstances and recognitive expression of its perceptual uptake. Kukla and Lance point out that perception's indispensable role in forming empirical concepts and justifying empirical knowledge requires experiential uptake: "The edifice of empirical knowledge, and with it our justified ability to make empirical assertions, depends upon there being chains of commitments and entitlements that terminate in *someone's* first-personal experiences. The necessary termination of empirical claims in experience means that whenever we make an empirical assertion, we are committing ourselves to someone having had an experience—a receptive encounter with concrete features of the world—that grounds this assertion" (Kukla and Lance 2009, 62). What Kukla and Lance call "observatives"—discursive performances expressing perceptual recognition—are not just declarative assertions, even when grammatically structured as declaratives. Their pragmatic force indicates not just an empirical content but taking responsibility for that content and licensing others to rely on it. The pragmatics of observation extend further, however. The conceptual and epistemic roles of observatives depend on recognition by others as well as observers' experiential encounters and their discursive expression. Observatives claim epistemic authority and fail if that authority is not acknowledged.[19]

As a second consideration, observatives extend beyond expressions of first-personal authority by picking out aspects of surrounding circumstances

---

19. Dependence of observative or testimonial authority on recognition by others has been widely discussed philosophically, from Hobbes (1994, chap. 10) on honoring speakers and Hume (1993, chap. 10) on assessing testimony to miracles through Shapin and Schaffer (1985, 55–72) on witnessing experimental demonstrations before the Royal Society and Fricker (2007) on epistemic injustice.

to invite shared perceptual uptake by others. Whether by word or gesture, people direct others' perceptual attention and recognitive abilities toward aspects of their partially shared circumstances. People only partially share surroundings with others, partly because organismic environments extend beyond their immediate surroundings in different ways and partly because sharing ostensive and recognitive uptake requires orientational reconfiguration (Rouse 2002, 248–54; chap. 7 below), and in part because other persons belong to one another's perceived environments. As a telling example, monozygotic twins growing up in the same household differ in the most salient feature of their practical/perceptual and developmental environments, since each encounters and engages with a different sibling. People do not merely inherit conceptual content from others' observation reports; they coordinate their postured engagements with circumstances through ostensive calls to take up one another's perceptual recognitions. Kukla rightly takes this point even further by regarding all discursive practices as ostensive.

> Discourse is not a tool for trading representations around; it is a set of second-personal practices that orient us toward a shared world. Telling is something we *do*; it is a special kind of comporting activity. . . . Practices, rather than representations, are what most paradigmatically have aboutness, although representations . . . can perfectly well have aboutness as equipment within such practices. And what gives them that aboutness is their function: insofar as the function of my practice is to call attention to something, to make it show up in a way that allows it to be grasped, then that thing is what it is about. (2017, 115)

Picking out aspects of the environment and directing others' attention toward them are critical in enabling high-fidelity social learning and active teaching. These aspects of interpersonal, world-articulative practice are widely recognized as enabling the iterated niche reconstruction that reproduces and builds on the material culture, complex skill sets, and practical interdependence of human ways of life (Laland 2017; Sterelny 2012a, 2021; Dor 2015).

Kukla and Lance thereby undermine shared assumptions driving philosophical disagreements over whether perception provides nonconceptual content or is conceptual "all the way down," even though they do not yet indicate the full import of that challenge.[20] Kukla and Lance identify as a mistaken assumption driving these disagreements that, "if what is absorbed in perception is not a full declarative judgment, but still engages our concepts, then it must somehow be a proto-declarative. . . . Once we cease to presuppose

20. Classic defenses and criticisms of the notion of nonconceptual content include Evans (1982), Peacocke (2001), Crane (1992), Davidson (1984), McDowell (1994), Noë (1999), and Schear (2013).

that everything that has any conceptual structure has the form of a declarative . . . we open new room for [a more satisfying account of the conceptual significance of perceptual receptivity]" (Kukla and Lance 2009, 72–73). Kukla and Lance (2009, 66–78) rightly diagnose some debilitating consequences of this widespread presumption. With that problematic commitment in hand, three different unsatisfactory responses have appeared as plausible alternatives. Davidson, Brandom, and Richard Rorty disconnect judgments from direct practical-perceptual accountability, exemplified in Davidson's claim that "only a belief can justify another belief" (2001, 141). Alternatively, McDowell ascribes propositional content to perception, while being unable to account adequately for which of the many possible contents to ascribe, or how perceptual receptivity differs from commitment to judgments (Kukla and Lance 2009, chap. 3). Finally, other philosophers (e.g., Evans 1982; Peacocke 1992; Crane 1992) instead posit "nonconceptual content" as intermediary between mere causal impact and fully propositional content, thereby only postponing the question of how receptive openness to the world contributes conceptual content to judgments.[21]

Kukla and Lance respond that observative utterances are conceptual but not propositional: "To observe is not just to inherit entitlement to a belief, but to recognize how things show up *to me*. I recognize what I see *with my concepts*, and hence such recognitions already bear articulate rational relations to the rest of the space of reasons, including beliefs. . . . But my observation is not *itself* the production of a propositional judgment, nor does my expression of this observation in an observative give voice to one" (2009, 74). Characteristically, observatives give discursive uptake to perception of *objects*: "A rabbit!" Applying a singular term invokes its possible inferential role in judgments without making any specific judgment. Kukla and Lance do not then draw the further step of recognizing that perception *of objects* is already conceptual, and is coconstitutive with a grasp of language and other discursive

---

21. I cannot develop these critical assessments in detail here. Brian Cantwell Smith (2019, 29n10) argues for nonconceptual content in a different sense than those mentioned. For Smith, nonconceptual content arises in "thoughts and judgments that . . . are not framed (in our minds) in anything like a discrete set of articulable concepts" (2019, 30). Smith refers to what cannot be framed in discrete concepts but still has a conceptual role in thoughts and judgments. Those thoughts and judgments are contextually situated, and ostension and indexical expressions play a conceptual role in those thoughts and judgments *together* with the situations in which they occur. Smith and I agree that those "registrations" (his term) or "articulations" (mine) are both part of the world and accountably of or about that world. The world outruns particular registration schemes or articulated domains, although that "excess" is only specifiable by further articulation.

practices. The underlying point is more widely recognized in philosophy of science. Nancy Cartwright notes that the canonical quantum mechanical model for the hydrogen atom is not a wave equation for an object in isolation, but instead a model of a two-body universe containing only a proton and an electron. There are no isolated objects: "Real hydrogen atoms appear in an environment, in a very cold tank, for example, or on a benzene molecule; and the effects of the environment must be modeled in the Hamiltonian. What we study instead is a hypothetically isolated atom" (1983, 137). Andrea Woody (2004) similarly notes that the mathematical model of an "ideal" gas, $PV = nRT$, has conceptual significance for chemistry that far outruns its limited descriptive or prescriptive accuracy because it models the conceptual basis for the chemistry of bulk materials as well as gases. Chemists analyze those substances as composed of individual atoms and molecules that canonically appear as such in the gas phase, although even gas behavior also depends on its surroundings such that the "ideal" gas model conceptually abstracts from omnipresent intermolecular forces and physical containment.[22]

Perceived objects are likewise conceptual abstractions from people's practical-perceptual openness to their surroundings. People perceive meaningfully configured circumstances, and picking out an object abstracts an aspect of those circumstances as reidentifiable in other settings. That conceptual articulation allows people to be both immersed in current circumstances and also oriented toward that setting's more limited interdependence with other concrete situations. The ability to pick out objects from their concrete entanglements, whether through ostensive indication in words or gestures or in judgments about them, orients one toward distal situations and other practical contexts in relation to current involvements. For Brian Cantwell Smith, an important lesson from the failures of early work on artificial intelligence, reinforced by subsequent successes with machine learning, is that

> developing appropriate registrations does not involve merely "taking in what arrives at our senses," but—no mean feat—developing a whole and integrated picture accountable to being in the world. It is not just a question of finding a registration scheme that "fits" the world in ways locally appropriate to the project at hand, but of relentless attunement to the fact that registration schemes necessarily impose non-innocent idealizations—inscribe boundaries, establish identities, privilege some regularities over others, ignore details,

22. Other work in science studies generalizes Cartwright's point, notably Barad's (2007) treatment of objects as objects-in-phenomena, and Mol's (2002) account of bodies and diseases as both multiply enacted and interactively unified.

and in general impose idealizations and do an inevitable amount of violence
to the sustaining underlying richness. (Smith 2019, 36)

That idealization and indicative directedness beyond current circumstances are
both exemplified and partially coconstituted in abilities to recognize words as
iterations of expressions used elsewhere. Word recognition and use gain their
significance from the interdependence of their differently situated deployments.

Brandom took one of several requisite steps toward more adequate recognition of this aspect of discursive practices by basing his explication of how
inferentially articulated judgments answer to the world on the inferential and
substitutional roles of singular terms. On Brandom's account, speakers undertake commitments that extend beyond and possibly contrary to judgments
they would explicitly acknowledge, through the truth-preserving intersubsti-
tutability of the singular terms they employ (1994, chaps. 6–8; 2000, chaps. 4–6).
Using singular terms undertakes commitments that extend to *any* expressions
appropriately intersubstitutable for those terms, even those one would incorrectly repudiate. Brandom thus initially identifies objects discursively by open-
ended groups of extensionally intersubstitutable singular terms. Where Brandom goes wrong is in abstracting these discursive performances from their
practical-perceptual entanglements. The relevant conceptual capacities involve
the complex practical-perceptual ability to recognize, respond to, and articulate
relations among different environmental circumstances by discerning objects
as reidentifiable aspects of those settings. That capacity arises in part from the
related ability to recognize and track as objects the iterative uses of linguistic
expressions such as singular terms in multiple discursive performances.

Implicit presumptions that nonlinguistic animals also perceive and track
objects has likely blocked recognizing perceptual recognition of objects as a
conceptual capacity whose evolution and development are linked to ostensive
coordination among speakers and the ability to understand and use singular
terms in observative expressions and judgments. An ecological-developmental
understanding of other animals' practical-perceptual abilities instead indicates that they perceive life-relevant Gibsonian affordances of their developmental and selective environments. That perceptual recognition is intimately
connected to what they can do in those circumstances and what matters to
their way of life (Gibson 1979; Chemero 2009, chaps. 7–8).[23] Many organisms

---

23. Chemero (2009, 170–79) discusses how animal "object exploration" experiments have
gone astray by not attending to how the objects used relate to the animal's capacities; experimental animals are not responding to an object as such, but to behavioral affordances of an
experimental situation as a whole.

have sophisticated, flexible practical/perceptual repertoires for sensitive, appropriate response to life-relevant variation in their circumstances (de Waal 2016). Such capacities evolved in response to other organisms' subversion of more direct and invariant responses to life-relevant environmental signals. These more sophisticated capacities for perceptual discrimination and behavioral flexibility typically turn on close attentiveness to multiple, possibly conflicting environmental features or signals (Sterelny 2003, chap. 2; Rouse 2015, chap. 3). As Sterelny said, "Agents with robust tracking—with the ability to use several cues either built-in or learned—have islands of resilience in their behavioral repertoire. The cues that control behavior have become flexible and intelligent. . . . Robust systems, like [simpler] detection systems, are [nevertheless] behavior-specific. Their function is to link the registration of a salient feature of the world to an appropriate response" (2003, 28–29). Organisms with robust, flexible repertoires are capable of complex, subtle responses to environmental variations, including tracking their own prior responses and their outcomes.[24] This flexibility and resilience are thereby all the more closely attuned to the practical import of those variations.[25] These developed capacities are then an impediment to organisms developing abilities to attend to identifications displaced from their current practical significance.[26] Similar situations can solicit relevantly similar responses without abilities or need to discern and identify objects as common to both.

Conceptual capacities allow understanding and behavior that is not proximally responsive to the organism's current situation. They allow response to other possible performances that make sense within a partially autonomous repertoire, such that the sense they make within the repertoire and their significance within proximate circumstances can diverge. Language is paradigmatic. What it is appropriate to say in a conversation, in a language, is shaped both by past encounters with uses of similar linguistic expressions and by the conversational context. Discursive response to current circumstances is

24. Organisms whose responses are directly linked to specific features or cues can also produce complex forms of behavior if those features and responses are chained together, but with more rigidity in sequences and cues which more flexible or resilient, multichannel repertoires circumvent.

25. The close coupling that enables organisms' attuned responsiveness to behavior-relevant circumstances has motivated dynamical systems models in enactivist and embodied approaches to the cognitive sciences (Chemero 2009, chaps. 2, 4–5; Anderson 2014, chaps. 5–6; Thompson 2007, chap. 3).

26. I have argued (Rouse 2015, chap. 3) that immature development of close, multifaceted attunement to current circumstances explained why the infant bonobo Kanzi could acquire an experimenter-facilitated protolanguage when his mother and other mature bonobos could not.

mediated by the utterances' place in linguistically constituted patterns. Emphasizing language as a partially autonomous repertoire enabling symbolic displacement from practical orientation to current circumstances should not be misconstrued, however. A linguistic repertoire is not composed of relations among words or other expressions isolated from their circumstances of use. On the contrary, niche-constructive conceptions require understanding the repertoire itself as a postured responsiveness to broader circumstances of word use. There is nevertheless a crucial difference between postures language speakers take up toward a conversation in situ and postures of other organisms toward environments in which postures and utterances by conspecifics or companion species are merely among many multifaceted aspects of current circumstances to which they are closely attuned and flexibly responsive.

The most plausible cases for object-identification by nonlinguistic animals are better understood in alternative terms. Consider predator-tracking, in which animals vulnerable to predation keep a predator in view at a safe distance. They are tracking not the predator as an object, however, but the security afforded by its visibility and distance. The predator visible at a distance and the predator dangerously close or not visible are not a single object differently related to its potential prey. Rather, these are dramatically different situations. What people would regard as the absence from view of predator *objects* is perceived by potential prey as a *situation* of heightened vulnerability.

The well-known example of vervet monkey warning cries (Cheney and Seyfarth 1990) should be similarly understood. Vervets undergo predation from snakes, leopards, and eagles, whose visible presence each evokes distinctive warning cries. These cries do not pick out kinds of predators, however, for they also indeterminately indicate a direction of danger and of relevant movements to safety. Moreover, they are not evoked solely by the visible presence of a predator but instead only in situations where other nearby vervets, especially close genetic relations, are manifestly unaware of the urgency of flight. Vervet calls do not indicate kinds of predator, a locus of danger, a direction of movement to safety, the kinship relations that prompt the calls, or some other *aspect* of the situation. Differences among such indications are not distinguished by vervets' situated responses. The calls respond to situations that confront the vervets holistically, as visceral, affectively fraught solicitations to call and flee. An exchange between Robert Cummins and John Haugeland instructively highlights this lack of articulation. Cummins argued that, "Predator recognition systems tolerate a lot of false positives in the interest of speed. But when a prairie dog dives in its hole in response to a shadow made by a child's kite, this is a *false* positive, a mistaken predator identification, not an accurate shadow detection" (Cummins 2002, 132). That

characterization presupposes an independent conceptualization and identification of predators as the behavior's target, and Haugeland countered that
such identification is underdetermined between object-detection and "a representation for some 'consumer'—a prairie-dog activity selector that takes it
as a cue to dive. . . . We could, with equal justice, attribute the 'error' (or design trade-off) to this activity selector. In other words, instead of saying that
prairie dogs are easily fooled, we say that they are skittish or easily spooked"
(Haugeland 2002, 140). People encounter vervets' and prairie dogs' predica-
ments under descriptions or as discursively-articulate observers, but the lives
of vervets and prairie dogs do not discriminate mistaken identifications from
mistaken responses or pick out objects as joint targets of both.[27]

In many cases, however, animals seem to track objects in their vicinity
over time through presence and absence. A dog penned in a yard may run
and bark alongside a passing car and continue to do so when the car disappears behind the house and reemerges on the other side.[28] Even setting aside
that the car's continuing olfactory and auditory manifestations are more salient for dogs than visual perception, this behavior provides no basis for attributing to the dog perceptual recognition of an object persisting in time
but only an ongoing practical-perceptual response to similarly life-relevant
circumstances that encompass the fenced constraints on pursuit. Whether
responding to the situation as threatening, an opportunity, or a disturbance,
dogs are attuned to practically significant circumstances as life-relevant situations. People differ in that our biological environments and bodily postures
are discursively articulated. Dogs are closely attuned to people's situated behavior and often recognize and differentially respond to vocal utterances. As
with vervet monkeys, however, they respond to those expressions as integral
to a situation that solicits some responses and discourages others. While their
behavior is flexibly responsive to multiple, conflicting solicitations, they do
not understand utterances as inferentially connected or their situation as discursively articulated and composed of reidentifiable objects. They instead encounter the situation holistically as affording or soliciting a visceral response.

The most striking examples of putative object-recognition by nonlinguistic animals involve complex social hierarchies. Cheney and Seyfarth (2007)
carefully documented complex patterns of social dominance and deference
in baboons' differentiated responses to other baboons in their group. As with

27. Haugeland (1998, chap. 10) develops a more extended argument for why other animals
do not perceive objects as such. My reasoning for this claim is differently grounded, but mostly
compatible with his.
28. I owe this example to Mark Okrent.

vervets, however, these behavioral variations track subtle differences in life-relevant social situations rather than individual baboons as objects or agents. Their behavioral variations are not just responsive to the proximity or absence of an individual in a social hierarchy but to a more inclusively configured situation, including the presence of other conspecifics to whom they have significant relationships, gradual shifts in the hierarchy and its stability, prospects for concealment of their behavior or others' distraction from it, and the affective significance of those behaviors. Whereas human observers only understand and respond to baboons' or vervets' situation by sequentially listing features or aspects distinguishable conceptually, baboons encounter a situation that is holistically configured by how it viscerally solicits a response, even if the attraction is internally conflicted.

Taking object-identification as a conceptual capacity linked to language acquisition does not minimize or denigrate the perceptual and cognitive capacities or behavioral sophistication of nonhuman animals. Flexible, sophisticated, perceptually sensitive, and practically appropriate animal behavior does not depend on perception and reidentification of objects and properties or relations. Discerning subtle, life-relevant differences in their environment and responding to complex environmental patterns is instead an evolved cognitive capacity that had to be overridden in hominin evolution to permit symbolic displacement and object reidentification (Sterelny 2002, chap. 2; Haugeland 1998, chap. 9; Rouse 2015, chaps. 3–4). These examples of sophisticated animal behavior thereby contrast to the different shape of human ways of life in discursively articulated environments. Human biological environments are discursively articulated by the differentiation of words and sentences as perceptual phenomena, postured patterns of response, and complex patterns of practice extended over time and space. Discursive articulation begins as ostensive coordination of people's differently postured orientations amid partially shared circumstances. Discursively developed humans—not infants—understand situations as already analyzable and recomposable concatenations of discernible and trackable objects with properties and relations that can be named and predicated.

The orientation of human bodily postures toward discursively articulated environments is an outcome of language evolution and each generation's lifelong developmental immersion in ostensively, grammatically, and inferentially articulated discursive practices. The environments humans inhabit are profoundly transformed by inferentially significant, iteratively recombined words conjoined with ostensive indications and perceptual reidentifications of objects and kinds of objects across varied situations. That evolutionary transformation carried both benefits and costs. The costs include partial suppression of subtle practical-perceptual discriminations of behaviorally relevant differences in people's circumstances not relevant to object-identification and predicative expression. As Haugeland pointed out,

> The number of "bits" of information in the input to a perceptual system is enormous compared to the number in a typical symbolic description. So a "visual transducer" that responds to a sleeping brown dog with some expression like, "Lo, a sleeping brown dog" has effected a huge data reduction. And this is usually regarded as a benefit, because, without such a reduction a *symbolic* system would be overwhelmed. But it is also a serious bottleneck in the system's ability to be in close touch with its environment. Organisms with perceptual systems not encumbered by such bottlenecks could have significant advantages in sensitivity and responsiveness. (1998, 220)

These aspects of language evolution and development have profound significance for philosophical understanding of language and human cognition. The classical philosophical problem of understanding predication and the unity of the proposition has long been recognized as a serious difficulty for philosophical analyses of language that start from reference to objects and sets of properties as "truth-makers" (Davidson 2005a). The preceding discussion generalizes that issue. Understanding language biologically is in significant part understanding people's symbolic displacement from current biological exigencies conjoined with conceptual articulation of those circumstances. That accomplishment, with its evolutionary advantages and costs, is presupposed by understanding objects as reidentifiable, trackable components of an environment and cannot explain it. Singular terms are only understood as such in relation to the "unsaturated predicates" (Frege, Davidson) or "sentence frames" (Brandom) into which they can be inserted or substituted. Objects can likewise only be encountered and understood as articulated components of situations or "phenomena" (Barad 2007), or more generally, of bodily involvement in the world from which ostending, naming, registering, and judging abstracts them (Smith 2019, 140–44). Other organisms' postured responsiveness to their environments exemplify more holistic involvements. The next chapter further explores how the evolved environmental pervasiveness and recurring developmental reconstruction of language and other conceptually articulated practices condition the reidentifiability of objects across spatiotemporally and virtually extended environments. The widespread philosophical presumption that an independent grasp of object-identity and reference then enables predication and symbolic displacement blocks adequate understanding of conceptual capacities generally, above and beyond the "problem" of understanding predication.

Discursive and Other Practices

A final aspect of the environmental embeddedness of discursive practices to consider is their two-dimensional normativity as practices. Discursive practices have their own constitutive issues and stakes—understanding and communication within mutually comprehensible linguistic repertoires are integral to a horizontally and vertically practice-differentiated way of life. These practices and performances also answer to many other normative concerns. People always hear or utter words and sentences in a larger practical/perceptual context even when it is configured to foreground and facilitate conversation.[29] It matters whether those conversations occur within a coffeehouse, a classroom, an office, or over a breakfast table and among family, friends, coworkers, antagonists, or strangers. Moreover, while one may think of conversations primarily as linguistic exchanges, they are more comprehensive interactions among people ranging from transitory micronegotiations to sustained interpersonal relationships.

Commonly, however, conversation and other less direct discursive exchanges take place amid other practices. Discursive practices play at least a tripartite role in those contexts. First, linguistic expressions are pervasive aspects of most situations people encounter. Human biological environments are verbal. Words are often salient within those environments, and people's postures are responsive to denotable and ostendable objects and discursively articulable categories. Categories in this sense extend beyond those groupings of objects, properties, relations, or situations which are explicitly marked by words to include those distinguished ostensively or by behavioral responses to those differences. People can "tell" differences for which they have no words (Haugeland 1998, chap. 13). Which categories to distinguish is a normative issue of which distinctions and groupings ought to be sustained in ongoing discursive practices rather than which ones are distinguished currently. Some current distinctions ought not to be sustained in subsequent practice, and some not (yet) discerned as categories should make a difference in ongoing discursive practices.

A second contribution of discursive practices to other practices is that the latter often incorporate utterances or exchanges of utterances among their performances, including explicitly performative utterances such as promises or apologies, conversations to coordinate activities and facilitate interpersonal

---

29. Written language's contextual character is more complicated, by normally separating situations in which expressions are written and read, and enabling reading them on different occasions.

relationships, or exchanges of information or orientation through queries, reports, explanations, or justifications. Vocative and recognitive utterances direct verbal and nonverbal performances toward specific audiences, acknowledge and respond to such directed performances, and point out relevant aspects of the conversants' partially shared circumstances.

A third contribution is central to the complex normativity of human ways of life. Other practices often incorporate expressive articulation of the aims, scope, and character of that practice and its place amid other practices. People often try to indicate or say what is at issue in apparent misalignments within a practice or among practices, and what is at stake in whether and how those issues are resolved. Often those efforts lead to further negotiation over how to continue or revise a practice to respond to those issues or resolve them. Talking about a practice is often integral to that practice, whether in seeking to recruit, train, and retain participants, direct and coordinate one another's performances, call attention to or work out conflicts over processes and goals, or critically assess performances and outcomes. In this articulative role within other practices, discursive performances are often closely aligned with and responsive to the situations, equipment, skills, and other performances integral to those practices. Even when a practice deploys no specialized vocabulary, understanding what is said often both presupposes and contributes to an understanding of those situated performances, such that neither the linguistic nor the nonlinguistic performances of the practice are intelligible apart from one another.[30]

Linguistic utterances could not play these diverse but integral roles in other practices if they did not belong to partially autonomous discursive practices. Discursive practices have their own constitutive ends—ways of making sense—at issue in those practices. The interdependent performances making up a language as a niche-constructed component of human environments are sustained over time only through the developmental reconstruction of that discursive repertoire in successive generations. People grow into the languages spoken around them and pick up on what it is for their expressions to make sense. They help reconstitute those practices as integral to the developmental environments of others in turn by sufficiently aligning their own sense making with that of others around them. Sustaining those alignments is a normative issue with the characteristic ambiguity between predictive and authoritative aspects of those constitutive normative concerns. The normative force sustaining the authority of those discursive alignments comes from

---

30. I have extensively discussed elsewhere how conceptual articulation is entangled with material interactions in niche constructive scientific practices (Rouse 2015, chaps. 6–10).

the place of discursive practices in an encompassing, practice-differentiated way of life—the expressive freedom it enables to make discursive sense and thereby to make sense *of* partially shared life-situations. Recognizing their interdependence with other practices and concerns does not thereby subordinate language to other issues, for which discursive articulation would then be merely instrumental. Taking up a language and engaging with one's environment discursively has its own constitutive ends that must be accommodated within other practices. Discursive practices thereby exemplify the two-dimensional normativity of a practice-differentiated way of life.

We saw earlier that Brandom modeled discursive practices as a *social* phenomenon that then could be connected to a causally or nomologically structured natural world through its supposed interfaces with perception and action. Mastering the communicative and justificatory norms that constitute discursive practices would then enable people to produce and comprehend novel descriptions and other claims about the world as the central and characteristic end of discursive practices. That ability in turn would allow people to undertake new intentions and engage in new practices that depend on the articulative and adjudicative abilities of language speakers. Brandom regards those capacities as what is distinctive about humans: "This expressive account of language, mind, and logic is an account of who *we* are. For it is an account of the sort of thing that constitutes itself as an expressive being—as a creature who makes explicit, and who makes itself explicit. We are sapients: rational, expressive—that is, discursive—beings. But we are more than rational expressive beings. We are also *logical, self*-expressive beings" (1994, 650). This proposed token-reflexive self-understanding as a rational and logical community leaves behind people's biological history as a lineage dependent on its earthly environments, which Brandom dismisses as an arbitrary distinction (1994, 3). He aspires instead to a more inclusive and less parochial sense of who we are which asks, "What would have to be true . . . of chimpanzees, dolphins, gaseous extraterrestrials, or digital computers (things in many ways quite different from the rest of us)—for them nonetheless to be correctly counted among us" (1994, 4)?

This chapter has argued that Brandom's account of people's discursive capacities failed in its own terms by not recognizing the significance of language as a practical-perceptual capacity of human animals. Linguistic expressions are embedded in people's selective and developmental environments and are taken up into interdependently developing bodily postures toward one another and the world. The chapter's primary aim, however, is ultimately constructive rather than critical. A naturecultural conception of practices as forms of biological niche construction also takes discursive practices as

integral and central to "who we are" as humans. For Brandom, our biological evolution as organisms is a contingent precondition for the emergence of
humans' rational and logical capacities, but is not integral to those capacities
themselves. A naturecultural conception instead integrates discursive practices into our biological environments and the ongoing, niche-constructive
coevolution of our lineage and the world we inhabit.

This inversion of Brandom's social-rationalist account of discursive practices not only accommodates his extensive technical apparatus, including the
expressive roles of logical and semantic vocabularies, although I do not work
through the details of this accommodation here.[31] A naturecultural conception also shares his aspiration to a less parochial understanding of human
ways of life. The next chapter takes up this alternative conception of discursive practices and their place within a naturecultural conception of who we
are. The two remaining chapters then work toward a conception of us and
our ways of life that would be less parochial in two further ways. The first
considers how a practice-differentiated way of life introduces new, normatively fraught differences *among* humans. These differentiations of power and
resistance matter to what is at issue and at stake in how we live. The concluding chapter similarly asks what is at issue in recognizing and responding both
to the historical contingency of our practices and to our interdependence
with the ways of life of other beings who are not rational and logical sapients.

31. Brandom (1994) provides the canonical exposition of this technical apparatus. I have
summarized more of Brandom's account (Rouse 2002, chap. 6) and shown how to accommodate
it within a more naturalistic account of language use, without specifically discussing niche construction. I have further explicated the niche constructive aspect of this naturalistic inversion of
Brandom (Rouse 2015, chaps. 3–5; 2018).

7

Discourse

Articulation

The previous chapter argued that a niche-constructive conception of language and other discursive practices turns Brandom's pragmatist model of language use inside out. Brandom construes language as a social practice among score-keepers who track and assess one another's shifting discursive commitments and entitlements. These performances and assessments would not just be moves in a self-contained game because they supposedly acquire objective *representational* significance, in two distinct steps. Discursive scorekeepers identify and adjudicate differences in their conceptual perspectives by holding one another's commitments accountable to "objects" defined as open-ended groupings of intersubstitutable singular terms. Those intralinguistic "objects" are then held accountable to causally efficacious objects in the world through speakers' judgments about one another's perceptual and practical reliability as observational reporters and agents. What people say and do is articulated by social-perspectival inferential relations, which then represent how things are in the world through their connections to people's perceptual reports and agential performances. Brandom's inferentialist semantics thus shares with most other semantic and literary theories a representational conception of performances "in" discursive practice as related to a world of facts and objects.[1] Brandom challenges that construal of his model: "It is wrong to contrast discursive practice with a world of facts and things *outside* it, modeled on the contrast between words and the things they refer to . . .

---

1. Davidson's (1980, 1984) semantics is not representational—semantic relations are intralinguistic but causally engaged, because semantically significant mental events are token-identical with physical events. Davidson thus shares a broader conception of the world's causal systematicity as external to rational, intralinguistic systematicity.

and to think of facts and the objects they involve as constraining linguistic practice from the outside—not because they do not constrain it but because of the mistaken picture of facts and objects as outside it" (1994, 332). Brandom does avoid some problematic aspects of other representational conceptions of language-world relations. He treats semantic concepts as expressive rather than substantive and insists that people's perceptual and practical interrelations with the world are integral to discursive practices. He still understands relations between words and things as representational interfaces *within* discursive practices, however, by treating words as already-discrete noises and marks, and objects as already-determinate components of the world. Perception and action then become interfaces between uses of words as noises and marks that form a "distinct and largely independent realm within the world" and objects as discrete components of causal relations between bodies and their environments (1994, 331).

A naturecultural, niche-constructive conception of language replaces representational relations between intralinguistic uses of words and (entities in) the world with recognition of how human developmental environments are discursively articulated.[2] Linguistic performances and their interrelations are inextricably part of the world, integral to human ways of life, and only meaningful and intelligible within more encompassing situations. Perception and action are not interfaces between intralinguistic expressions and "nonlinguistic" entities or states of affairs, whether as connections between words and things or sentences and their "truth-makers." What people say and hear is instead embedded amid postured bodily interaction with their surroundings. Living within that discursive environment does enable extensive representational *uses* of linguistic expressions. As they grow into language, people can talk and think about and act toward situations that are spatially, temporally, or imaginatively displaced from surrounding circumstances without needing causally effective contact with those displaced situations.[3] What enables those representational uses, however, is how the continuing reproduction of language and other conceptual practices already articulates human developmental environments.

Discursive practices allow people to make sense of the practices they engage in, the circumstances in which they are situated, and those practices'

2. The parentheses indicate that Brandom treats linguistic representation as a holistic feature of discursive practices via their inferential mediation rather than directly assigning representational relations to linguistic types or tokens.

3. Bickerton (2014, chap. 7) indicates why "growth into language" better describes human development than "language acquisition." Smith (2019, 9–16) discusses the (causal) "noneffectiveness" of semantic relations.

horizontal and vertical interdependence with other practices by making discursive sense. The ability to make discursive sense of one's situation and one's life and the concern to do so now characterize human ways of life. The autopoietic tradition has used "sense making" in a more expansive way that encompasses how the bodily processes of any organism configure its circumstances as a meaningful environment: "Adaptivity (Di Paolo 2005) is what enables living bodies to distinguish a situation as a risk or an opportunity, to tell the difference between good and better, bad and worse. . . . Sense-making is the capacity of an autonomous system to adaptively regulate its operation and its relation to the environment depending on the virtual consequences for its own viability as a form of life. Being a sense-maker implies an ongoing (often imperfect and variable) tuning to the world and a readiness for action" (DiPaolo et al. 2018, 32–33). Organismic sense making in this broader usage includes people making sense of the world discursively. Human developmental environments are discursive, and people regulate what they do partially in response to the discursive sense they make of those environments. Assimilating discursive sense making to organism-environment interdependence more generally, however, overlooks the significant transformation of human ways of life by their two-dimensional normativity and the discursive articulation of a world of objects and facts.

The only-partial autonomy of discursive practices and their constitutive concerns situates those practices within an ongoing biological way of life. It also enables people's postured engagement with that discursive environment to incorporate symbolic displacement from their proximate circumstances. Enabling symbolic displacement was an important selective consideration in the evolution of language, but the environments people then inhabit are reconfigured by the discursive practices pervasive within them.[4] This chapter develops an overview of human ways of life and the discursive articulation of the developmental environments they encompass.

The previous chapter initiated this reconception with one aspect of how language and other expressive practices are embedded in the world in materially situated performances. Languages and other conceptual repertoires only exist through their continuing reproduction in vocal utterances, expressive gestures, or readable markings, and the bodily skills exercised in doing so. Brandom and many others overlook this key point by treating words as discrete noises or marks that comprise a body of "specifically linguistic facts (thought of as a class of facts about words)" that could have been different

---

4. Bickerton (2009, 2014), Dor (2016), Dor and Jablonka (2000), and Deacon (1997) each treats symbolic displacement as the central challenge for language evolution.

while otherwise leaving the world unchanged (Brandom 1994, 331).[5] There are indeed many languages and people could have grown up amid one language rather than another, with different words. One only encounters those words, however, within an ongoing discursive practice that incorporates concrete engagement with the world.

The environmental embeddedness of words is multifaceted. Vocal utterances incorporate the speaker's posture and expressiveness, situated by the (intended) audience's presence and bodily postures amid partially shared circumstances.[6] Understanding language use and other conceptual practices as articulative emphasizes that utterances in language are always integral features of more complex environing circumstances. What is said or what it is about may be the primary focus of a speaker's or hearer's orientation and attention, but that orientation is shaped by its more encompassing local situation. People's bodily postured, situated involvement in high-bandwidth environmental intra-actions is the naturecultural alternative to the role usually assigned to perception and action as more limited interfaces between thought or language and "the world" as an imaginary abstraction from which discursive practices have been excised.

Linguistic performances are situated in the world in other ways that matter to their conceptually articulative significance. First, these expressive performances are only linguistic as concretely situated within a language as a practice integral to people's developmental environments. The sense in which they are "concretely" situated in language was concisely expressed by Ruth Millikan: "The phenomenon of public language emerges not as a set of abstract objects, but as a real sort of stuff in the real world, neither abstract nor arbitrarily constructed by the theorist [consisting] of actual utterances and scripts forming crisscrossing lineages" (2005, 38). Utterances are used and understood in context as acquiring significance and intelligibility from other concrete, situated, past and possible future uses of recognizably iterated expressions.[7] People take themselves to be using words that have been used

5. Ebbs (2009, chap. 4) criticizes a similar conception of words in Davidson rather than Brandom but from a different philosophical orientation than mine.

6. Written language and machine-computational states may seem to differ from speech, because writing is not as ephemeral and machine-computational structures are multiply realizable. They are nevertheless written, read, or realized in specific circumstances, and writing or programming belongs to larger patterns whose character, orientation, and provenance matter to what is said to whom.

7. Davidson (1986) famously denied that audiences interpret speakers' utterances via shared grasp of a structured language. He cited a passage readily intelligible to English speakers although its expressions were mostly semantically or graphemically anomalous or not English

before in different circumstances, and are reusable and intelligible in part through interlocutors' familiarity with some prior uses. Inferential relations among those expressions are not disembodied, but instead carry gestural, intonational, affective, and other contextual significance. Talk of utterances as concretely situated in a language is thus an alternative to widespread conceptions of natural languages as formally structured by governing rules or norms. Structural and formal relations can be discerned within linguistic practices, but the direction of fit is from structure to practice, and important aspects of meaning and understanding are obscured by ignoring their concrete embodiment. People sometimes attempt to revise linguistic practice to accord with formal structures, but if they succeed those revisions cannot correct the practice but only change it, because the structures proposed are accountable to the practice and not vice versa.

The two aspects of the worldly embedding of linguistic expressions discussed so far—in concrete, postured, bodily engagements with extended situations, and in languages as crisscrossing lineages of situated uses of iterable expressions—work together to constitute a third. A communicative repertoire situated only in these first two ways would be more complex than vervet monkey cries but not different in kind. For conceptually articulated discursive practices, it matters that these uses also occur amid many other practices with only limited and partial forms of interdependence. Such practice-differentiated uses of the same linguistic expressions across diverse local contexts gives those expressions dual significance. Linguistic expressions have both a rich, context-specific role in local contexts of use and a more limited role in establishing and maintaining coordination and interdependence across contexts.[8] Discursive practices allow expressions intimately involved in one concrete context to have more limited significance elsewhere. That dual contextual role becomes more than just an appropriation or a causal influence if the two settings remain interdependent and mutually accountable for making sense together. Lance similarly emphasizes how linguistic expressions mediate among practices and vice versa: "To be a person—a language user, an intentional being, a truth claimer—is to operate in a system of mutually interdependent subpractices, held together by incompatibility

---

words at all. Davidson's objection does not apply here, however, since his examples do rely on "crisscrossing lineages" of actual utterances and scripts, as do the wide range of figurative, playful, erroneous, proper-nominative, or other idiosyncratic word uses that concerned him.

8. Boundaries between high-bandwidth interdependence within a practice and limited, partial interfaces between practices are relative differences among "levels" of interdependence rather than differences in kind; see chapter 4 above.

norms. One must be part of a social system such that it is a general norm that certain performances in one context cannot be jointly entitled alongside performances in other contexts, regardless of the differing practical goals and internal norms of the two contexts" (2017, 179). Lance talks about incompatibility *norms*, whereas I take the compatibility of subpractices and how people make sense of them together to be *at issue* in how those practices are undertaken.[9] We agree on the central point, however, that the body-postural interdependence between persons and their environments, the discursive articulation of the practices making up those environments, and the normative accountability of the resulting postured performances work together.

Although linguistic expressions are integral to how practices, subpractices, and their performances are interdependent, they work in concert with other traffic among practical settings. Words alone do not constitute and sustain low-bandwidth interfaces among practices whose "internally" situated performances are more extensively and intimately interdependent with one another. People and materials move among practices and settings in concert with their discursive coordination. Individual people also take up those settings differently, with different significance in their lives; to that extent, they inhabit differently configured environments. The ostensive coordination of people's bodily postures and the reiteration of linguistic expressions amid interdependent practical contexts enables discerning objects, properties, and relations, and reidentifying them across those different configurations. Brandom took relations between intersubstitutable singular terms and causally constituted objects as interfaces between language as a social-normative practice and the world as a Given causal nexus.[10] I look to more complex patterns of normative interdependence. Practice-differentiated ways of life are partly enabled and sustained by the reidentifiability of objects and their conceptual determinations across practical settings, even while those objects are also more complexly entangled amid those settings. These local entanglements only become manifest and intelligible as involving more or less discrete objects, properties, and relations through the involvement of linguistic and other conceptual expressions. These engagements occur in people's movements among different practical settings, and their ostensive indications,

9. See chapter 5 above for this reconception of normative concerns as only anaphorically specifiable.

10. "Given" is capitalized here because the role of objects in Brandom's and other accounts of representational "relations" between language and world exemplify what Sellars (1997) criticized as "the Myth of the Given," in trying to ground normative concerns on nonnormative determinacy.

reidentifications, and critical assessments within and across those settings.[11] That is why Brian Cantwell Smith called "objects, properties and relations the long-distance trucks and interstate highway systems of intentional, normative life, . . . undeniably essential to the overall integration of life's practices . . . [by] packaging up objects for portability and long-distance travel [such that] they are thereby insulated from the inexpressibly fine-grained richness of particular indigenous life" (Smith 2015, 222).[12]

This chapter sequentially considers three ways discursive performances are embedded—in concrete situations, in a language or other expressive practice, and across multiple, concretely situating practices. Living a practice-diversified way of life involves not only knowing one's way around in intimate, multifaceted engagement with particular concrete situations. It also involves situating those practical involvements with respect to other practices taken up vertically at different times or places. We recognize how and when to "move" among practices and when to take account of their dependence on other practices.[13] The discursive concern to make sense of that interdependence helps articulate and coordinate people's practice-differentiated environments and ways of life, but sense making involves adjustments in both directions. Understanding discursive practices requires recognizing how people's utterances and the words employed have richly entangled significance in specific contexts while also answering to more limited connections to their uses and to other expressions in other practical contexts. In familiar settings, people readily and often seamlessly negotiate among dealings with words, with objects and their properties as recognizable and transportable across practical contexts, and with those practical situations as sites of tasks at hand. People's postured involvements nevertheless also extend beyond those familiar settings, with a more indefinite openness to their place within a more complex, practice-differentiated way of life.

11. Objects are "more or less" discrete since boundaries between objects and contexts or among properties and alternatives, and classificatory boundaries between kinds of objects, properties, and relations are rarely sharp, often requiring normative adjudication. On objects, see Smith (1996) and Scheman (2011, chap. 8); for kinds, see Bowker and Star (1999).

12. Smith is quoting his own unpublished manuscript in challenging how computer scientific and other theoretical accounts of intentionality are built around these "long-distance trucks and highways" without recognizing their ineliminable involvement in and responsibility to concretely situated involvements.

13. Sometimes physical movements are required—leaving home for work or a restaurant, or going from a bedroom to the bathroom or kitchen. Sometimes we shift orientation and practical involvement while remaining physically in place. Either way, immersed, postured involvement in situated activity incorporates a practical ability to resituate oneself in other practical contexts.

Practical Immersion

As living organisms, people always take up bodily postures grounded in and responsive to concrete practical situations. Such situations extend beyond direct causal contact, although some causally situated grounding and responsiveness is always involved. Human ways of life enable engaging with distal circumstances out of sight or reach or temporally or modally displaced from that bodily grounding. What matters instead is the multifaceted, high-bandwidth interdependence among bodily postures, conceptual repertoires, and the meaningful environments to which people are practically and perceptually responsive. Any organism is engaged with and responsive to an environment configured by significance for its way of life. Human lives are distinctive partly by how our discursively articulated environments extend spatially, temporally, imaginatively, and conceptually. Those capacities nevertheless also remain grounded in bodily postured, skillful negotiation of more immediate but still practically configured settings. Moreover, those postures and skills fit together to sustain practical, strategic repertoires to engage with and respond to more complex or distal settings.

Those bodily skills and repertoires are developmentally attuned by life-long interactions with different situations, through characteristic postures and movements among them. People acquire different capacities for standing, sitting, walking, driving, dancing, or climbing, but also speaking, listening, and reading in language(s) and vocabularies, through close interaction with terrains, furnishings, equipment, and practical-discursive interlocutors. They—we—take up differently postured movements when moving through familiar spaces with definite intent or through unfamiliar places prepared for more or less unexpected encounters. Those postures are often focused and directed through parts of the body, but rely on whole-body-postured orientations, whether in throwing, reaching, grasping, writing or drawing, speaking or listening, visually focusing at varied distances, thinking, or daydreaming. Auditory uptake and vocal responsiveness to conversational, reportorial, interrogative, and other discursive interactions are integrated into people's bodily postures and repertoires. People similarly develop perceptual attunement to other practically and affectively salient aspects of their surroundings, often through ostensive molding of one another's orientation and response. That broadly perceptual attunement includes proprioceptive and emotional responsiveness to one's own bodily activity integrated within every postured orientation. Humans have evolved extraordinary developmental plasticity and flexibility in bodily-postural repertoires, but that variability is hardly unlimited or unconstrained by developmental reinforcement or atrophy.

People's developmental attunement to patterns of discursive practice enables close engagement with things-in-concrete-settings alongside recognition of their more limited significance in other settings. Voting for a candidate, a motion, or an award takes place in concrete situations—raising one's hand, standing up, or saying "aye" at the right time; marking and turning in a ballot at a polling place; clicking an online form; moving to one part of a room at a caucus; and so forth. People also tabulate votes originally recorded in different ways to produce numerical totals indifferent to their concrete expression. In some cases, people also recombine such transportable significations in new practical contexts, for example by mapping or subdividing them and performing operations on numerical data or the spatial arrays of maps. Moreover, those practices sometimes enable novel, concrete, skillful knowing one's way about the recombinatory practices that disconnect those reidentifiable significations from their previous situated involvements. Statistical analysts, for example, engage in high-bandwidth interactions with different sets of data without concrete familiarity with the specific contexts where they were generated. Boundaries between bodily repertoires for high-bandwidth engagement with concrete, practical settings and more limited, low-bandwidth interdependence among practices are not fixed in advance.

High-bandwidth practical-perceptual embedding of discursive expressions and performances in concrete situations replaces traditional appeals to perceptual interfaces between linguistic representations and objects represented. In the latter context, Sellars (1997) instructively reworked empiricist distinctions between observation reports and theoretical statements graspable only inferentially. He replaced substantive distinctions between observable and unobservable entities with a methodological distinction between modes of conceptual authority. Sellars argued that what matters for epistemically relevant observational authority is conjoining grasp of the inferential significance of a concept with abilities to identify its applications noninferentially. For example, physicists *observe* particles below the threshold of visibility if they grasp the relevant concepts and identify their instances noninferentially from their signatures on a detector. The distinction between people's bodily-postural immersion in a concrete context and discursive engagement with other situations is similarly methodological. As a parallel example, practical grasp of a terrain from walking through it extensively differs from understanding acquired only from maps, but one can also acquire concrete, bodily postured familiarity with navigating different kinds of maps and coordinating them with terrain. New practices with their own concrete, situated know-how can thereby arise at interfaces among concrete, practical domains. Some higher-level concrete involvements are "theoretical" analyses of more concrete situations—students working with pedagogically structured problem sets acquire concrete familiarity with physical settings or chemical reactions *as* mathematically articulable—but they also include second-order practices such as international currency exchange, appellate legal work, sports management, government regulatory policymaking and enforcement, or statistical analysis of concrete enumerations. In such cases, concrete grasp of these activities and settings often abstracts from first-order practices of spending money, conducting trials, playing a sport, or collecting data.

Consider first how people's available and familiar modes of transport configure their bodily mobility and accessible circumstances. Cars, bicycles, wheelchairs, horses, motorbikes, or public transit are not simply discrete objects or even tools but instead configure people's repertoires for postured engagement with their surroundings. These ways of moving not only solicit, impose, or assimilate different bodily postures, skills, and physical demands. They also work in close concert with roadways, trails and tracks, stairs and curb cuts, and with places to get on or off, park or store, and feed, fuel, or repair those vehicles. People's accessible, inhabited modes of transport shape their surroundings as patterns of accessibility, salience, sociality, bodily vulnerability, and weather-sensitivity.

These concrete configurations of mobility and accessibility also embody or express other salient aspects of people's postures and capacities: social and economic standing, agential autonomy and dependence, aesthetic sensibility, or age-related stages of life. These entanglements in concrete forms of life nevertheless also have more limited involvements with other practices. People do not encounter cars or bicycles only as extensions of their mobility that configure the accessibility and familiarity of their circumstances. Vehicles are also bought and sold; damaged, stolen, or towed; flaunted or hidden away; licensed and taxed. Involvement in these other practical settings only partially engages a vehicle's embeddedness in people's situated bodily repertoires and spatial orientation. Vehicles are also talked about and referred to. Those discursive performances belong to and engage with the vehicles' entanglements in concrete ways of life but also enable and depend on how the words used are iterable and detachable in other contexts where vehicles matter in more limited ways.

Postured engagement with other people is also integral to and salient within the concrete situations in which people's lives are occasionally, periodically, or sequentially immersed. People now differentiate, name, describe, enact, and respond to many context-dependent roles. In doing so, they deploy classificatory vocabularies for kinship, occupations, gender and sexuality, body types, employment, friendship, citizenship, political affiliation, race and ethnicity, forms of worship, cultural identity, cooperative and competitive gameplay, and more. We often describe these roles abstractly but inhabit and respond to them concretely, and their discursive articulation draws on both forms of manifestation. These roles are not just relations among individual persons or groups, for they extend into the world as marking and marked by other situations. People encounter others not only face-to-face but by recognizing and responding to their possessions, attire, practical commitments, other personal relationships, interests and desires, and bodily styles and routines. Those interactive responses are shaped by their situations as well by one another's postures and performances. People encounter and respond to others differently on the street, at work or in class, in a bar or on playing fields, on an airplane or in a lifeboat, or in other situations arising in vertically differentiated lives. People also inhabit and respond to their own circumstances as configured by how other people navigate those settings.

In those concretely situated relationships, people's postures and performances rely on and play off the expected comportment of others. As chapter 5 emphasized, such expectations fall between predictions of what others will do and normative demands for which noncompliance would be sanctioned. Counting on a friend's support when in distress or passing the ball to a teammate on a fastbreak is not just predicting their behavior or attributing obligations to them. People comport themselves in concrete, temporally extended entanglements with one another's postured repertoires for engaging the world. These concrete engagements often incorporate and draw on their discursive expression, such as "being there for me when I needed her support" or "filling a lane on the break." These expressions have uses in other contexts detached from that concrete interdependence, but one has only a limited grasp of what they mean without some grip on how they are concretely embedded. Empirical predictions and normative ascriptions are sometimes extracted from those concretely embedded involvements, through further inferential relations to their situated discursive articulation. A steadfast friend or successful fast-breaker who could not express those engagements discursively has only an inarticulate but still articulable involvement in those situations and relationships.

Ásta's (2018) insightful account of how people are differentiated in particular interactive contexts begins with how two travelers encounter venues in San Francisco: "[Rebecca] Solnit experiences herself as Western in Chinatown, as white in Bayview, as straight and female in the Castro. [Guillermo]

Gómez-Peña is mistaken for a tourist from Argentina in Chinatown, at the Bollywood Café he is 'the wrong kind of brown,' in the Castro he is an older gay man, and in the financial district he is nobody" (Ásta 2018, 1, citing Solnit 2010). These concretely situated differences are consequential. They constrain and enable people's behavior and action in those settings through other people's situated responses to their presence, posture, and performance (Ásta 2018, 44). These differences are also notably practice-dependent ("context"-dependent in Ásta's terms). With gender categories, for example, "not only is gender deeply context dependent when it comes to historical periods and geographical locations, but the same geographical location and time period can allow for radically different contexts, so that a person may count as of a certain gender in some contexts and not others. . . . In some contexts, it is the person's perceived role in biological reproduction [that matters], in others it is the role in societal organization of various kinds, sexual engagement, bodily presentation, preparation of food at family gatherings, and so on" (Ásta 2018, 74). Ásta also rightly notes that consequential categorizations need not depend on the availability of descriptive terms for them, even though discursive articulation is often integrally involved (2018, 52).

I reframe Ásta's treatment of such "categorial" differentiations in practices. For Ásta, people's *"features are socially significant in a context in which people taken to have that feature get conferred onto them a social status"* that then "generates constraints on and enablements to a person's behavior and actions" (2018, 3, original emphasis, 44). Where Ásta refers to social significance in context, I talk about significance for the niche-constructive practices that make up people's inhabited environments. Ásta starts from "base" features people have, and then asks what *social* significance is added to those features by other people taking them to have that feature and responding accordingly. One problem is that what is a feature and whose responses matter are not specifiable apart from the practices that confer consequential differentiations. The point is not to equate features with responses to them, since features can be misrecognized; rather, the practices people participate in shape what count as relevant features in that context, while the interdependence of recognizable features and significant responses to them shape practices as contexts for recognition or misrecognition. Haugeland characterized this mutually constitutive relation as a distinction between "two fundamentally different sorts of pattern recognition. On the one hand, there is recognizing an integral present pattern [or practice] from the outside—*outer recognition* we could call it. On the other hand, there is recognizing a global pattern from the inside, by recognizing whether what is present, the current element, fits the pattern—which would, by contrast, be *inner recognition*. The first is telling whether something (a pattern [or practice]) is *there*; the second is telling whether what's there *belongs* to a pattern [or practice]" (1998, 285). While distinct, these two forms of pattern recognition are *mutually* constitutive. Practical significance, feature-articulation, and their discursive explication are thus interdependently coupled in ongoing practices rather than hierarchically related as Ásta suggests.

Reframing Ásta's conferralist conception of social categories parallels my larger strategy of turning Brandom's account of discursive practices inside out. Although Ásta does not refer to Brandom, her conferralism about social categories works analogously to his social-phenomenalist expressivism in explicating conceptual normativity. Brandom "offers an account of the practical attitude of taking something to be correct-according-to-a-practice, and then explains the status of being correct-according-to-a-practice by appeal to those attitudes" (1994, 25). He then addresses how correctness-according-to-a-practice is accountable to objects and their properties through the "messy retail business" of sorting out which conceptual contents *ought* to be ascribed in which circumstances. Ásta's analysis similarly moves from *taking* someone to have a feature to the status those takings confer within a practice. She does not then explicitly undertake a parallel to the messy retail business of assessing whether those statuses are appropriate, but her analysis aims to enable such assessment, as a social justice project. That aspiration is to bring the significance various practices confer on people's features into accord with whatever significance they ought to have (if any) "over and above the constraints and enablements that come with simply possessing the property" (Ásta 2018, 33). I am arguing instead that the objects or features, how they are taken as present, and the status and significance thereby accruing to them are reciprocally entangled rather than unidirectionally constituted. The "messy retail business" of sorting out what significance ought to accrue to how people are responded to in different contexts is the process of holding these local contexts accountable to normative concerns that apply across contexts, such as social justice or respect for cultural difference.

Engagements with people and things are not the only sites where discursive articulation draws on concrete immersion in practical settings. People's understanding of activities or processes and the words used to describe them are also often concretely situated and mutually articulative. As one example, Searle (1983, 145–48) highlights many different situations where the English verb "open" is used. These are telling cases of how word uses and situations inform one another through people's concrete familiarity with the relevant situated activities. Competent speakers of the language readily grasp on first encounter the different ways in which someone opens a door, their eyes, a wall, a book, a wound, a computer file, a meeting, a restaurant, and so forth, because grasping a word is an open-ended bodily skill at engaging in an ongoing discursive practice. People who can walk, throw, or point do not repeat the same motions each time, but instead adapt their postures to their settings; people's grasp of words is similarly extended flexibly to new cases or circumstances. People with linguistically coevolved and developed neural organization acquire the ability to use words such as "open" in different concrete contexts of use. They similarly develop abilities to extend the semantic significance of those words into other contexts, in part through ostensive and discursive interaction with others using the same words.[14]

I conclude with an example of people's concrete situational involvements that figures prominently in recent work on social ontology. Money has been diversely described as an abstract entity (Dennett 2013, 104), a social institution in equilibrium (Guala 2016, chap. 3), a "massive fantasy" (Searle 2010, 201), or a form of social positioning (Lawson 2019, chaps. 5–6), among others. Much can be learned from these conceptions, but they each mostly overlook the concrete entanglement of money in locally situated practices. Kukla vividly characterizes how money is taken up in people's postured bodily involvements in different circumstances:

> Money is caught up and intertwined almost maximally robustly in a huge number of our concrete practices. How much of it we have determines what we wear and eat, where we live, and so forth. An enormous number of our daily actions are directed toward getting it, calculating out how much of it we have in our pocket, spending it, and so on. Meanwhile, how we interact with and respond to one another, at an intricately embodied level, is shaped in all kinds of ways by how much money we perceive one another as having and how much money we have and have had in the past and expect to have in the future. . . . None of this is at all abstract. (Kukla 2018, 12–13).

Money could not acquire these complexly entangled worldly involvements, however, unless they were also implicated by more distal interactions with other comparably concrete forms of bodily dependence on monetary exchange and accumulation. Among the latter practices are the printing or coining of currency, the authorization of interest or discount rates, legislative determinations of legal tender, and assessment of taxes payable in a designated currency.

14. Searle accounts for the semantics of "open" differently. I do not address his analysis in particular, because my reasoning for a different understanding of language turns on the overall explanatory fecundity of a niche constructive conception of discursive practices rather than analysis of specific examples.

Money also shows up cross-culturally in different concrete practical involvements. Michael Taussig described some unusual ways of engaging with money in peasant communities in Colombia.

> According to the belief in *el bautizo del billete* (baptism of the bill) in the southern Cauca Valley, the godparent-to-be conceals a peso note in his or her hand during the baptism of the child by the Catholic priest. The peso bill is thus believed to be baptized instead of the child. When this now baptized bill enters into general monetary circulation, it is believed that the bill will continually return to its owner, with interest, enriching the owner and impoverishing the other parties to the deals transacted by the owner of the bill. The owner is now the god-parent of the peso bill. (1980, 126)

Taussig recognizes that this belief is very strange and that unbelievers will find it utterly unwarranted, but it informs many concrete bodily engagements with peso notes and other persons by parents, priests, godparents, and shopkeepers. These activities include shopkeepers' efforts to avoid the anticipated consequences of accepting baptized bills, and parents' and priests' precautions and admonitions against such illicit baptisms, alongside some godparents' efforts to subvert those ceremonies. Taussig invites readers to juxtapose this way of trying to activate money as interest-bearing capital to other people's more widespread engagement with banks, investments, and capital growth to defamiliarize those forms of commodity fetishism. I make a different point. People can exchange peso notes for sale goods, talk about money and interest, and cope with fluctuations in their available wealth, goods, and reliance on sales or purchases, while each also understands those exchanges as having very different concrete significance.[15] Elsewhere, traders at international currency or commodity exchanges and officials at national banks regularly have "at their fingertips" the situated ability to intervene in larger monetary flows. These mediated interventions engage with money abstracted from the concrete embodiments Kukla and Taussig described. They are nevertheless consequentially interdependent with how money is entangled throughout people's lives. Meanwhile, economists construct and apply theoretical models that track these abstract monetary movements quantitatively, which both subsume and suppress the qualitative diversity of their concrete entanglements in multiple practices.

---

15. Peter Galison (1996, 803–4) uses Taussig's work to illustrate "trading zones" that coordinate different local contexts of action and belief. Brandom's social-rationalist inferentialism emphasizes how people talk about the same objects while holding very different beliefs about them, but he only considers interrelations among discursive "scores" and not interactions among people's concretely postured dealings with one another.

In learning to deal with different concrete settings, people also acquire open-ended facility with words used there. That concrete ability to hear, recognize, and use words prepares them to make connected sense of their uses in other settings, often with guidance and correction by others. Learning a language is not just acquiring a verbal repertoire. Second language acquisition may begin that way, but only via connections to discursive practices one already inhabits. One then may grow into that language by gradually taking up concretely situated uses without translation. That skill is extended by picking up on related uses of its words in settings where only some aspects of past use make sense. Facility in a language conjoins concretely grounded verbal capacities with abilities to recognize and use words across contexts. That recognition still draws on some concretely situated interconnections, even while making new sense of them to accord with uses elsewhere. People differ in their prior familiarity with words, but the ambiguities are not debilitating; in adding words to a working vocabulary they can track uses across contexts by drawing on their assimilation of prior situated uses of other terms. Understanding another speaker is not interpreting an idiolect, but instead assimilates their word use within a public practice that outruns anyone's situated uptake. The intralinguistic facility thereby acquired allows further use of those words to configure imaginary settings remote from anyone's actual encounters.

Sellars (1997) influentially argued that perceptual *experience* extends beyond differential responsiveness to the presence or absence of stimuli or impressions. It conjoins a richer competence over contextual constancy and variation with an inferential facility with related concepts. Kukla and Lance expand on Sellars by recognizing experience as articulated not only discursively and inferentially, but also by perceptual activity: "Observations are the activities through which we engaged with the elements of the world, the complex transitions from the competent and interactive moving about of our body—focusing eyes, picking up things, all the rest—to normative output. . . . Perceptual episodes are different from judgments, but not because they are less conceptually articulated or inferentially fecund. Rather, they *do* something different: they take up, acknowledge, or recognize the normative significance of worldly events and objects" (2009, 78). For Kukla and Lance, perceptual activities play dual roles in discursive practice. They both allow people to connect with one another as situated in a publicly shared world, and they provide multifaceted access to worldly objects as an independent tribunal to which judgments and actions are accountable.

A naturecultural conception of discursive practices expands on both lines of argument. Perceptual activities enable inferentially articulated discursive practices to gain concrete "empirical content" and interactive intentional directedness. These activities are situated amid practical immersion in con-crete life activities. People do occasionally take up roles that are primarily ob-servational, but observing is usually integral to other engagements. Someone who was only an observer would have no experience in the sense that matters for understanding how discursive practices articulate the world people in-habit. The word "experience" is sometimes used to talk about first-personal uptake, and sometimes in a more encompassing sense. This broader sense of experience involves more extensive practical-perceptual engagement with concrete situations, as when a job advertisement requests three years of ex-perience. The discursive articulation of practices requires participants who both grasp the inferential articulation of conceptual relations which Sellars rightly emphasized and can also bring those inferences to bear in concrete practical immersion with one another in the world.

That dual conceptual articulation of practical-perceptual situatedness is not an interface between bodily engagement with the world and inferential relations among sentences or propositions. Those inferential relations must also be (re)integrated amid concrete, situated involvements. The need for such integration is highlighted by its occasional failure, when people recog-nize and acknowledge inferential relations without reintegration into their lives. As an illustration, Lance asks that we "imagine Frank, who is so com-pletely in love with his partner John that he cannot truly embrace the fact that John is a cheating, lying philanderer who is treating him contemptuously. One day, he might be confronted with copious evidence by a third friend, and in the face of pressure, accept—as a sort of theoretical proposition—that yes, John is cheating. . . . In such a case, it is natural to say, 'yes, I know you are right, but I just can't bring myself to believe it'" (2017, 289). Someone whose inferential acknowledgments were always or mostly insulated from concretely situated uptake, however, would not be engaging in inference or discursive practices. That recognition parallels in the other direction Sellars's claim that perceptual impressions without inferential significance would not constitute experience. Conceptual relations that lack sufficient concrete em-bodied significance would also not be inferences. Discursive practices involve perceptually accessible bodily performances that are also dually situated amid concrete practical involvements in the world and their partially detachable, inferential significance in other practical contexts.

Discursive Practices

We now consider how discursive practices are situated amid concrete worldly involvements that contribute to articulated conceptual understanding. An important aspect of the situated materiality of linguistic performances is communicative. Utterances always have a vocative directedness that might be explicitly marked by proper names, descriptions, hypothetical constructions ("if you are not wearing your seat belt, please buckle it now"), or merely indicated by gesture or context (Kukla and Lance 2009, chap. 6–7). Assertions are normally directed toward someone or some group in particular even though anyone can then take them up. Grammar may be expressed in generic rules, but grammatical constructions vary observably and sometimes intentionally to situate speakers or audiences socially. Derek Bickerton reports that twenty Guyanese students asked to translate an English sentence into Creolese generated thirteen different versions of its opening words, "I am sitting." When asked who might actually use those expressions, however, they readily identified a "most probable speaker" for each—for example, "an illiterate female rice-grower in Berbice" or "a middle-aged black stevedore on the Georgetown docks" (Bickerton 2008, 22). Speakers situate themselves, and may also establish rapport or more aloof social distance by using different grammatical constructions or vocabulary.

Some utterances are pure vocatives that only pick out an intended audience and solicit a response: greetings, roll calls, or an interpellating "Hey, you!" Interplay between vocatives and acknowledgments is central to discursive practices as socially articulative. The conversations they shape are integral to interpersonal relationships and also open onto other aspects of a situation. Requests, orders, promises, reports, warnings, advice, and a plethora of other such performances integrate speech acts with other activities, engaging prior events and subsequent responses to what was said conversationally. Those addressed may fulfill a request or order, rely on a promise, act on a report or warning, or instead refuse, challenge, discount, or ignore what was directed or indicated. In either case, speech acts and their uptake sustain discursive interdependence between what is said and what is done.

Utterances not only locate speakers and audiences among differentiated identities, relationships, roles, standings, or situations. Conversations institute, extend, challenge, or revise those positionings.[16] Their instrumental roles in sharing information, making plans, or coordinating activities emerge against that background. People spend time and hang out with others just to talk. They connect with others by telling stories about themselves or others, expressing and responding to opinions, making jokes, describing or imagin-

16. The literature on conversation as a social practice is enormous. Classic sociological studies include Sacks (1995), Goffman (1983, 1959), and Garfinkel (1967). Di Paolo et al. (2018) give sustained attention to the bodily aspects of conversation.

ing other situations, or commenting on shared circumstances or past and projected-future interactions. Conversations have their own harmonious or discordant emotional attunement, which is also affectively shaped by participants' underlying moods and expressed emotions. While conversation may seem to consist in extended verbal exchanges, the relationships it establishes and sustains integrally depend on their settings: it makes a difference whether a conversation takes place at work, around the dinner table, on a street corner, in a bar, or in bed. Conversations are also set amid other personal relationships: what happens in a conversation and how it matters depends in part on participants' relationships with others. Whom they know and care about, who knows them or will encounter them in other settings, and how they fit amid networks of social roles and standings affects how people interact, and how those roles, relationships, and standings change over time. Interpersonal relationships are marked by what people say, have said, would say, and do not say to others, including what they let slip or hold back and how they consequently open up to others or close themselves off.[17] Intimacy or estrangement conditions people's self-presentation and engagement with others and also helps shape people's developing sense of self as interactively situated.

Sustaining and differentiating relationships and interpersonal roles or standings is but one aspect of the discursive articulation of human ways of life, however. Conversational interactions ostensively redirect participants' postured engagement with their surroundings toward distinct entities, features, or relations. In common use, "ostension" may seem limited to directing others' perceptual orientation and attention by pointing, turning one's gaze, cocking an ear, or otherwise making visibly directed gestures. Even in this narrow sense, ostension requires complex attentiveness to one another's postures as pointing beyond themselves. Ostension only succeeds if the one addressed attends to and recognizes the other's posture as directing her elsewhere and picks up the intended reorientation. Those who gesture ostensively must attend not only to their target but also to others' postured orientations and any divergences from the intended redirection. The result, when successful, does not just direct someone's attention to aspects of the world, but also directs people toward one another *as* mutually directed and focused. We saw in chapter 3 that bodily postures are always selectively grounded and directed within an environment. Ostension

17. Sharon Traweek, an anthropologist studying high-energy physics communities, reports that one physicist discussing how information-sharing defines socially significant boundaries in scientific work said, "One thing we never tell anyone is . . . ," thereby simultaneously offering partial, temporary admission to professional relationships while also barring her as not "anyone" to whom such admission would count (Traweek 1988, 14).

carries that directedness further by picking out features or aspects of only-partially shared surroundings. Redirecting another's perceptual-practical orientation requires implicitly shifting orientations: what is in front of me may be behind you, what is to my right is to your left, and what is near me may be far from you, with those horizontal differences typically anchored by a shared vertical orientation of human bodies. The requisite reorientations of posture and location work together with other shifts in salience or significance due to differences in discursive familiarity or social location, involvement, or standing.[18] Ostensive triangulation among people's postures and their environmental directedness transforms postured orientations within an environment into interactive directedness toward objects, features, or relations. Those familiar with Brandom (1994, chap. 8) on the social-perspectival articulation of "objects" as open-ended groups of intersubstitutable singular terms will recognize the ostensive articulation of objects, features, and relations as generalizing his line of argument. What Brandom describes as "perspectivally shifted conceptual content" from others' assertions is a special case of ostensive convergence, in which people coordinate what is perceptually salient for them by indicating it discursively to others differently situated. This assimilation of conceptually articulated content within people's practical-perceptual engagement with one another in partially shared circumstances further exemplifies what it means to turn Brandom's social-rationalist inferentialism inside out as a naturalcultural rather than social activity.

Ostension is thus not confined to gestured direction and perceptual uptake. It only succeeds by drawing on contextual cues, practical-perceptual skills, linguistic and other social patterns, prior interactions, and more or less common features of human bodily postures and environmental salience and significance. People use words together with expressive gestures to direct attention to an animal, a color, a juxtaposition of things, a normative concern, or an object's location or proximity, and much more. Kukla also emphasizes the possible open-endedness of ostensive reorientation.

> Successful ostension results in grasping or seeing what is ostended in a way that is undistorted and enables practical coping with it, and this will be embodied. But there is no need for us to assume any simple sensory contact as the medium of this grasping. . . . Wittgenstein's main example of ostension is of how to go on following a rule. . . . The ostender tries to get the addressee to attend to a norm of behavior in a way that involves practical grasp of how to continue it. Regardless of any puzzles concerning how we manage to do this,

18. I have more extensively discussed perspectival shifts required to share conceptual contents elsewhere (Rouse 2002, 217–23).

we are routinely successful at this kind of ostension. Hence we can indicate things that are nothing like mid-size physical objects. We can ostend states of affairs, moral complications, responsibilities, abstract trends, and things with any number of other metaphysical characters. (Kukla 2017, 107).

In this context, Lakoff and Johnson's (1980) account of the systematicity of orientational and ontological metaphors exemplifies how practical-perceptual ostensive reorientations can be deployed and built on to facilitate more complex or abstract ostensions.

Ostension triangulates among people's bodily orientations within mutually interdependent environments to pick out objects, properties, and relations at their intersections. Each person has a postured bodily directedness within and toward an environment in which other persons and their postured directedness are typically salient. Ostension is more complexly directed within that environment, triangulating among one's own practical-perceptual orientation, another's postured directedness, and what she is directed toward. That triangulation contributes to conceptual articulation of the world but only in conjunction with other aspects of the practices now making up human environments. Grasping the various resources that can be deployed to redirect others' postured orientation ostensively shows few constraints on these interactions that pick out entities and their properties or relations as standing out from an environmental background.

The most basic form of discursive articulation is anaphoric. Ostensive redirection is always a situated, occasioned performance. For that redirection to sustain a reorientation, whether for conversational interaction or other joint projects, people must at least implicitly direct themselves toward a previous ostensive performance and what it indicated. As Brandom noted in commenting on Hegel:

> Deictic tokenings as such are unrepeatable in the sense of being unique, datable occurrences. But to be cognitively significant, what they point out, notice, or register must be repeatably available—for instance, to appear in the premises of inferences, embedded as the antecedent of a conditional used to draw hypothetical consequences, and embedded inside a negation so that its denial can at least be contemplated. Demonstratives have the potential to make a cognitive difference, to do some cognitive work, only insofar as they can be picked up semantically by other expressions, typically pronouns, which do not function demonstratively. Deixis presupposes anaphora. (Brandom 2019, 124–25)

Anaphoric inheritance from previous token performances requires tracking and marking an ordinal sequencing of performances, distinguishing what was done or indicated then from what is happening now, and from when some other related performance occurred or will occur. From a broadly evolutionary perspective it is unsurprising that marking tense (when something occurs), modality (whether the occurrence is actual, hypothetical, or future-possible), and aspect (whether an indicated situation or occurrence is completed, continuing, or anticipated) are built into grammars.[19] These distinctions belong to the "semantic envelope" of language (Dor and Jablonka 2000), the basic structure of how languages organize and direct what can be said and what those sayings are about or for. As part of people's normal developmental environments, utterances that systematically mark such differentiations extend and order how speakers can direct themselves toward one another and their mutually encompassing circumstances.

The differentiation of words as semantically significant elements of discursive performances involves a different anaphoric inheritance from verbal developmental environments. Words are used as tokens of expressive types, but that type-token distinction is embedded in anaphoric relations among current, past, and possible future uses of recognizably iterated expressions. What matters is not objective similarity or dissimilarity among utterances— uses of an expression on different occasions or by different speakers vary considerably as sonic patterns or visible marks—but rather mutual recognition among speakers and hearers that an expression is repeatable and inherits semantic significance from *its* uses on prior occasions. Subsequent uses partially transform that significance when added to the stock of recognized iterations. Prior uses of words carry dual contextuality, however. Words occur in ordered combination with others. Part of recognizing and acquiring a word is grasping how it can combine with which other words, including in novel intelligible combinations. Combinations of word use are also situated in a temporally extended environmental context, however. Word uses are conditioned by past as well as current social and material circumstances, including which aspects of those circumstances were salient, which aspects those word uses made salient, the affective significance of those occasioned uses, what the speaker was doing or attempting with those words, and how the circumstances resemble or differ from prior uses. Speakers of a language overlap only minimally in their direct perceptual encounters with and implicit recollections of word use, but linguistic practices provide dense networks of partial overlaps among multiple speakers. Moreover, speakers and listeners mostly do not assimilate prior exposure to word use through explicit

19. For a brief, nontechnical discussion of tense-modality-aspect systems from an evolutionary orientation, see Bickerton (2014, 158–61).

recollection and comparison. Exposure to and use of particular words is instead embedded in bodily postures within current circumstances. That postured orientation shapes dispositions to use those words, inclinations for how to understand and respond to others' uses, and encompassing patterns of perceptual and practical salience.

Anaphoric uptake of prior ostensive performances and word use also has more widely recognized articulative significance as classificatory. In acquiring and using words in a language, people also inherit and sustain classifications of their surroundings into kinds of entity, property, or relation, along with these and other kinds of kinds. Acquisition of a working vocabulary registers, codifies, and builds on patterns of differential perceptual and practical responsiveness to circumstances. Geoffrey Bowker and Leigh Star point out: "To classify is human. Not all classifications take formal shape or are standardized in commercial or bureaucratic products. We all spend large parts of our days doing classification work, often tacitly, and we make up and use a range of ad hoc classifications to do so. We sort dirty dishes from clean, white laundry from colorfast, important email to be answered from e-junk" (1999, 1–2). Even behaviorally embedded classification requires ability to tell differences and recognize relevant similarities, and to sustain distinctions with sufficient clarity and effect to matter in practice. Such classifications become explicit when marked verbally, and public when people hold one another accountable for their uses of classificatory expressions, or for reconstructing or adjusting situations to align better with classificatory practices. The significance of classificatory performances as conjoined material and behavioral niche construction becomes especially evident wherever such practices are partially standardized in material or discursive infrastructure.[20]

The classificatory role of practices also indicates especially clearly how discursive articulations are embedded in concrete situations, while also establishing and sustaining more limited, partially disentangled interdependence across practices. The adjudication of boundaries among categories, their application in practice, and the significance of the differences they express and sustain are concretely embedded in particular settings. People and their skills also move among settings, and verbal expressions of distinctions, standards and skills often travel with them, sometimes in concert with things classified. Discursive articulation thereby brings classificatory differentiations into more complex and far-reaching inferential relations, usually

---

20. Bowker and Star (1999, 6–10, 33–38), Edwards (2010, 8–25), Edwards et al. (2007), Hughes (1983), and Bakke (2016, xi–xxx) discuss infrastructure as pervasive in and integral to many overlapping domains of practice.

requiring mutual adjustment of contexts and categories. Some properties are incompatible, with the presence of one excluding others, while other groups of properties or relations support symmetric or asymmetric entailments, or vary independently. Objects and kinds then stand out as loci of entailments, compatibilities and incompatibilities, and identifications: the same object can be encountered on multiple occasions, and identified in different situations; mutually exclusive properties or relations cannot coexist in the same object(s) at the same time but can hold for different objects at the same time or the same object at different times; compatible properties or relations can combine in the same object(s).

Discursive practices thus articulate the world, allowing new patterns to show up by introducing words, uses of words, and the skills and performances they enable. Ostensive redirection allows triangulation among different environmentally directed postures to pick out objects, properties, and relations from those ongoing involvements. The ostensive conjoining of words and gestures works in either direction: sometimes people use presumptively shared ostensive salience to introduce words or refine their uses; sometimes prior familiarity with word use helps make other aspects of their surroundings stand out recognizably; and sometimes reliance oscillates between words and situations. Availability of and familiarity with situated verbal utterances partially stabilize those articulations and thereby change how people perceive, track, and respond to situations. Having a vocabulary to register and track differences among kinds of entities and families of properties and relations allows those features to stand out from more complex involvements and connects different circumstances.

Discursive Construction

Linguistic utterances are salient aspects of people's developmental environments, and they are also assimilated into their expressive capacities. Languages are thereby niche constructive, as forms of behavior whose salient presence and uptake in people's developmental environments help reproduce behavioral descendants in subsequent generations. Linguistic expressions do their work in part by complex embeddedness in postured bodily engagement with concrete situations, while also belonging to linguistic practices with more limited inferential connections to differently situated uses of those expressions in other practices. That dual contextualization articulates the world by bringing out aspects of people's concrete engagements in the world as encounters with objects, properties, and relations reidentifiable across practical contexts. The incompatibilities and other inferential relations among those situated discursive performances then require ongoing adjustments all around to bring those practices and discursive performances back into compatible alignment.

I have so far emphasized horizontally differentiated practices as the contexts that condition the reidentifiability of objects and inferential relations among linguistic constructions. That articulative role is enhanced and extended by discursive practices that introduce novel constructions within or conjoined with language. Those discursive constructions enable further distinctions and connections across contexts, which can then be reintegrated into people's situated activities to transform their practical-perceptual and ostensive capacities. Counting is a familiar example. The availability and mastery of sequences of number words or symbols and the practice of counting enables enumeration of things in the world and a more differentiated responsiveness to quantitative relations as numerical differences. Adding these canonical sequences to people's verbal repertoires lets new patterns of similarity and difference emerge across practical contexts. Further constructive activities and practices built on enumeration—arithmetic, algebras, the real and complex numbers, and mathematical objects and algorithmic or computational operations more generally—allow movement back and forth between self-contained operations in these constructed domains and connections to other practices.

These discursive constructions are not limited to verbal and symbolic practices. The establishment and stabilization of unit measures discernible with rulers, scales, thermometers, or timers, conjoined with counting, adding, and other discursive constructions, then enumerates continuous differences in length, weight, volume, temperature, or temporal duration. Introducing, stabilizing, and standardizing units of measure with the requisite skills and practices of measuring and assessing their performance and outcome are often complex achievements. Once rudimentary forms of such discursive practices are in place as familiar and reliable aspects of human developmental environments, they also enable more complex forms of discursive articulation.[21] The invention and use of calendars and clocks meshed with new practices and discursive articulations to similar effect. The division and organization of temporal durations into hours of the day, days of the week, weeks of the month, months of the year, and numerically or cyclically differentiated years was one outcome. These niche-constructive practices enable new ways to organize,

21. Alder (2002, chap. 5) discusses shifts from units of measure attuned to locally relevant differences to measures transportable and reusable in distal contexts. Latour (1987, chap. 6) discusses metrology more generally.

track, and give significance to the duration and sequencing of events and activities. As Brian Cantwell Smith notes, "clock faces do not come labeled in advance by God, like plant slips at a nursery, identified with a white plastic tag. . . . The problem is particularly acute for time itself, especially the periodic cycle of hours, minutes, and seconds . . . [due to] the incestuous fact that clocks themselves are probably largely responsible for the temporal registrations (hours, minutes, seconds, etc.) of the times they represent" (1988, 10).[22] Adding these and other stabilized discursive-practical performances to people's postured and ostensively triangulated environments is both transformative and enabling. Discursive articulation changes people's environments and how they can and do respond with a vast array of new perceptual discriminations, practical capacities, affective attunements, and mutually interactive and interdependent performances. The distinctive trajectory of human evolution was enabled by iterative cycles of discursive niche construction. Assimilating these transformations of human developmental environments allows for new elements and practices, while also creating new developmental processes that produce selection pressures and developmental advantages for adaptation to these recursively transformed environments.

These discursive constructions similarly integrate postured bodily embeddedness in concrete practical-perceptual relations into more extensive forms of counting, tracking, measuring, and comparing things via newly articulated properties or relations. People now develop amid discursive environments replete with widely deployed units of measure, but they usually acquire visceral grasp of the multifaceted significance of those differences only within limited ranges. Most people who grow up with English systems of measures, for example, do not rely on a measurement scale to grasp the length of a yard, the height of a six-footer, differences in body weight between 150 and 200 pounds, a mile's walking or running distance, fifty "highway miles," or ambient temperatures of 50 or 80 degrees Fahrenheit. Within specific ranges and tolerances, people develop an intuitive, bodily-postural facility with enumerations of lengths, weights, or durations, whereas measurements outside those ranges or in different units only make sense via scale-based calculations and comparisons. That visceral grasp of familiar measures makes the outcome of the calculations concretely intelligible, even though it also matters to

22. Mumford (1934, 14–15), Landes (1983, 72–78, 285–287), and Galison (2003, 13–14, 84–98) discuss clocks' constructive role in articulating times and temporal durations. Landes and Galison emphasize that construction and coordination of timing was driven by and responsive to other practices where timing mattered, notably the coordination of railroad schedules.

one's visceral grasp of familiar measurements that those units are indefinitely extendable on standardized scales.

Distinctions between people's viscerally grasped facility with concrete situations, and their merely inferential, calculative, or other discursively extended understanding of objects and properties in other practical contexts, have no fixed boundary. People with developed know-how in one system of measures can acquire another through extended practical immersion, and with sufficient practice and concern they can also develop similar immersive grasp of more esoteric measures or unusual ranges or tolerances within familiar systems. Most people acquire a similarly visceral, flexible knowing their way around their geographic neighborhoods and often-traveled routes on roads, paths, or public transit systems, which they may also situate more abstractly in relation to locations on a map. Others, however, can acquire comparable familiarity with relationships and distances depicted on maps and their significance for point-to-point travel, while lacking concrete grasp of what they would encounter or negotiate at the particular points mapped.

Scientific practices exemplify how discursive and other material forms of niche construction work together to articulate novel conceptual relationships.[23] New conceptual domains become accessible to scientific reasoning via experimental systems that let intelligible patterns stand out clearly. Conceptual relationships are definable by the regimented behavior of carefully prepared materials and instruments in experimental settings. These conceptual relationships can then be extended beyond controlled settings, either by adjusting the conceptual relations for previously excluded effects (friction, air resistance, chemical impurities, lethal mutants, inelastic collisions, developmental noise, and other "confounding" factors), or isolating or simplifying settings elsewhere to resemble laboratory conditions (Latour 1983). Many classic experimental practices helped bring whole conceptual domains into the scientific space of reasons: air pumps for gas behavior (Shapin and Schaffer 1985); pure-bred lines, hybridization, and back crosses with breeding stocks to distinguish genetics from development (Kohler 1994; Müller-Wille and Rheinberger 2012, chap. 6); voltaic cells and copper wires for electrical currents and electromagnetism; systems of particle sources, targets, and detectors for subatomic physics (Galison 1997); purification, isolation, and weighing of chemical substances; ultracentrifuges and electron microscopes for cell biology to correlate structure and biochemical function (Bechtel 1993,

23. I discuss the niche constructive role of experimental systems, which articulate scientific domains by mediating application of theoretical concepts and families of models to other situations in the world (Rouse 2015, chap. 7–10).

2005; Rheinberger 1995); work with alcohols and their derivatives for organic chemistry (Klein 2003); and more. These examples also illuminate how widespread uses of scientific concepts in diverse practical contexts gain empirical grounding through concretely situated practical know-how rather than at simpler observational interfaces. Scientific concepts have their empirical home in practical-contextual grasp of instruments, protocols, prepared materials, and theoretical models amid a general laboratory ethos of cleanliness, standardized procedures, and careful record keeping (Haugeland 1998, chap. 9). These practices nevertheless vary in different laboratory settings.

Two prominent and extensive groups of niche-constructive transformations conclude this sketch of human developmental and selective environments as discursively articulated by human capacities for ostensive triangulation, symbolic displacement, and their anaphorically interrelated expression. The first group consists of proximately intralinguistic practices. They include material inferences, metaphoric complexes, logical constructions, and narratives, along with writing and reading. Material inference is integral to language use: one does not understand what a linguistic performance says or does without some grasp of the inferential significance of the words used.[24] Material inferential relations also incorporate metaphoric complexes that significantly expand people's expressive capacities (Lakoff and Johnson 1980). The densely interconnected inferential relationships embedded in word use and sentential constructions are central to the articulative and expressive capacities of language. Linguistic utterances are never freestanding, but instead implicate an open-ended range of inferentially interconnected performances, both in specific contexts of use and in understanding how to extend and modify those uses in other settings. Uses of words or phrases in new settings, including novel linguistic contexts, also partially transform their anaphoric dependence on past patterns of use: all discursive niche construction is reconstructive. Mastery of logical expressions lets speakers make the relationships implicit in concept use and material inference available for criticism or justification. People similarly develop abilities to recognize, appreciate, and respond to rationality and making sense as constitutive ends at issue in what those conceptual relations express.

Narrative is also integral to discursively articulated human environments. Storytelling is pervasive in human life, from everyday recounting of sequential events to widely shared familial, mythological, historical, moral, or literary

24. Sellars (2007, chap. 1–2) and Brandom (1994, chap. 2, section 4; and 2000, chap. 1) distinguish material inference (the inferential significance of conceptual contents) from formal-logical inferences.

narratives. The familiarity and availability of many stories, people's ability to place events within a story, and how the stories people tell and hear bear on how they live are salient aspects of human life and discursive sense making. Storytelling enables understanding human lives as narrative sequences from birth to death, marked by milestones and transitions among stages, with turning points and moments of triumph or disaster, loss and recovery, setbacks and comebacks, and concern with death's finality. Storytelling also constitutes human communities united or divided by shared or contested histories. Telling and understanding stories enables people to live within overlapping and conflicting stories and situate actions within narrative fields.[25]

The emergence of written language and its subsequent transformations by technologies and practices of printing or electronic storage and transmission introduced different kinds of intralinguistic environment. The tempting claim that writing first enabled persistent and decontextualized linguistic expressions is not quite right on either count. Oral traditions made some linguistic performances reliably persistent, while the vagaries of manuscript preservation and copying limit the stability of written texts, and rapidly obsolescent technologies of storage and retrieval render electronic preservation more fragile. Written texts are also only partially decontextualized from their production, since provenance and internal features of the text typically constrain authorship and projected readerships. Written and printed texts nevertheless importantly shifted how people engage discursive environments.[26] Written records enable more complex forms of economic activity and bureaucratic regulation and control, which transform many other practices.[27] They enable novel, more widely distributed forms of pedagogy and social learning. Writing facilitates constructing more complex and lengthy linguistic performances and their preservation and uptake beyond the circumstances of their production or continuing reproduction.[28] Written language

25. Whether stories are lived as well as told is controversial in philosophy of history and of action (Fay 1996, chap. 9; Carr 1986). I have argued (Rouse 1990; 1996, 162–73) that people situate their lives within a field of narrative significance rather than giving them unified narrative form. Roth (2020) astutely discusses narratives as explanatory.

26. Goody (1977) and Eisenstein (1979) are classic discussions of the transformative significance of writing and printing, respectively, as is Johns (1998) on the book format.

27. Some work in social ontology (Searle 2010; Guala 2016) emphasizes the collective creation of token institutions such as nation-states, corporations, clubs, and the like. That emphasis presupposes practices of written documentation, storage, retrieval, and normative authorization. See also Scott (2017, 139–49).

28. Although the *Iliad, Odyssey, Beowulf, Gilgamesh,* and other epic narratives originated and were long sustained orally, only those preserved in writing outlived their original communities.

also transformed human cognitive capacities. Oral traditions developed people's mnemonic capacities, materially facilitated by rhythm, rhyme, musical accompaniment, and extensive repetition. Written texts partially displaced mnemonic skill but also brought forth practices and skills of logical and rhetorical criticism through juxtaposition and comparison of what was said on different occasions.[29] Literacy is a novel, powerful skill requiring new pedagogical practices to maintain and disseminate. The resulting stratification of skills affects people's social position and prospects while constituting dyslexia and illiteracy as novel forms of disability. The development and expansion of written and print cultures and the associated transformations of people's developed cognitive capacities are superb examples of how conjoined material and behavioral niche construction transform human ways of life and cognitive development. Human developmental environments have likely not encompassed writing and reading long enough to enable genetic assimilation. These practices have nevertheless been reproduced and transformed over hundreds of generations, maintained and expanded in scope and complexity, adapted to different linguistic traditions, extended to almost all human societies, taken up by increasing proportions of most human populations, and integrated into almost every aspect of human life.

Although language is a preeminent evolutionary novelty in the hominin lineage, it was not isolated from other forms of conceptually articulated niche construction. The second part of this concluding sketch of people's discursive environments considers nonlinguistic practices and capacities that contribute to conceptually articulated ways of life. Among these practices are the making, understanding, and deployment of images, mimetic or other bodily expressions, musical performances, games, or rituals. They also include the pervasive deployment of equipment, distinguishable from mere tools by systematic interconnections within equipmental complexes. Tool use is instrumental deployment of something to bring about an intended effect. Tools become equipment when used in concert to bring about intended effects that require their appropriate use together. Equipment is also typically produced and made available together for recurrent, organized use, often amid practices with differentiated roles and skills.[30]

29. See Goody (1977). Plato's Socratic dialogues are written constructions of oral encounters between the mnemonic skills of traditional oral cultures and the cognitive capacities for critical assessment enabled by juxtaposing written texts.

30. Okrent (2018, chap. 5 and pp. 217–23) similarly distinguishes tools from "toolkits" but to somewhat different ends.

Several features of these practices and performances combine to mark them as *conceptually* articulative. First is their symbolically displaced significance. Although they always occur in specific circumstances, their use cannot be adequately understood simply as responses to those circumstances or as indicative of what to do there. These performances are expressive and communicative even while usually lacking determinate *propositional* content. The resulting capacities for expression and understanding usually outrun what can be readily said in words. As Daniel Dor and Eva Jablonka suggested, "many of the messages which turn out to be difficult to communicate through language seem to be very well suited for communication through other means of communication: we can *mime* or *dance* them, use *facial expressions* and *body language* to express them, *paint* and *draw* them, write and play *music*, prepare *charts* and *tables*, write *mathematical formulae*, screen *movies* and *videos*, and so on" (2000, 40). Subtle differences in mood or emotion that are difficult to describe in words can be brought out and recognized in musical performances or dramatic characterizations, as but one of many examples of how nonverbal expressions are conceptually articulative. These nonverbal articulations can then be taken up anaphorically to enrich discursive capacities.

A second aspect of these expressive repertoires is the practices that enable and condition articulative and communicative roles. Undertaking and understanding performances with those repertoires requires grasping their anaphoric interdependence with prior and possible future performances, as maintaining and revising ongoing traditions of expression, communication, or enactment (e.g., Risjord 2014 on jazz improvisation and chap. 4 above). People extend and develop their expressive and recognitive capacities in these conceptually articulative repertoires by undertaking performances against the background of others, as differential repetitions, transformations, or implicit commentaries. People now develop distinctive capacities in environments replete with many different expressive or enactive practices. These practices depend on recruiting and training new generations of participants who can recognize, fulfill, and revise their constitutive ends, and take up those practices amid their other activities and concerns. Their articulative and communicative character is closely bound with their two-dimensional normativity. They are open to assessment in one dimension for their appropriateness, excellence, originality, or success *as* linguistic utterances, musical performances, dramatic impersonations, visual depictions, moves in a game, or skillful use of equipment, and thus as continuations and transformations of a practice. Those performances and practices are also open to assessment for their place within and effects on lives, relationships, careers, institutions, communities, ways of life, or their biological lineage. Such conceptual practices do not just enable continuation and reproduction of human ways of life. They often embody a sense of having made people's lives worth living, as but one indication of the more complex normative space that encompasses not just whether but how human lives and lineage continue.

Power

Understanding Power

The practice-differentiation and discursive articulation of human ways of life situate people in different developmental environments. Which practices they participate in or depend on, which roles are available to them, which skills and capacities they could acquire, and which courses of action they can take up depend on these environmental differences. Compared to other organisms, humans exhibit great phenotypic plasticity. People inhabit different geographic locations and material circumstances, living very different lives and often confronting different kinds and degrees of challenge or threat to their well-being. Their developmental trajectories are partly channeled by other people they encounter, those people's responses to their own postures, performances, and prior developmental outcomes, and how their situated actions align or misalign in cooperatively interdependent ways. Extensive migration among environments has not homogenized human lives across that geographic range but instead often enhanced the horizontal diversification of human ways of life within particular geographic regions. One consequence of this diversification is the recognized linguistic, cultural, institutional, and socioeconomic differences within practice-based ways of life.

Human developmental plasticity accords with that geographic and practice-differentiated environmental diversity. Infants transported from one environment to another early in life typically develop within the range of normal variation among those around them. Even migrants to different environments later in life often successfully adapt to and accommodate living in different settings, amid other patterns of practice.[1] The diversification of linguistic environments

---

1. Differential treatment of migrants as outsiders or inferiors complicates the plasticity of human development in novel environments, whether predicated on visible indicators of race or

is illustrative. Human infants normally learn any language spoken around them during their early years of life. Although their developmental trajectories are shaped and constrained by those initial encounters, people still often learn languages first encountered later in life with varying degrees of fluency and discernible differences in performance.

The diversity of interdependent practices not only leads to historical and geographic differences within human ways of life. People are often differently positioned within the practices they encounter and take up, which are also differently positioned in relation to other practices. We have so far discussed the niche-constructive evolution of a practice-differentiated way of life and how people's postures, orientations, abilities, and normative concerns are affected *in general* by the practices amid which they develop. Humans never develop and live within a practice-based way of life generally, however, but only in particular locations amid specific practices. Those practices and people's roles within them are consequential for their developmental trajectories, postures, and orientations, the issues and choices they confront, and the developmental and life outcomes at stake in resolving those issues. People differ in what they can undertake and accomplish on their own or in concert with others, in the practices and roles within practices open to them, in how their choices and actions affect and are affected by others, in their vulnerability to or protection from harms or constraints, and in their abilities to understand, criticize, or change their situation.[2]

Differences in people's capacities, constraints, and vulnerabilities are often discussed in terms of power. In everyday parlance, power seems a ubiquitous but diverse social phenomenon. Some people or institutions can act in ways that effectively serve their interests or concerns while others' capacities and opportunities are more constrained. Those differences are often linked. Some people's actions and capacities dominate or oppress others, limiting their capabilities and prospects and leading them to act against their own interests and desires or to further the prospects and capacities of those more powerfully situated. Social theorists' efforts to analyze power and criticize its workings and effects have nevertheless encountered serious difficulties, despite

---

ethnicity, linguistic divides, other residual differences in customs, or sexual or gendered identities or expressions. The normative concerns raised by such treatment are a central theme in this chapter.

2. I use "action" here in a broad sense including not only overt performances, but also capacities to act, perceptions and belief formation, affective and orectic responses to situations, and personal affiliations.

widespread recognition of the phenomena of domination or social control and the resulting differences in people's capacities and vulnerabilities.[3]

The diversity of power attributions may put the coherence of the concept in question. "Power" often designates both control of or constraint on others' actions ("power-over") and abilities to bring about normatively significant effects ("power-to") (Wartenberg 1990, chap. 1). Force or violence may directly establish and maintain power or may only indirectly or subjunctively sustain credible coercive threats. Power over others may be achieved and sustained through persuasion or indirect influence without reliance on force, but that influence may instead result from ideological distortion or censorship to which force or coercion contribute (Lukes 2005; Wartenberg 1990, chap. 5). Some theorists construe these exercises of power over others as constraining agents' free or autonomous self-determination. One influential tradition prominently insists that the powers that thereby disempower some people, while enabling others to effect significant outcomes, have multiple "faces" (Isaac 1987; Lukes 2005). Other theorists argue that self and agency are always conditioned by social constraints. The political theorist Clarissa Hayward, for example, called for "de-facing" power by recognizing how constraints circulate throughout social life and differentially shape everyone's abilities and prospects without localization in some agents' power over others (Lukes 2005; Hayward 2000; Lukes and Hayward 2008). Power is further "de-faced" by constraints on action built into the material arrangements of neighborhoods, schools, clinics, prisons, workplaces, or markets (Foucault 1977, 1978, 2003; Hayward 2000). Power over others is most often identified with imposition of harmful or unwanted consequences, but power can also be constructive or liberating. Overall, ascriptions of power in the diverse forms of political leadership, socially enforced constraints, ideological distortion, the epistemic authority of experts, or capacities for instrumental effectiveness, force, or violence may seem to conflate altogether different phenomena or at best display family resemblances (Haugaard 2010).

Theoretical disagreements over how power is located and targeted are complemented by disputes over what power is. One tradition emphasizes power as a consensual or communicative outcome of concerted agency or collective intentionality, which constitutes and sustains rights, duties, obligations, permissions, and other forms of authority (Hobbes 1994; Arendt 1965, 1970; Searle 2010, chap. 7). An alternative tradition that encompasses debates

---

3. Wartenberg (1990, 1992), Hayward (2000), and Menge (2015, 2018, 2019, 2020) usefully review competing conceptions of power in social theory and the issues shaping those controversies.

over the "faces" of power understands it as instruments or instrumental capacities for powerful agents to control or redirect the actions of those less
powerful (e.g., Dahl 1958; Bachrach and Baratz 1962; Lukes 2005; Gaventa
1980). Social-structural conceptions locate power in relatively enduring institutions, relationships, or boundaries, whether they confer power on agents in
particular social locations (Isaac 1987; Barnes 1988) or shape the boundaries
of possible action for all agents (Hayward 2000). Dynamic conceptions of
power instead treat those mediating boundaries or networks as reproduced
and sustained interactively within temporally extended fields or alignments
of accommodation and resistance (Foucault 1977, 1978, 2003; Wartenberg
1990). Some construe the relative stability of those fields as a constitutive pretense or fiction (Menge 2015, 2020).

These divergent conceptions of power share common ground in understanding power and resistance as social phenomena. Understanding humans
as social-discursive animals developing in practice-differentiated environments allows for a more expansive naturecultural conception of power. That
conception is informed by constructive and critical engagement with some
aspects of Thomas Wartenberg's, Clarissa Hayward's, and Torsten Menge's
social theories of power, with Michel Foucault also prominently in the background.[4] Its primary basis, however, is recognition that development amid
the two-dimensional normativity and discursive articulation of practice-
differentiated environments differentially shapes people's bodily postures,
capacities, and prospects.

Power and Practices

The evolution of a niche-reconstructive, practice-differentiated way of life
mostly enhanced and refined people's collective ability to satisfy basic biological needs, while also enabling new kinds of cooperative activity and
achievement through that mutual interdependence. Those new capabilities
were nevertheless accompanied by new constraints on and threats to how
people live. All organisms are dependent on and constrained by their biological environments. People, however, now live in closely coupled interdependence with one another, in practices they take up together, and they must
adjust what they do to how those practices are sustained and reproduced.
People develop different capabilities and also encounter different constraints
because they take up or have access to different practices and are differently

4. I attend to Wartenberg, Hayward, Menge, and Foucault as the most important and insightful recent social theories of power, but I do not defend that critical assessment here.

placed within them. Those differences are further shaped by the categories by which others understand and respond to them (Hacking 2002; Haslanger 2012; Ásta 2018). People often find themselves excluded from or subordinated within some practices by how participants there or elsewhere categorize and respond to their involvement.

Practice-related differences in human capacities and constraints are grounded in people's neediness, vulnerability, and lack of self-sufficiency as living animals.[5] Social theories of power almost invariably but mistakenly take this fundamental aspect of human biology as given rather than integral to their own theoretical concerns in two initial respects. From one direction, maintenance of bodily functioning for people's survival and reproduction is always at issue in human lives, even when it may seem reliably assured by ongoing, stable practices. Force and coercion are not limited to imposing or threatening bodily harm or deprivation, but vulnerability and neediness play a prominent role in coercive power that leads subordinate agents to act to benefit others more favorably situated (Wartenberg 1990, 101–4). The significance of biological vulnerability in constituting power is clearest for those most severely constrained. Many people are rendered more vulnerable to hunger, disease, overwork, violence, environmental hazards, or premature death by how others' situated actions and capabilities align within practices. People care about their lives and prospects, and power often works through their affective response to these vulnerabilities. Those who encounter more severe forms of domination are usually well aware that their lives—both whether they continue living and what lives they can lead—are at stake in how those practices shape what they can do.

Power also depends on human animality in a further way that encompasses the capacities of more powerfully situated agents. Most conceptions of social power recognize that people's ability to affect the capacities or action-outcomes of others depends on whether other people act in accord with them. People's individual capacities to affect others are limited and those limitations are magnified in cooperative, practice-differentiated environments. The normative force of power-constitutive situations is grounded by people's biological interdependence. Any capacities to affect or control what others do and to achieve whatever goods people aspire to through that control depend on supportive alignment with what others do in relevant material settings. In the other direction, many powerful constraints work by excluding or limiting

5. People's vulnerability and neediness were long recognized as background to social and political theory, and as conditioning social power in book 1 of Hobbes's *Leviathan* (1994), but they were rarely foregrounded as integral to social practices or institutions.

people's participation in some practices or blocking their reliance on them. People's power-inflected abilities to satisfy basic biological needs, circumvent hazards or threats, and fulfill other normative concerns are mediated by practices in which they are interdependent with what others do in niche-constructed circumstances.

In our vulnerability, neediness, and affective response to circumstances, humans are not so different from other organisms whose bodily postures engage their developmental and selective environments in part through evolved affective responsiveness to the resulting patterned changes in their bodies (Prinz 2004). People's dependence on others to limit their vulnerability and provide for their needs are also to some extent continuous with those of other social animals. The many forms of evolved interdependence in other animals include herd behaviors to protect against predation; social learning and parental or kin-selective provision of resources; dominance hierarchies and "mock combats" enabling group cohesion, reproduction, and avoidance of internecine violence; animal communication systems as adaptive contributions to meeting basic biological needs (Hauser 1996); and biological needs for bodily interaction and sustained relationships with conspecifics in social animals. Here again, however, what notably differentiates human ways of life from those of other organisms is the practice-differentiated complexity and multiplicity of the developmental and selective environments with which the species coevolved. Even those targeted by power-constitutive alignments often find that the harms or deprivations threatened or imposed on them take more complex forms mediated by the capacities and normative concerns enabled by and expressed in their practice-differentiated environment.

Conceptions of social power and practice-based approaches to social theory thus often intersect in recognizing that the practices people participate in or depend on condition their lives in significant ways. Hayward, for example, argues that analysis of power should "focus attention on political mechanisms that comprise relevant practices, as well as the institutions that sustain and govern these practices" (2000, 38). Practices and institutions set boundaries to action, define ends and standards, and determine and distribute normative standings. Wartenberg's (1990, 72–74) practice-based conception of power focuses on how social alignments shape people's "action-environment" as a *field* of actions and possible actions, with explicit analogy to an electromagnetic or gravitational field. The field is composed of agents and their capacities, but interrelations among those capacities are not decomposable because of their field-dependence. People's actions and capabilities are enabled and constrained in many ways, but what characterizes *power* relations is their mediation by the field and the boundaries of practices where those actions are situated.

That field of possible alternative courses of action incorporates but is not limited to agents' understanding and evaluation of those alternatives. Understanding the significance of power within human ways of life depends on recognizing the distinctive, evolved biological character of human action-environments. Many organisms evolve and develop flexible repertoires for appropriate responsiveness to conflicting solicitations from surrounding circumstances. Human environments composed of horizontally and vertically differentiated practices establish temporally extended alternatives that reach well beyond how to respond to current circumstances. They yield significantly different developmental trajectories and resulting ways of life, even though linked by extended chains of interdependence. These more complex, differentiated action-environments are discursively articulated, extending, enriching, and sometimes constraining people's understanding and evaluation of alternative courses of action. Human environments are also two-dimensionally normative. What is at issue in how people respond to this field of alternatives is not only their survival and biological reproduction, but also the many normative concerns constituted and enabled by the practices composing those extended environments. What matters to the practical-perceptual configuration of people's postured responsiveness to their action-environment is not only what to do now and whether that response lets them survive and reproduce but *how* they live overall. The latter includes which practices they take up, how they develop as persons, the normative significance of those practices and their places within them, and how those practices and their own developmental possibilities change over time.

Many social theories of power nevertheless rely on an attenuated recognition of how people are situated and thereby enabled or constrained. Wartenberg, for example, only considers how the actions of some agents are oriented by and coordinated with those of dominant agents: "A situated power relationship between two social agents is thus constituted by the presence of peripheral social agents in the form of a *social alignment* if and only if . . . the coordinated practices of these social agents [are] comprehensive enough that the social agent facing the alignment encounters that alignment as having control over certain things that she might either need or desire" (1990, 150, original emphasis). He distinguishes people's social action-environment from their physical environment, contrasting what is socially possible for an agent to what is physically possible to show that the social possibilities are what matters. That reasoning omits consideration of what is biologically possible for an organism whose bodily capacities develop through interaction with a material environment incorporating other environmentally interactive human bodies. What supposedly matters for understanding power is only the "action-alternatives" shaped by agents' understanding and evaluation of their situation (1990, 80–84), considered apart from how those alternatives are realized physically or biologically.

This conception of how people's action-environment is constrained is too narrow in at least two ways. First, the niche-constructive constitution of human action-environments incorporates people's material circumstances and their ongoing, iterative reconfiguration. People develop within physically and biologically transformed environments that enable cooperative interaction and channel the possibilities for constraint or control embedded in those practices. Equipment, infrastructure, and other material components of those social alignments often significantly shape whether and how their aligned activities and capacities bear on another agent's needs or desires and capacities to fulfill them (Rouse 1996b, chap. 7; 2005b; Bowker and Star 1999; Star 2015). The significance of material infrastructure in reconfiguring embodied agents' capacities and constraints figured prominently in Foucault's work on power (1977, 1978, 2003) and stands out especially clearly through work in disability studies (Kafer 2013; Crosby 2016).

Biologically motivated physical niche construction on a large scale was often integral to the establishment and effectiveness of powerful state institutions for transportation systems, irrigation, the enclosure or terracing of land, and military control and defense of populations. The domestication, cultivation, distribution, and use of specific crops for food or clothing has similarly helped establish and maintain associated forms of economic capacity and political domination. Prominent examples include how wheat and barley cultivation both enabled and required state control of labor,[6] how domesticated and feral animals extended settler colonialism in climatic "neo-Europes" (Crosby 1986, chap. 8), and development of plantation systems of chattel slavery, indentured service, or other labor controls for the cultivation and global exchange of sugar, cotton, or bananas for consumption elsewhere (Mintz 1985; Williams 1994; Schoen 2009; Baptist 2014; Soluri 2005). The contingent

---

6. James Scott (2017) argues that the intermittent emergence of centralized state power and domination in the Tigris-Euphrates valley and elsewhere depended on grain-based agricultural economies. The short, observable harvest periods of wheat and barley and the ability to store them with minimal loss enabled state surveillance and control of agricultural labor and reliable cumulative extraction of food resources to sustain a state. Graeber and Wengrow (2021) defend more varied and contested relations between agricultural cultivation and centralized state power than Scott suggests, but they agree that questions of power and resistance are central to the history of agricultural practices.

availability of accessible deposits of coal and oil dramatically affected the economic and political possibilities and prospects of people throughout the world (Pomeranz 2000; Mitchell 2011). The niche-constructive design, construction, and uses of factories, mines, prisons, and organized military violence along with the arrangement of housing, transportation, educational facilities, ports, markets, and other material reconstructions of the environment both facilitate and depend on practices organized around them. Those material transformations are integral to the practical alignments that enable some interdependent achievements and impose other coercive and ideational constraints. Not only does conjoining material and behavioral niche construction impose specific forms of constraint and collective productivity within the varied human environments constituted by naturecultural practices. The regulation of access to what is thereby achieved or produced is also accomplished through practices linking the material infrastructure of manufacture, transport, distribution, and policing with specific ways of organizing and connecting different human activities.

Miranda Fricker (2007, 11–13) emphasizes a second widespread constriction of social-theoretical conceptions of action-environments shaped by power relations, in a criticism also central to Hayward's (2000) insistence on "defacing" power. Power has often been localized in individual agents' power to dominate or control others' actions. Wartenberg expands that conception by recognizing the mediation of one agent's power over another by more extensive social alignments, but his account still analyzes power as passing through such "power dyads." That constriction limits power to orientations of a social field around a dominant agent's actions and capacities, obscuring how agents may be more diffusely constrained by the practices they and others take up together.[7] The power differences constituted by a practice-differentiated way of life are not limited to how powerful agents constrain the powerless but instead concern how the practices people undertake shape the possibilities and prospects of all participants. Wartenberg rightly recognizes that these more diffuse power relations shape relationships among individuals in that context; his preeminent example is pervasive forms of sex discrimination that unequally empower men within marriage even when both partners seek a more egalitarian relationship. The primary issue in such cases is not the personal

7. Fricker describes nonagential forms of power as "structural," but that term suggests too static a conception and seems to separate social structure from individual agents' doings altogether. Constraints on people's actions and abilities can be dispersed throughout their environments yet still dynamically produced by agents' ongoing, situated interactions. See discussion below of dynamic conceptions of power.

domination enabled by alignments around sex differences, however, but the more pervasive constraints that enable and enforce that personal domination, give it a more extensive, "impersonal" form, and constrain the possibilities for interpersonal relationships.[8]

These two neglected aspects of power work together. Some constraints are built into material environments and enhance the effects of or limit the needs for active involvement of other agents. Elizabeth Anderson (2010) has shown how racial domination in the United States exemplifies that conjoined effect. White supremacy is sustained in significant part by racial segregation, produced and sustained by coupling social policies and aligned attitudes and performances with the material infrastructure of neighborhoods, highways, public transport, and service provision. Together, they limit African Americans' access to employment, adequate nutrition, health care, personal safety, educational opportunities, and influential social networks. Hayward similarly argued that the globalization and local deindustrialization of economic activity and the materialized arrangements of transportation, housing, and political boundaries shape power relations in school systems (2000, 3, 37). Anthony Hatch's (2016, 2019) studies of the racialized construction of "metabolic syndrome" and the pharmacological implementation of mass incarceration exemplify how materialized racial domination works biologically and medically in concert with physical infrastructure.

Discursive practices have an extensive role in sustaining differences in people's capabilities and prospects by affecting how they understand their action-environment. These effects are also difficult to localize in specific relations of domination. Ideological distortion and informational limitation or falsification are systemic constraints on discursive articulation (Haslanger 2021). Fricker (2007, chap. 7) points out how conceptual limitations function as distributed discursive constraints through widespread inability to make some forms of domination intelligible to oneself or others. Here the problem does not primarily result from influential utterances by particular people. Rather, some conceptual contents and inferential connections are familiar and readily available in discursive interactions; others are difficult to express in ways that can be recognized and acknowledged by anyone. Marilyn Frye called attention to difficulties in recognizing the systematic and pervasive character

---

8. Marx famously introduced a parallel distinction as a *historical* transition from relations of personal domination in feudal economies to capitalist social relations of distributed, impersonal class domination by owners of capital over those reliant on their own labor (Marx 1973, 163-65). I am arguing that "impersonal" practices of male privilege and sexual oppression (Frye 1983) have *synchronic* effects that shape interpersonal relationships.

of oppression: "It is now possible to grasp one of the reasons why oppression can be hard to see and recognize: one can study the elements of an oppressive structure with great care and some good will without seeing the structure as a whole, and hence without seeing or being able to understand that one is looking at a cage and that there are people there who are caged, whose motion and mobility are restricted, whose lives are shaped and reduced" (1983, 5). Fricker similarly argues that the concept of sexual harassment coalesced diverse constraints and diminutions of women's public lives into a coherent pattern whose harmfulness could then be more readily recognized, understood, and countered. People's action-environments are not just composed of specific alignments among agents and what they do but incorporate the material and discursive circumstances in which those agents encounter one another, interact, and assess their situations. Those circumstances are shaped by ongoing, situated interactions among those agents, which are often at work without being focused on or by particular personal relationships. Other potent examples come from shifting conceptual relations among aspects of sex, gender, and sexuality. These transformations include identification and recognition of intersex, transgender, and other nonbinary or queer modes of material and discursive bodily interaction amid heteronormative discursive constraints that left no intelligible place for those possibilities. Work in critical race theory has articulated similarly pervasive but heretofore obscured forms of anti-Black racism.[9] In such cases the discursive transformations typically occur in concert with other materialized forms of niche-constructive practice.[10] Such examples appear more clearly in retrospect after the articulation of new discursive possibilities, often in concert with new practices that enact as well as express previously occluded possibilities.

Wartenberg and Menge (2015, 2019, 2021) emphasize another aspect of the field-structure within which human agency is situated as temporally extended and dynamic. Wartenberg introduces the temporality of agency and power by arguing that the field of power relations is constituted by diachronic

---

9. Mills (1997), Hartman (1997), and Coates (2015) are illustrative examples of a more extensive literature, drawn from different genres disclosing often-hidden or obscured manifestations of anti-Black racism that require conceptual clarification.

10. Butler's (1989, 1993) classic accounts of gender and sex embed them in performative practices encompassing bodily movement, clothing and other accoutrements, and their material settings. Butler is often read as emphasizing only linguistic performance, but their conception of discursive performativity is more inclusive. It *materializes* gender in discursively articulated ways, best read as forms of material-discursive niche construction. Barad (2007, chap. 5) further develops and critically extends this reading of Butler, without explicit characterization as biological niche construction.

relationships among agents, and not just their synchronic interactions: "power exists primarily in social relationships and not in isolated exercises. . . . When relationships rather than events become the focus of a theory of power, it becomes clear that power is not a piece of property that can simply be possessed by its owner" (1990, 165). Their temporal character arises from how relationships and the practices encompassing them are both the basis for action and its product. These relationships and practices only exist through having been produced, reproduced, and transformed by the very actions dependent on them. Power thus does not just constrain physical movements of people's bodies but affects bodies in *action*. Power is integrated with agents' freedom and consideration of what to do in response to constraints and opportunities, opening up "a whole field of responses, reactions, results and possible inventions" (Foucault 1982, 220). Power relations at most constrain or redirect that freedom, attaching different anticipated or likely consequences and significance to possible courses of action. That reconfiguration does not compel specific outcomes but instead changes the environments in which agents act and develop their capacities and normative concerns.

It then makes a difference to the working of power whether those affected accede to or resist its impositions. In particular, their responses in concert with others may affect the aligned agencies and circumstances that originally mediated that constraining power. As Wartenberg noted, "The *present* actions of the dominant agent count on the *future* actions of the aligned agents being similar to their *past* actions. But this faith in a future whose path can be charted entails that the dominant agent not act in a way that challenges the allegiance of his aligned agents" (1990, 170). These temporally extended connections between powerful agencies and their mediating alignments also provide targets for resistance, which need not confront powerful agents directly. Resistance to power may instead undermine the aligned agencies or change the circumstances that would otherwise have sustained those powers' effects. Menge concludes that

> While an emphasis on the social bases of power may seem uncontroversial, it appears to be at odds with the robust character that we usually associate with power. Power is generally understood as a robust causal capacity to affect others' actions in a wide range of cases, most importantly in the face of resistance from other agents . . . However, social relations are not static; they involve ongoing relationships that must be continuously reproduced. . . . Since a social alignment is constituted by the mutually responsive actions of the aligned agents, its reproduction is subject to continuous negotiation and struggle. Thus, a dynamic view makes power seem rather precarious or at least . . . more fluid. (2020, 3)

A dynamic conception of power does not deny that some forms of power and domination remain relatively stable over multiple generations or longer. Prominent examples include subordination of women, settler colonialism and its partial extermination of many indigenous populations, colonial and neocolonial domination of the Global South, associated forms of racial classification and domination, differential capacities accruing to holders of capital and those dependent on their own alienated labor or excluded from meaningful economic participation, dominance of heteronormative sexual practices and discourses, and intersectional effects of many coexisting power relations. Dynamic conceptions of power instead highlight that these relatively stable patterns of power and domination are maintained through practices that must be reproduced in the face of changing resistance and practical realignments. Even relatively long-standing and consequential patterns of difference and constraint have not been monolithic or invariant but vary in form and intensity locally and over time.

Recognition that power relations are constituted by people's practice-differentiated biological interdependence also challenges a distinction central to many social theories of power. The concept of power in ordinary English use is ambiguous between power *to* accomplish outcomes and power *over* other agents, roughly equivalent to how *Kraft* and *Macht* in German are translated as "power." The more widespread approaches take power-over others as the primary issue for a social theory of power. Hannah Pitkin advanced one prominent reason for that priority: "One man may have power over another or others, and that sort of power is relational, though it is not a relationship. But he may have power to do or accomplish something by himself, and that power is not relational at all; it may involve other people if what he has power to do is a social or political action, but it need not" (Pitkin, 1972, 277). On that basis, power over others would be a coherent focus for social theory, whereas power to would be more heterogeneous.

Pitkin's reasoning is instructively problematic. As Davidson (1980, essay 11) influentially noted, actions can be described differently. Many philosophers thinking about action start from nonrelational descriptions, whether as voluntary movements of one's own body, efforts at such movements, or their foreseeable or intended consequences (Wilson and Shpall 2016). In understanding people's powers to act, however, what matters for social theory broadly construed are only actions and abilities under descriptions that express normatively significant contributions to people's lives and ways of life.[11]

---

11. I speak of social theory "broadly construed" to encompass a naturecultural reconception of human ways of life.

Brandom brings out the relevant difference in characterizing Hegel's discussion of action.

> Whereas theories of action of the sort epitomized by Davidson's find their paradigmatic actions in momentary, punctiform events such as flipping a switch or letting go of a rope, the paradigms of the actions Hegel addresses are to be found rather in complex, extended processes such as writing a book or properly burying a slain brother. . . . An action succeeds in this sense if the consequential descriptions that are true of it include the purpose whose achievement is the endorsed end in the service of which all the other elements of the intention-plan function as means. An action fails in this sense if, although some things are done intentionally, i.e., as part of the plan, the purpose is not achieved, because the means adopted do not have the consequences envisaged. (Brandom 2019, 400, 402)

On this conception of action as purposive, even when more atomistic movements and performances are voluntary, they are only *actions* when understood as contributions to normative concerns that matter in one's life or how people live together. In these respects, however, people's capacities and achievements almost invariably depend on supportive alignment in practices and the specific powers-to enabled by those practices. Moreover, a central reason for attending to power over others is to criticize people's domination or oppression of others. Domination or oppression would only be matters of concern, however, if they blocked people from capacities and achievements—powers to—that are normatively significant.

An alternative tradition in social theory has mostly taken an opposing tack, giving priority to what can be achieved through concerted action as "deontic powers" conferred by collective recognition or acknowledgment (Hobbes 1994; Arendt 1965, 1970, 1958, 199–207; Searle 2010, chap. 7). These theorists prominently argue that even power over others and capacities for force or violence depend on capacities for concerted action and collective recognition. To that extent, this consensualist tradition has some common ground with the practice-based account of power developed here. The consensualist tradition has nevertheless been insufficiently sensitive to how concerted action and collective recognition or acknowledgment of deontic powers are often predicated on domination or oppression.[12] Attending to the biological character of a practice-differentiated way of life shows more clearly

---

12. I do not develop that criticism here. Consensual theories of power often differ in *how* they diminish the significance of power as domination. Hobbes notoriously treats the collective institution of a commonwealth as conditioning nearly all other significant human capabilities, thereby overriding concerns over how it is achieved and sustained. Arendt and Searle each

why power to achieve normatively significant ends and power over others should be considered in tandem.

This consideration starts by recognizing that power relations do not simply enable and constrain what agents do. They also affect their development as agents and persons. Foucault prominently called attention to processes of "subjection," or subject formation, in which agents change by accommodating themselves to the practices they take up and taking advantage of whatever limited opportunities those practices make available (1978, 1982). Wartenberg succinctly summarizes this theme from Foucault.

> Subjection is the creation of the human agent as having desires that are adopted as his own as a result of his interaction with a power structure over which he has no control. . . . The presence of power relationships causes human beings to make choices that determine the sorts of skills and abilities they will come to have. As a result, since the formation of skills and abilities is a fundamental aspect of the constitution of character, human beings come to be the sorts of beings that they are as a result of the presence of power relationships. (1990, 160)

Foucault was in part aiming to understand how power relations shape agents themselves without needing to contrast ideological domination that distorts people's self-understanding to an imaginary conception of their undistorted "true self" freed from power.

Understanding how power-differentiating practices shape agents themselves nevertheless starts from recognizing that people are not born agents with capacities to act, but are instead helpless and in need of extended developmental support and guidance. People become agents at all only by depending on others to sustain their lives and enable their development of central human capacities. Parents or teachers must exercise power over children to enable their skillful, discursive, and affective capacities for agency and critical assessment and their full participation in various practices. Power relations between teachers and students have repeatedly been prominent examples in social as well as educational theory (e.g., Freire 1970; Giroux 2020; Isaac 1987; Wartenberg 1990; Hayward 2000), but human developmental dependence enters into social theories of power in different ways. Some theorists separate the establishment and exercise of power by or over mature adults from people's development and education into mature human agents; some distinguish power-over as domination from power-over as transformation or

---

analytically distinguishes concerted or consensual achievements of power from control of others by force, deprivation, or violence.

empowerment (Hartsock 1984; Wartenberg 1990); and some do extend criticisms of domination into the education of children (Giroux 2020).

Each of these approaches nevertheless acknowledges the Janus-faced character of power (Allen, Forst, and Haugaard 2014, 9) by distinguishing criticism of domination from recognition of how developmental dependence on others constructively enables people's capacities for agency, judgment, and autonomy. Wartenberg's version of this strategy distinguishes "transformative" power over others from domination, paternalism, or Foucauldian subjection. His concern to understand transformative power over others arose from considering the constructive but fraught role of parental nurturing in human development,[13] but Wartenberg's approach is instructive in its apparent breadth. His conception of transformative power encompasses not only parents' and educators' power over children but also therapists' and advisers' power over adults. He aims to vindicate transformative power over others while also attending to "[its] dangerous nature . . . , because [when] one person is in the power of another, he lacks the normal means of control over his fate and thus can be made into a pawn of the more powerful person's desires" (1990, 213). That vindication stems from a supposedly self-effacing role for genuinely transformative power, such that "the dominant agent's aim is not simply to act for the benefit of the subordinate agent. . . . [but] a use of power that seeks to bring about its own obsolescence by means of the empowerment of the subordinate agent" (1990, 184).

Wartenberg's conception of transformative power supposedly excludes consideration of power-to, but he implicitly imagines a nonrelational power to shape one's own actions and life course rationally and voluntarily as people's most important power-to. He takes power over others as transformative when enabling development or enhancement of autonomous self-determination, and as dominating, subjectivizing, or paternalistic when other agents' socially aligned power over others compromises autonomy. Power over others can therefore be constructively transformative but only as temporary means to individual autonomy and self-determination. Wartenberg is not alone in theorizing power-over by valorizing human freedom, as we shall see. That analysis overlooks the many transformative relationships that are not self-effacing and should not be. These include love, friendship, political engagement, coauthorship,[14] religious community, and other cooperative projects whose ends are not yet determinate.

13. Wartenberg's concern with nurturing arose from constructive critical engagement with feminist discussions of care and mothering in the specific context of parenting practices in which women bear disproportionate burdens and responsibilities. Both the historically specific context and broader issues about human dependence and development are relevant here.

14. The acknowledgments sections of books ritualistically, partially, but accurately acknowledge that all authorship is coauthorship.

Wartenberg is right that transformative relationships involve vulnerability and trust because people's lives are partly in others' hands. He also rightly attends to whether relationships are noncoercively initiated and revocable, but a life that avoided or escaped all relationships of vulnerability and dependence would be debilitated and impoverished. Ongoing transformative interdependence with others is enabling. Even exiting a mutually transformative relationship is not free as he suggests because the transformations it brings about are forms of agential interdependence rather than autonomy.[15] People sometimes choose not to continue particular relationships, but having been involved remains transformative. People become different kinds of agents with new capacities, limitations, and normative concerns through sustained forms of interdependence.

The problem is more general, however, extending beyond transformative relationships among individuals. Wartenberg treats dependence on others and a need for trust as if they were deficiencies people must trust others to help them overcome. Biological development would then have a normative telos of mature adulthood in which people become freely self-determining agents. The coevolution of neotenous dependence, ratcheted niche construction, and practice-based ways of life thoroughly undermines that conception. People do not develop into autonomous, self-determining agents through submission to others' transformative power. They instead develop specific postures, orientations, affects, skills, and involvements as one kind of agent rather than another, enabling capacities and goods not achievable otherwise and hindering others counterfactually open to them. In doing so, however, they are subjected to and subject themselves to new forms of constraint. Aging also characteristically brings on new or renewed forms of dependence. Biological development is not a process culminating in mature adulthood abstractly conceived as autonomous agency but a lifelong process of people becoming who they are through vertically differentiated participation in a complex, horizontally differentiated way of life. The practices amid which people find themselves throughout their lives enable some forms of agency, self-development, interdependence, and self-understanding and occlude others.

Power and Normativity

Situating the concept of power within a niche-constructive conception of people's evolved, practice-differentiated ways of life reinforces some themes from social theories of power and expands on others. The horizontal differentiation

---

15. Lance (2015) discusses these relationships as transformative but not self-effacing.

of interdependent practices and the vertical differentiation of people's lives among those practices both enable new capacities and possibilities and also constrain and sometimes threaten how people live. We can now ask how it matters to think about power and difference as a naturecultural aspect of people's niche-constructive, practice-differentiated ways of life. Social theories of power are rightly guided by the concept's normative significance. Steven Lukes and Clarissa Hayward initiate a dialogue about their disagreements by highlighting their "shared conviction that analyzing power relations is an inherently evaluative and critical enterprise, one to which questions of freedom, domination, and hierarchy are—and should be—central" (Lukes and Hayward 2008, 5). Wartenberg similarly emphasizes the normative orientation of the concept in social theory: "The concept of social power plays a fundamental role in describing and evaluating social inequities. Just as the concept of justice plays a significant role in legitimating social institutions, the concept of power plays a fundamental role in their critique. . . . The goal of this study . . . is to provide a theoretical conception of social power that will enable social theory to make use of this concept as a fundamental one in articulating a critique of social relationships, practices, and institutions" (1990, 5). Even Robert Dahl's (1958, 1961) and Nelson Polsby's (1960, 1968, 1980) empiricist, behavioral analyses of the concept which initiated debate over the "faces" of power in political science were motivated to defend a pluralism of power relations in American democracy against earlier criticisms of domination by a "power elite" (Mills 1956; Hunter 1953).

A more expansive conception of power initially complicates its critical role. These complications are already partly evident in commitments taken over from social theory. A naturecultural conception of power as pervasive in a practice-differentiated way of life undermines one familiar strategy for critical assessment of power over others as domination. Hayward (2000, chap. 6) argues that the critical aspirations of many social theories of "power-with-a-face" are problematically framed by contrast to freedom from social constraint. The normative concern animating those criticisms of domination is for a negative freedom enabling people to choose their own plan of life freely and develop their capacities and life activities without constraints imposed by others. Recognition of the horizontal and vertical differentiation of practice-based ways of life supports Hayward, Foucault, and other social theorists of power in dismissing that normative aspiration as both chimerical and un-self-critically coercive. Hayward argues that "character formation itself is to some extent unavoidably coercive; what a given individual hopes to achieve and become ('her plan of life') is always, in part, the product of actions by others that enable *and* constrain possible 'beings and doings'; the processes through which people arrive at more or less stable and coherent senses of who they are and to what they aspire, although these enable them to act in ways that they and others value, also always close off possibilities in ways they do not choose" (2000, 165).[16] This aspiration to free, unconstrained personal autonomy is not only illusory. Any such conception of autonomous agency is also problematically coercive even in its own terms, by building particular conceptions of normal human development and individual autonomy into various practices while placing them beyond critical appraisal or effective resistance.

Similar problems afflict efforts to focus the critical assessment of power and domination on powerful agents who can rightly be held responsible for intended constraints on others' capacities and prospects. Many aspects of human life clearly do call for such criticism and redress. A practice-based way of life differentially imposes constraints and limits on all, however, many of which are diffuse or built into niche-constructed transformations of people's material and discursive developmental environments. As Hayward concluded, "some [forms of constraint] are the unplanned net effect of the actions of multiple actors who could not—not through their individual choices, not through their coordinated efforts—control and direct the outcomes that, together, their actions produce. . . . The normative project of criticizing power relations, and identifying and evaluating alternatives, is badly served by an approach that excludes from analysis *a priori* [these] set(s) of significant and inegalitarian social constraints" (Hayward and Lukes 2008, 9). Moreover, if criticism and transformation of "significant and inegalitarian social constraints" were based on holding those who were responsible for those constraints accountable for their remediation, then any effort to address more diffuse forms of domination would require intentionally imposing further constraints on those who benefit from power relations for which they are not personally responsible. Such impositions may or may not be appropriate but not so on those grounds.

A common response recognizes that power is Janus-faced, so that illegitimate domination or subjection of others needs to be distinguished from legitimate, transformative empowerment. This distinction is often drawn by appeal to sophisticated conceptions of freedom, as in Wartenberg's conception of transformative power as a self-effacing empowerment of those it targets. I argued above that Wartenberg's conception of autonomous agency and freedom from dependence on others did not recognize the importance and

---

16. Hayward frames this passage as suppositions that she endorses, as do I. The quotation here removes "suppose" from the beginning of each claim, in accord with her endorsement.

ineliminability of people's lifelong biological development. People's coupled interdependence in mutually transformative relationships and practices is not and should not be self-effacing. Hayward's analysis nevertheless instructively generalizes this difficulty in Wartenberg's project. Her comparative ethnography of two school systems aimed to show that even educational practices striving to empower students, encourage their autonomy, and provide them some collective control over their educational program also excluded and dominated those students who do not embrace or conform to their orienting conceptions of excellence and self-direction. As Hayward concludes, "Every possible matrix of social boundaries to human action (every practice, for instance) creates a class of actions (and their agents) or attributes (and their bearers) that exceed the norms and other mechanisms of power that comprise it" (2000, 173). Wartenberg's conception of transformative power as self-effacing promotion of subordinate agents' autonomy, if taken as criterial for differentiating legitimate transformative power from problematic forms of domination, subjection or paternalism, would similarly undermine those relationships and practices that bring about lasting forms of trust, vulnerability, and mutual dependence.

Hayward responds by proposing a conception of *political* freedom that would promote both inclusion of all in shaping the norms and boundaries of practices in which they participate and an openness of those norms and boundaries to contestation, criticism, and transgression. In proposing these "criteria for distinguishing better from worse forms of power" (2000, 7), Hayward claims common ground with critics ascribing a normatively confused and implicitly conservative view of political change to Foucault.[17] Hayward nevertheless also recognizes that the conception of political freedom she proposes cannot be insulated from criticism by or even subordination to other normative concerns: "Clearly promoting freedom [in this sense] is a goal that can compete and at times conflict with other valued social ends, such as efficiency, security, or community, . . . requir[ing] attention to conflicts between the democratic commitments that inform this critique of domination, and the non-, or even anti-democratic commitments constitutive of some highly valued social practices" (Hayward 2000, 176). On my reading, her unwillingness to specify further criteria for how such normative conflicts ought to be resolved shares Foucault's rejection of a standpoint of normative sovereignty which was the target of these classic criticisms (Rouse 2005b). I

---

17. Hayward (2000) and Hayward and Lukes (2008) approvingly cite Habermas (1986, 1987, chaps. 9–10), Fraser (1981), Walzer (1983), and Taylor (1984) as sources for criticism of Foucault's supposed normative "neglect."

do not make this claim to criticize Hayward for inconsistency or to rehash familiar arguments for and against Foucault. My aim is instead to indicate how a naturecultural conception of power and normativity as pervasive in niche-constructive practices poses and responds to these issues differently.

   This naturecultural reconception of practices, power, normativity, and political criticism begins by recognizing one further complication to how a practice-differentiated way of life brings about power-constitutive differences among people's capabilities and prospects. The dynamic alignments of performances and circumstances that constitute practices do not just enable the *effectiveness* of power relations in shaping people's actions, capacities, or prospects and affecting their development *as* agents. They also help constitute the normative *concerns* with respect to which those constraints can be significant, effective or ineffective, and open to criticism. The evolved, two-dimensional normativity of practice-differentiated ways of life brings many normative concerns into play. The achievement, recognition, and assessment of these capacities, skills, goods, virtues, proprieties, and other normative concerns are enabled by cooperative interdependence within and among the practices making up human developmental environments. At the ground level, those normative concerns express the goods or excellence character-istic of the practices, and are often only fully recognizable to participants in those practices or in other practices dependent in more limited ways on how those practices are carried out and sustained. Those capacities and concerns arise within environments that incorporate the continuing reproduction of the practices in which their constitutive goods, proprieties, or virtues are rec-ognizable and at issue.

   Practice-constitutive normative concerns nevertheless also open a broad-ened second dimension of critical assessment. This second dimension emerges in how those concerns fit together within the vertically differentiated lives of persons engaged in multiple practices and the mutual interdependence among horizontally differentiated practices. Many integrative normative con-cerns emerge here, ranging from personal fulfillment, integrity, or well-being to social justice, political freedom, or communal identities. Recognizing that discursive practices are integral to a practice-differentiated way of life clari-fies how the practices making up people's developmental environments also help shape and articulate normative concerns that make sense to people and guide, express, or justify their vertically differentiated involvement in those practices.

   In other organismic lineages, phenotypically plastic adaptation to di-verse environments may expand a lineage's geographic range and buffer it against extinction from local environmental changes. More dramatic cases lead to speciation. In the human lineage, the evolved, coupled interdependence of performances and practices instead both enables and brings about a more diversified, cooperatively interdependent way of life. If taken in isolation, a horizontally differentiated practice or interdependent groups of practices would not further complicate the lineage's biological normativity. By themselves, occupations, rituals, and other interdependent performances—religious worship, artistic production and consumption, scientific inquiry, games or sports, languages, modes of travel, and the extraordinary variety of other practices now part of human ways of life—elicit no basis for assessment beyond their success or failure in attracting and retaining participants who sustain the practice and take its constitutive normative concerns as authoritative. As Lance noted:

> Whatever the complexity of a local practice considered *in isolation*, its norms exist only as articulations of our engagement with that practice itself. In one sense we can distinguish between the rules of the game now (say that this part of the field is in-bounds) and discussions of what the rules should be (during a timeout we consider whether it really should be out of bounds because the field is just too large, or too dangerous). But however complex the reason-giving of such revisionary moves within the practice, they are still within the practice. We are still engaging in our game-practice, governed by our lived sense of its point, living out our roles and statuses within it. (2015, 299, emphasis added)[18]

Practices never do exist by themselves, however, due to their constitutive interdependence with other practices. Even those practices directly connected with human subsistence still typically must engage with other patterns of activity, both in participants' vertically differentiated lives and in their interactions with other practices around them.

The ongoing adjustments between the predictive and authoritative senses of normativity and its temporality also show how power and normativity are coconstituting. People can only engage in practices through which they fulfill biological needs and live human lives by relying predictively on normative concerns guiding those practices and their relations to other practices. Participating in ongoing practices implicitly anticipates others continuing to act so as to let their performances make sense and sustain those practices' constitutive, normatively constituted ends. That predictive sense of practical normativity arises from familiarity with the practices in which people developed their postured orientation within the world and acquired the practical capacities those strategic orientations enable. Those prior patterns did not

18. Chapter 5 above suggests speaking of normative issues or concerns rather than norms.

simply consist in then-present performances, however; they were oriented by and toward normative concerns at issue in the practice's subsequent continuation. In taking up and continuing that prior pattern, current participants are in turn acting for the sake of what the practice is by adopting an authoritative normative orientation toward what is at stake in its continuation. A practice *is* an anaphorically constituted continuation of what *it* has been toward what is at stake in *its* future evolution in response to what is now at issue in *its* practitioners' current situations.

This temporal pattern exhibits a mutually constitutive relation between power and normative issues and stakes. People act out of concern for what is at stake in the practices they take up. An essential component of those stakes for living organisms is their ability to sustain their lives by meeting their biological needs. In a practice-differentiated way of life, however, what is called for by their situation to sustain their lives in this way is not simply determined by their evolution and development as organisms in that lineage.[19] The normative concerns for the sake of which people participate in practices are two-dimensional. People participate in practices in different roles and vertical combinations. The normative concerns at issue and at stake in those practices then become part of the lives they maintain through that participation. What those concerns are, however, and how they bear on their life prospects, are partially shaped by how other people's involvements align with one another in circumstances they inhabit together. If I am a teacher or a nurse, whether and how I can sustain a life in these practices and what such a life could be depends on others' situated performances. How those performances align reconfigures what concerns are *actually* at issue and at stake in teaching or nursing here and now. Teachers can only teach the students they have, amid and in response to the educational practices and institutions around them, except to the extent that their efforts align to effect change. Similarly, nurses can only take up a concern to provide professional and ethical care for patients with the patients they encounter in settings accessible to them. What one does as a teacher, nurse, parent, or in any other practice, and hence what it could mean to practice teaching, nursing, or parenting, is constitutively linked to how others understand and enact supportive roles. Whether someone can sustain herself in a life she would accept, responsive to concerns shaped by her developed, postured normative orientation, is always

19. Dennett (2020) emphasizes that what matters is not causal determination of people's actions by prior events, but the extent and locus of *control* of the outcome. The degrees of freedom enabled by complex, interdependent causal relations among performances in a practice-differentiated way of life dramatically reduce control of the normative development of practices.

at issue and at stake in how she engages in or abandons various practices. Whether and how those practices remain possible is in turn shaped by how others respond to their shifting alignments.

The normative temporality with which practices evolve and people develop in their midst is recapitulated in the interdependence among those practices and how people's participation must accommodate other practices in their lives. How practices contribute to and depend on one another has a similar temporality and associated interdependence of power and normative concerns. Recall the example in chapter 4 of the complex development of volleyball as a multifaceted practice whose rules evolve in response to changing issues and stakes in playing the game. Whether and how one engages in such a practice depends in significant part on the normative concerns at issue—the kinds of activity, skill, achievement, camaraderie, competitive engagement, recognition, and more that it enables—and how those concerns are worked out interactively with others as teammates, opponents, coaches, spectators, and others involved. Many of these concerns arise in multifaceted, close engagement in playing the game. The practice may also acquire other more limited interactions, however. My team may no longer be able to play with our opposite hitter, Neil, if a medical laboratory determines that he violated league rules against performance-enhancing drugs or tested positive for COVID-19.[20] Interaction between volleyball and laboratory practices to analyze blood samples or nasal swabs is limited in scope and "intimacy" compared to players' multifaceted involvement in the game.

Initiating and sustaining that interaction between athletic and laboratory practices nevertheless shifts the normative concerns at issue in both. On one side, the game is no longer just volleyball but organized intercollegiate or professional volleyball, with players now subject to medical surveillance and dependent on the resources needed to retain their sponsoring organizations. Other aspects of the game shift accordingly—who can play; with what role in their lives; for how long, where, and when; what rules and other concerns govern play; and how the game aligns or conflicts with other practices and concerns. The laboratory in turn no longer just diagnoses patients' health to support their medical treatment but also certifies their medical standing for other purposes. That normative shift can affect the laboratory practices and standards themselves and rearrange the alignments that provide its sustaining resources. For example, the assessments of inductive or epistemic risk that determine the significance of false positives and false negatives (Douglas 2000; Biddle and Kukla 2017) take different shape for certifying athletic

20. I am indebted to Lance (2015) from which this example is adapted.

eligibility than for diagnosis of disease or addiction, and the negotiation and implementation of those epistemic standards is affected in turn by realignments of interested parties.

The second dimension of normativity is not confined to limited interdependence among particular practices, however. These local interrelations—between athletic leagues and medical laboratories; religious traditions and food preparation; product marketing and legal proscription and prosecution of fraud; rural school calendars and harvest seasons; and the like—are indeed pervasive in practice-differentiated ways of life. The normative concerns at issue in particular practices nevertheless also encounter more comprehensive normative concerns in belonging to vertically differentiated lives and horizontally differentiated ways of life. Those integrative concerns have similar temporal entanglements of their predictive and authoritative senses, and of normativity and power. Consider first the normative interdependence implicitly at work in how practices are interrelated in people's lives. The temporality of normativity is embedded in people's postured orientation toward their surroundings. People live vertically differentiated lives that involve them in multiple practices—kin relationships, occupations, food provision and consumption, friendships, sexual relationships, recreational activities, waste excretion, citizenship, domestic life, religion, and so forth—but at any given time must orient themselves toward some practices rather than others. People may be more or (usually much) less reflective about their lives as temporal wholes, but their postured orientation embodies a normatively ordered temporal sequencing: now it is "time for" lunch, then work, interspersed with breaks, followed by exercise, dinner with family, a political meeting, return home, and preparation for sleep.[21] This normative sequencing also occurs at multiple levels, since people's participation in practices responds to the significance of culturally varied life stages from infancy to old age and the intermediate spanning and sequencing of longer-term practical commitments.

Different normative concerns and their affective import overlap and interpenetrate in shaping these temporal sequencings. On day-to-day scales, these concerns may include familiarity and reliability of routines, coordination with others, circadian and metabolic bodily urgency, responsibilities to and for others, accommodation to solar and clock-regulated circumstances, comparative importance, dependence of one task upon others, and many more. On more extended scales, cultural and personal narrative fields intersect with

21. Heidegger (1962, sects. 79–81) characterized this temporal ordering of activity-defined temporal spans ("times for x") that are normatively significant and publicly accountable as "world-time."

developmental changes in bodies and publicly significant events to guide
what matters now rather than earlier or later in a life. All of these normative
orderings are interspersed with the all-too-ordinary occurrence of extraordinary events realigning the practices people inhabit. War, famine, economic
boom or collapse, epidemics and infestations, political regime changes and
shifts in cultural normativity, individual violence or other injury, catastrophic
weather, or tectonic shifts often mark associated patterns of change or continuity whose normative significance reverberates through interlinked networks of practices.

The normative temporal ordering of lives and relationships also implicates
power relations. The practices available for people to take up impose temporally constituted demands and constraints. These practices evolve through
conflicts among those demands and how people accept or resist the resulting
solicitations as authoritative for their own responses. As one example, influential cultural narratives of people's life course now impose heteronormative,
ableist, or racialized constraints on many people's lives. These normative concerns and expectations are embedded in multiple practices but also encounter resistance from those whose lives enact "queer times" or "crip times" that
do not align with dominant patterns of development and reproduction (Edelman 2004; Halberstam 2005; Kafer 2013, chaps. 2–3). In the United States,
racially differentiating practices of mass incarceration, health care provision,
employment access, and provision for child care and support together suggest racialized analogs to the reconfigured normative temporality of queer
or crip times. Racially subordinated people are often blocked from access to
opportunities widely accepted as normatively significant or as consequential
stages in a life.[22] Race, gender, sexuality, and disability are not independent
practical and discursive articulations, so their temporal ordering of and constraints on people's vertically differentiated lives are also intersectional.

Negotiation of alignments and conflicts among horizontally differentiated practices also introduces further normative concerns bearing primarily
on the interdependence and conflicting alignments among those practices.
These overarching, integrative normative concerns—social justice, economic
growth, freedom, ecological sustainability, political legitimacy and its limits,
scientific credibility, new forms of social solidarity or political division, and
more—can in turn align or misalign with normative concerns arising within

22. I do not know of explicit analyses of how racialized oppression affects the normative
temporality of people's life course, but the issues parallel those posed by sexuality or disability.
Anderson (2010) brings out some of these themes in passing when emphasizing the political,
cultural, and psychological costs of racial segregation.

particular practices and the temporal ordering of people's lives and relationships. As Lance noted, "There is a deeper sense in which the normativity
of a practice is transformed by the institution of a broader meta-normative
network of interconnected practices. Such instituted complexity brings with
it . . . the possibility of a new sort of rational critique, namely one not essentially tied to one's lived commitment to the practice in question" (2015,
299).[23] This no-longer-new form of criticism is rooted in the interdependence
of discursive practices and the other practices they articulate. People do not
just mutely adjust what they do in response to shifting alignments of others'
performances and their circumstances. Practices are discursively articulated,
and people talk about the normative concerns animating practices, reason
about their constitutive (mis)alignments and how to respond, and call one
another to account for how they take up, continue, or resist practices involving them together. These discursive practices of reasoned assessment transform agents themselves along with what they do, both in reconstructing a
practice from within and critically assessing the issues and stakes in alignments among practices. People's developed bodily postures and the unfolding
of their lives do not merely consist in temporally ordered integration of vertically differentiated lives. People are discursive sense-makers, and the sense
we make of what we do and how we live is integral to coordination of our
actions with one another and with otherwise disparate aspects of our lives.

Discursive practices have their own constitutive ends, including a power
to make sense of people's immersive involvement in some practices and those
practices' interdependence with others. Concepts and activities with a richly
interconnected sense in one practice acquire more limited uses and implications from how they matter in other contexts. Discursive practices enable the
expressive and inferential articulation and adjudication of how those uses
bear on one another. Misalignments between practices and the discursive
sense participants make of them can also call for adjustment. Sometimes
people adjust their performances and practices to align better with whether
and how they make sense discursively. Sometimes they instead adjust what
they say when and what inferential connections they draw to accord better
with the temporally ordered interdependence of other performances and
practices. Sometimes they learn to live with that discord.

Scientific practices and discursive articulation of their aims and normative concerns illustrate how complex realignments of concepts and practices
can generate adjustments all around. Logical empiricist philosophers and

---

23. Lance's undeveloped conception of "meta-normativity" is roughly equivalent to normative two-dimensionality.

scientists at mid-twentieth century explicitly specified normative concerns for empirically meaningful and logically sound scientific understanding. Those explications were gradually recognized as partly divergent both from contemporary scientists' immersive grasp of those concerns in their own research and pedagogy and from historical explication of the normative significance of earlier scientific practices and achievements.[24] These controversies changed both how science was done and how people talked about it. Such adjustments are not just up to individual participants, however, because their own performances and the discursive sense they make of them also depend on circumstances, the responses of other participants, and how those participants in turn coordinate and make discursive sense of what they do together.

People's developing capacities for critical reflection and transformation of themselves and the practices in which they participate are not separate from those practices. How people describe and assess what they do, advocate and act on behalf of concerns they find authoritative, and call one another to account for issues arising from their performances, situations, and other practices all play significant roles in how practices continue and change. Sometimes these integrative concerns are expressed and focused by other practices. Political activism, scientific inquiry, philosophical analysis, moral reflection, or religious worship often raise and articulate normative issues primarily directed at how other practices fit together, but such integrative activities also answer to their own normative concerns. The latter concerns include fitting those integrative practices into their participants' vertically differentiated lives, the practices' ability to recruit and retain new participants, their accountability to and relevance and authority for the practices they aim to assess, alongside conceptual issues arising in their own discursive sense making.[25] All of these practices are constrained and enabled by evolving alignments of people's performances and circumstances within larger networks of practices, which shape their normative significance.

Discursive practices are thus an important site where power and normativity are entangled. They conjoin the material and socially-articulated settings in which people use words and act on their verbal understanding with

24. Hacking (1983, part 1) and Zammito (2004) are instructive overviews of this complex history.

25. Lenoir (1997, chap. 2) instructively considers how these conjoined concerns affect transformation of scientific research programs into disciplines. Discipline formation depends on developing material and conceptual interdependence with other practices alongside the conceptual and practical issues internal to research programs. These discipline-constitutive alignments enable recruitment of and employment for students; provide requisite financial, institutional, and material support for research; and enhance the significance of their results.

the vocabularies, inferential relations, and normative concerns available and intelligible to participants. Understanding how normative concerns at issue within other practices are discursively articulated and enacted or critically transformed requires attention to what is enabled and constrained in both respects. Through such materialized and situated discursive articulations, people encounter the two-dimensional biological normativity of human ways of life. What is at issue and at stake in the evolution of this practice-based way of life is not only whether it continues but what possible alternative life courses it encompasses, how individual human organisms develop and fare, who they become individually and collectively, and what degrees of freedom are thereby enabled. Organisms in other lineages also develop different temperaments, behaviors, or lives in more or less accommodating environmental situations, including social positioning of individual animals. These differences are nevertheless cumulative effects of ongoing interaction with local developmental environments. Those local environments in human ways of life encompass alternative, temporally extended, practice-differentiated developmental trajectories. These trajectories are at issue and at stake in interactions among horizontally differentiated possibilities and discursive articulation of how those differences matter. The normative force of those alternative ways of living, individually and collectively, arises from how people understand the course of their lives as at issue and at stake in how they go on. The lives people can lead are dependent on the practices available to them, the sense those practices can make, and how other people's lives and actions reshape their possibilities. In all of these respects, power and normativity cannot be disentangled.

The Expressive Significance of "Power"

Social theorizing is itself an integrative discursive practice aiming to articulate and express normative concerns at issue in many other practices. Social theories of power are concerned to work out and assess the normative issues raised by how individual actions and practices differentially enhance or constrain what people can do and how they can live. To that extent, they typically have two components: analysis of what power is or how it operates, and assessment of what is at issue and at stake normatively in how those power relations or effects are reproduced or transformed. We can now understand better what difference it makes to situate the concept of power and its critical deployment within a conception of human ways of life which conjoins people's biological lives and lineage as animals with the centrality and complexity of people's situated, interactive dependence on one another.

The preceding sections argued that power relations—how people's lives and activities are enabled and constrained by the conjoined activities of others in partially shared circumstances—are grounded in the vulnerability and practice-differentiated interdependence of people as living organisms and the resulting two-dimensional normativity of human ways of life. People's lives are goal-directed processes. They aim both to sustain their lives and fulfill other normative concerns that animate their participation in multiple practices as aspects of a life in supportive alignment with other practices. Their ability to fulfill those goals depends on closely coupled interdependence with the postures and performances of others in partially shared circumstances. Power is not a robust causal capacity of agents to enact their goals or control the actions of others, because what outcomes are achieved and how they matter depend on circumstances that include the prior development and subsequent responses of others. Those others likewise act for goals whose fulfillment depends on situated, supportive alignment with one another. Wartenberg and Menge thus rightly argue that agents' causal capacities depend on mediation by a field of other actions and capacities; Hayward and Foucault rightly emphasize that this field both enables and constrains the possibilities of *all* human agents, not just the powerless; and Wartenberg and Menge rightly emphasize the temporally extended dynamics of that field as people's action-environments. I am arguing that people's action-environment is the coevolved, niche-constructed *biological* environment for their lifelong development as bodily agents. Their postures and capacities are shaped by and responsive to normative concerns enabled and discursively articulated by that environmental interdependence.

Starting from a similar conception of the dynamic, temporally extended interdependence of people's actions, Menge (2020) draws a provocative conclusion about the significance of power attributions. He argues that in taking other agents as powerful, people misrepresent that dynamic interdependence as a stable causal structure but thereby also help constitute the power they misrepresent by responding to others *as if* they had reliable causal capacities: "To act meaningfully in a complex social world, I have to assume that there are relatively stable structures that make the effects of my actions and those of others predictable. . . . Power is never simply present; rather it essentially involves a projection of stability into the future. Power is not a static feature, and yet it always depends on fictional expectations about its continuing stability and robustness" (2021, 18). Menge is right to ask what work is done by implicitly responding to or explicitly attributing power relations, but he misconstrues these attributions as misrepresentations of *causal* capacities of an agent. In doing so, he leaves out the normative dimension of power relations.

Power attributions do not merely describe or ascribe causal capacities and their consequences within a complex environment. They express the normative significance of situated causal capacities in how those capacities or their exercise matter to the concerns at issue in a practice or nexus of practices. Menge is right that power attributions have predictive significance but only through the constitutive ambiguity of the predictive and authoritative senses of normativity, which always function together. Predictively recognizable patterns of an ongoing practice do not determine how the practice will continue but only show how it would continue if undertaken "in the same way." The authoritative sense of normativity then concerns how it should continue—what one should do—in light of that predictive understanding and the normative concerns at issue and at stake in continuing the practice and its encompassing practice-differentiated way of life.

The expressive role of power attributions thus parallels the expressive roles Brandom ascribed to logical, semantic, and propositional-attitude attributions (1994, 498–99; 2000, chap. 1).[26] These latter conceptual repertoires let people say and reason about what they are doing when making inferences, holding others' utterances accountable to circumstances, and taking them as discursively committed or entitled. The concept of power similarly enables people to talk about and assess how causal effects of situated activities and capacities have normative significance for one another's situations. It matters here that causality is not a single relation but takes diverse forms, including standing conditions as well as triggers, hasteners, inhibitors, and many others (Cartwright 2003; Hitchcock 2003). Which of those causal relations matter in understanding people's situations is not separable from their normative significance. Attributions of power thus *do* concern the material, causal constraints and supports for what people can and cannot do even though they do not denote specific causal mechanisms or ascribe robust causal capacities. A defining feature of power is instead that the causal effects of people's actions and capacities are mediated by practices—dynamic alignments of situated performances that embody normative concerns. What people do and can do only has effects that matter normatively within ongoing practices. Those effects encompass and conjoin what those actions and capacities enable (power-to) and what they constrain or impose (power-over). To ascribe power relations is to talk about how the causal-material significance of what

26. I elsewhere argued, partly in response to prominent criticisms of Foucault on power, that power has an expressive role paralleling those Brandom ascribes to logical, semantic, or propositional-attitude ascriptions (Rouse 2002, 259–60, 359–60; 2005b, 108–19).

people do or can do reconfigures what is at issue and at stake in one another's situations and lives together.

This conception does not thereby treat *ascriptions* of power as normatively neutral. On the contrary, use of that concept to understand causal reconfigurations of the normative significance of people's actions and utterances builds in normative assessment. That nonneutrality is evident in many prominent theories of power. For Wartenberg (1990), what distinguishes domination or subjection from transformation is only whether one agent's power over another aids their autonomous self-determination. For Hayward (2000), what matters is the political freedom to help shape the normative concerns embodied in practices and to criticize or transgress them without severe consequences. For Hobbes (1994) or Machiavelli (1988), by contrast, what is at issue in understanding power is not how to limit its impositions but how to enable its effective, sovereign exercise amid ongoing social conflict.

The recognition that the concept of power plays an expressive role is a second-order determination, however, according to which the normative significance of the term's first-order application is always at issue in its uses. *Using* the concept always takes a stand on those issues and stakes and the substantive normative concerns they embody, but the concept itself does not depend on the normative concerns for which it is deployed. In this respect, this conception of power parallels propositional-attitude ascriptions and expressive or prosentential conceptions of truth (Grover, Camp, and Belnap 1975; Brandom 1994, chap. 5). Attributing discursive commitments and entitlements to others always undertakes commitments of one's own. Attributions of truth draw on normative concerns for which considerations would legitimately warrant those attributions. Metalevel *analysis* of truth's expressive role nevertheless does not directly determine substantive considerations of warrant for the ground-level claims to which it applies. It only rules out those alternative conceptions of truth that do embed substantive epistemic norms, and thus implicitly challenges ground-level claims whose justification partly depends on a substantive conception of truth. I am similarly arguing that understanding the expressive role of the concept of power rules out substantive conceptions of power as robust causal capacities of agents or institutions.[27]

The normative concerns invoked by attributions of power are at issue not only in social theory but in the practice-differentiated ways of life those

---

27. Lance (1997) argues that anaphoric conceptions of truth should be adopted generally, even by those who start from alternative semantic orientations. I would similarly argue that this expressive role of the concept of power should be common ground within social theory broadly construed.

attributions aim to reconfigure. Attributions and assessments of power aim to exercise the discursive power to make sense of various practices and the issues raised by how they (ought to) fit together. What is at issue in such efforts is not only what sense can be made of those practices and their normative ends, but also the extent and character of people's collective power to bring their practices into accord with the sense they make. Brandom concluded that his "expressive account of language, mind, and logic is an account of who *we* are. . . . We are sapients: rational, expressive—that is, discursive beings. . . . But we are also *logical, self-expressive beings*" (1994, 650). Incorporating power within this expressive conception of discursive practices is part of turning Brandom's account of discursive practices inside out. People are animals whose practice-differentiated ways of life make us two-dimensionally normative beings whose lives constitute and answer to an extraordinary range of interrelated normative concerns. Brandom rightly emphasizes that we are also reflective, self-critical beings. We make discursive sense of how those concerns are realized in our ways of life, and we often seek to bring our lives and practices into accord with those critical assessments. How those capacities for rationality, critical assessment, and reflective transformation of our lives fit into our ways of life is at stake in whether and how practices and our practice-differentiated way of life develop over time. The next chapter concludes the book with two considerations for how it matters to situate those critical-transformative capacities within people's biological lives and lineage.

Finitude

Conceptual Finitude

The previous chapter concluded with some distinctive capacities (powers-to) enabled by discursive practices within a practice-differentiated way of life. These capacities begin with abilities to look beyond the exigency of current circumstances through conceptual displacement. Discursive practices let people make sense of aspects of their environment and express the sense they make in words or other modes. Moreover, they often allow making sense of what is at issue in misalignments within or among practices, by showing how aspects of people's practical orientations do not make sense together. That capacity then enables critical reflection on how diverse normative concerns bear on one another. Misalignments between discursive expressions and the situations they address have a dual direction of fit, however. Sometimes speakers revise what they (would) say to accord better with what they are talking about or directed toward; sometimes they change what they do or reconstruct their surroundings to accord with the sense they make of practices they take up or depend on. Finally, that critical capacity sometimes enables collective transformation of practices or their circumstances guided by those critical assessments, or vice versa.

These distinctive human capacities helped motivate conceptual bifurcations of human ways of life between human biological nature and diverse social worlds. This book aims to overcome that bifurcation without denying the biological distinctiveness of two-dimensional normativity, discursive articulation, and how they inform the niche-constructive transformation of human biological environments. Discursive practices differ from communicative expression by other animals in their conceptual displacement from surrounding circumstances and their place within a cooperative, practice-differentiated way of life. Some organisms evolved sophisticated, flexible responsiveness to partially conflicting environmental solicitations. Nonhuman animal communication nevertheless only affords further indications of what to do in those animals' circumstances. Those communicative behaviors normally persist when their expression and uptake enhance the lineage's prospects for survival and reproduction. Human discursive performances undoubtedly originated as a communicative repertoire sustained by selection pressures for their acquisition and use but gradually acquired a place within patterns of interdependent performance with their own constitutive goal-directedness. The resulting two-dimensional normativity not only enabled accountability to multiple normative concerns but also contributed to ongoing, interactive adjudication and revision.

Human capacities for discursive articulation and reasoned adjudication nevertheless are evolved and developing biological phenomena, sustained and reproduced via phenotypic plasticity, material and behavioral niche construction, and partial genetic assimilation. Attentiveness to and concern for discursive interaction evolved and develops amid people's verbally expressive performances as neotenous social primates and occurs within the interdependent practices they take up in vertically differentiated lives. These practices are also enabled by people's more or less cooperative interdependence with the horizontally differentiated practices taken up and sustained by others. Discursive performances have their own constitutive ends in complementary ways of making sense, phonetically, grammatically, semantically, pragmatically, and affectively. They nevertheless also belong to other practices as making sense of those practices and of people's practice-differentiated environments. People's interpersonal relationships and roles in practices are also partly constituted by engaging one another conversationally, and categorizing and thereby making sense of those relationships.

This chapter takes up two further aspects of how it matters to understand people's capacities for discursive sense making, cooperative interdependence, and a practice-differentiated way of life as biological phenomena. Kant's insistence on the finitude of human cognitive and discursive capacities provides informative background to these two issues. Kant (1998) understood human cognitive abilities as normative phenomena of thinking and reasoning according to concepts as rules. These capacities are finite for Kant in being dependent on and answerable to what is receptively given in sensuous intuition. Rule-governed conceptual manipulations in isolation could determine no conceptual contents and thus make no determinate sense. Kant's well-known conclusion—"concepts without intuitions are empty, intuitions without concepts are blind" (1998, 193–94)—embodied this dual conception of empirically constrained rational normativity. He implicitly discerned the two-dimensional normativity of conceptual content, answering both to rational norms of inference and classificatory norms for structuring a sensuous manifold conceptually. These two aspects of conceptual normativity determine judgments as jointly answerable to norms of meaningful combination (objective *purport*) and objective correctness.

Kant conceived the reasoned application of concepts to sensuous intuition as finite in a further consequential sense, although rarely thought of as such. For Kant, human capacities to apply concepts and make judgments according to rational norms depend on an unknowable capacity for autonomous determination of judgments *from* reason, and not merely *according* to reasoned norms. If people thought or said only what they were causally inclined to think rather than freely judged as correct, these expressions would not have conceptual content and would not be *thoughts*. Kant's account of theoretical understanding in *Critique of Pure Reason* thus belongs within the scope of people's presumed noumenal capacity for rational self-determination addressed in his moral philosophy (1993). People cannot *know* whether they autonomously determine their judgments according to rational norms of conceptualization and inference rather than merely following heteronomous inclinations. Kant concluded that rational normativity was not a theoretically knowable capacity of humans but only a practical obligation to which agents ought to hold themselves accountable.

Although this book does not work out an account of conceptual normativity, its line of descent through Hegel, Heidegger, Sellars, Davidson, and Brandom identifies Kantian conceptual norms with inferential roles in public discursive practices. It also revises that tradition in at least two significant respects. First, the normative accountability of individual discursive performances comes from the constitutive temporality of practices, which conjoins predictive and authoritative modes of assessing and sustaining practices over time. What it is to answer to reason and make sense discursively is not already settled but is at issue in our discursive practices and practice-differentiated way of life. A naturecultural account of conceptual understanding thereby endorses Kant's focus on the prospective orientation of people's critical capacities. It relocates that prospective orientation in discursive practices rather than individual cognition, however, and does so without grounding that orientation in a distinction between appearances and things in themselves. Second, people's developing practical-perceptual postures are grounded in and directed toward their biological environments; bodily responsiveness to a practice-differentiated environment replaces sensuous intuition or causal reliability as the locus of the empirical dependence and accountability of discursive practices.

This revisionist conception of discursive practices generates two further aspects of the finite dependence of human capacities for rational sense making and reason-guided transformation of their ways of life. Katharine Withy characterizes the first aspect as people's finite *situatedness*, and with that, a consequent finitude of normative accountability: "As entities who make sense of things, we are delivered over to the things that we make sense of and to specific ways of making sense of them. We are thrown into particular situations, and this means that we are given over to particular things to make sense of (and not others), and particular ways of doing so (and not others). We are always in some situation that provides the content or material for our sense making and in doing so limits or constrains it" (Withy 2011, 63).[1] This chapter's first section considers how this situational dependence of people's capacities for discursive sense making and reasoned transformation of their way of life shapes the normative accountability of that way of life and the practices it encompasses.

The book concludes with a further aspect of human finitude implicit throughout the book, which comes to the fore now in considering limits to our capacities for reasoned, discursively guided transformation of our practices and practice-differentiated way of life. Discursive adjudication of conflicts and misalignments within and among practices only takes place among humans. People reason with one another about the concerns at issue in how practices develop and affect one another and work with or against one another to shape continuation of those practices. People who communicate and reason discursively with one another are nevertheless not the only agents involved in sustaining and transforming human ways of life. Understanding the normative accountability of a practice-differentiated way of life requires taking account of its more encompassing interdependence with other organisms and their overlapping environmental intra-action. Whether and how people can work out and sustain that way of life depends on finding ways to live together with myriad other organisms whom we cannot engage discursively.

Objective Accountability as Situated Transcendence

The traditional separation of causality from rational, discursive normativity has repeatedly raised concerns that language might enable an unconstrained expressive freedom, leaving it bereft of content and normative authority. This concern, often expressed as a need for *objective* accountability of what

---

1. Withy also examines more expansive aspects of human finitude that cannot be considered here.

people say and do, has taken two characteristic forms. A first worry concerns how concepts and claims are accountable to what they purport to be about.[2] Conceptual contents and their inferential interconnections, including their accountability to perceptually acquired but conceptually formulated judgments, might exhibit a self-reinforcing mutual coherence floating free of answerability to experience or to causal relations among entities in the world. That empirical disconnection would render discursive expression empty of content. The second worry concerns how to resolve conceptual or empirical disagreements among speakers and agents occupying different conceptual standpoints or perspectives on their situated interactions, including different ethical or political commitments. It also encompasses a concern that powerful agents might forcefully impose or ideologically promote a self-serving conceptual understanding that blocks or diverts critical challenges to their authority on their reasoned merits.

Understanding discursive practices and their normativity as forms of biological niche construction shows why the generality of those concerns is misplaced. Objective accountability in both respects is a localizable issue that answers to the place of discursive practices in more encompassing practical contexts. The first form of the worry is resolved by how a niche-constructive conception turns Brandom's social-rationalist inferentialism inside out. The practical-perceptual development of linguistic abilities through scaffolded initiation into ongoing discursive practices prevents that concern from arising in traditional skeptical forms. Linguistic expressions and their uses are concrete, materially situated aspects of the world and only become meaningful and justified in being caught up in people's bodily postured immersion in concrete worldly situations. Those expressions also have a second dimension of normative accountability through their use in other practices that are interdependent in more limited ways with those concrete engagements. Misalignments arise among the ways people talk, the significance of those discursive performances in other settings, and their concrete interactions with one another and the situations they talk about. In those situations, however, as Brandom himself rightly noted:[3]

2. Prominent formulations of this issue include Kant's (1998) transcendental conditions for the objective purport of concepts, Sellars's (1997) attacks on the Myth of the Given, and McDowell's (1994) differentiation of rational justification of a claim or action from its causal exculpation. Chapter 6 discussed how Brandom's social-rationalist model of language as discursive scorekeeping addresses this concern. See also Rouse (2015, chap. 5).

3. Brandom clearly expresses how the future-directed temporality of discursive normativity resolves skeptical concerns about the objective accountability of discursive practices. He nevertheless incorrectly applies this resolution to his *model* of the game of giving and asking

Sorting out who should be counted as correct, whose claims and applications of concepts should be treated as authoritative, is a messy retail business of assessing the comparative authority of competing evidential and inferential claims. Such authority as precipitates out of this process derives from what various interlocutors say rather than from who says it; no perspective is authoritative as such. There is only the actual practice of sorting out who has the better reason in particular cases. . . . That issue is adjudicated differently from different points of view, and although these are not all of equal worth, there is no bird's-eye view above the fray of competing claims from which those that deserve to prevail can be identified, nor from which necessary and sufficient conditions for such deserts can be formulated. (1994, 601)

No one can guarantee freedom from epistemic error or conceptual incoherence, but the worldly accountability and corrigibility of discursive performances is secured by the two-dimensional normativity of discursive practices. Language use belongs to and answers to the world as part of a developing and evolving way of life dependent on sustaining itself amid changing circumstances.

The horizontal differentiation of a practice-based way of life, in which practices are partly constituted by normative concerns that grip their participants, nevertheless complicates resolution of the second worry about possibly irreconcilable normative standpoints. The normative heterogeneity of a practice-differentiated way of life and the entanglement of power and normativity may seem to undercut objective resolution of people's conflicting life-commitments and normative concerns. Why are these diverse normative concerns, including those aiming to adjudicate among practices, not merely goals that some people happen to pursue de facto and others do not? People commit and develop their lives through having encountered some ongoing practices rather than others in their affective development within complex environments. Considered in isolation, the normative concerns internal to practices seem little different from the one-dimensional normativity of other organisms. The lives of cats or vervet monkeys are internally goal directed and can be assessed accordingly. They can succeed or fail in sustaining those lives and helping continue that lineage as a way of life, but that outcome provides no basis for assessing whether such a lineage should continue or should evolve in one way rather than another. The worry is that assessing

_____

for reasons (see chap. 7 above; Rouse 2002, chaps. 6–7; Rouse 2015, chap. 5). The model lets discursive practices answer to material circumstances only through scorekeepers' *judgments* of their perceptual and practical reliability as observers and agents, rather than directly to causal involvements.

conflicts within human ways of life might also answer to no further normative authority.

This concern arises in both dimensions of normativity. The normative
concerns constitutive of a practice are often at issue in its continuation. The
mutual alignment of some participants' performances and capabilities may
compel others either to yield to these dominant orientations or abandon participation. Such alignments may provide no basis to justify one way to continue the practice over another, apart from instrumental considerations of
how to sustain it at all. The latter considerations do not determine whether
the practice *should* be sustained in that form. Such effective alignments may
then seem to constitute power without normative accountability, as people's
cooperation with others shapes possible roles and performances with no further normative accountability than to continuation of a power-inflected pattern of interdependent performance.

Similar difficulties might seem to arise for critical assessment of how
practices work together. Practices may depend on other practices in specific
ways—employers may not be able to hire workers unless adequate care for
their dependents is available, or workers may have to distort their lives to
incorporate both remunerative work and provision of care for others; sports
competitions may depend on other practices for medical testing or to certify
participants' age, gender, citizenship, or academic standing, or such certifica-
tion may be imposed by others; scientific researchers may need suppliers of
instruments or prepared materials or employers for their students, but those
needs may not matter to others who might provide such support only in ways
that would compromise scientific understanding. Practices also answer to
normative concerns affecting their configuration together. Concerns about
pervasive racial or sexual inequality or oppression, environmental sustainability or habitat preservation, protection of privacy, information access, or
food security may call for adjustments to many practices, which would in
turn affect their own normative concerns or their fulfillment. In both dimensions the question is how to understand the demands practices impose on
one another and their participants or the concerns arising from conjoined
practices as genuinely normative rather than merely effective impositions
by powerfully aligned agents. This apparent difficulty has often motivated
quests for a sovereign normative authority apart from particular conflicting
concerns, perhaps at a metalevel to enable the rational adjudication of those
conflicts from a "standpoint" of aperspectival objectivity (Kukla 2006; Rouse
1996a, 2005b).

Understanding the exercise of critical capacities as an ongoing, repro-
ducible niche-constructive accomplishment shows why such aperspectival,

sovereign legitimation is not necessary for genuinely normative critical assessment and transformation of practices or practice-diversified ways of life. The normative accountability to one another of people's situated performances is grounded in the biological normativity of the human lineage. The niche-constructive history of that lineage not only substantially transformed its material surroundings and the behavioral postures and capacities of persons sharing that biological heritage. It also constituted new forms of normative accountability through its two-dimensional articulation. That development provides the basis for understanding how the constitutive normativity of a practice-differentiated way of life transcends parochial limitations of particular traditions and perspectives or even the evolutionary, niche-constructive history of the species.

To understand this normative transcendence, we need to think through the significance of that two-dimensionality, its temporal openness, and its biological basis. The biological grounding of people's normative concerns makes important contributions to understanding the more complex forms of assessment that emerged in the human lineage. First, human lives and ways of life are processes whose success or failure depends on their environments. That dependence makes the outcome of people's interactions with one another and their partially shared surroundings consequential. How people can live and whether they continue to live are at stake in those interactions, which are both enabled and constrained by their circumstances. How practices develop and whether they can be sustained or improved is not a free-floating consideration, but is accountable to the circumstances in which they develop and how those practices can accommodate their capacities and limitations. That environment is not simply a given constraint, however, because of the evolutionary significance of niche reconstruction. People's environmental dependence is immensely complicated by interdependence with one another. Sterelny (2002, chap. 2) pointed out that organisms whose developmental and selective environments encompass other organisms are vulnerable to deception and manipulation, producing selection pressures for more flexible developmental and behavioral responses. The evolution of a more complex, temporally extended, practice-differentiated interdependence further complicates that issue, as does the discursive articulation and continuing critical assessment of practices. Working through those complications nevertheless shows how they enable the mutual normative accountability of differently situated normative standpoints.

A first complication is that people are not only vulnerable to one another, but increasingly dependent on one another's performances in ongoing patterns of practice. Other people and what they do are influential components of human developmental environments. This complex interdependence explains why power is not simply a stable causal capacity of agents or agential alignments. Exploitative or oppressive social alignments remain targets for resistance or circumvention. Relatively stable forms of domination or oppression can sometimes persist for generations but not without continuing effort, under changing conditions, facing both new and persistent forms of resistance. New participants in oppressively aligned practices would have to be recruited and retained without further challenging or transforming those arrangements.

The discursive articulation of practices provides a second, crucial complication. People do not merely continue, change, or abandon participation in a practice due to their own affective and reasoned assessment. People can express what is at issue and at stake in practices, call others to account for their participation in or response to a practice, and articulate and reason about the normative concerns at issue and at stake in those circumstances. These capacities provide a more complex check on how practices develop. Discursive practices also build and sustain relationships among participants—friendships and commitments, along with mutually affirming recognition of goods, skills, or virtues, and difficulties, obstacles, or objections—making the assessment of participation more than an impersonal, instrumental matter of assessing outcomes and weighing benefits. These embodied, affective, and discursive relationships, along with skills, postures, affective orientations, and motivating concerns developed in or by a practice, transform the *people* involved. First-personal, reflective assessment of who I am (in each case) and how I might develop my life always answer in part to who *we* are, how *we* would develop together or apart, and who or what belongs among "us."

A further complication of assessment and transformation of practices is the need for any single life to accommodate multiple practices and relationships and for any one practice to accommodate many participants with differing practical and normative involvements. People confront genuine conflicts among the normative concerns of multiple practices and how they are addressed. Sometimes practices need to change to accommodate other aspects of participants' lives; sometimes those practices instead impose demands on participants to change their lives to accommodate participation in the practice; sometimes part of the issue is who can, should, or would continue participation or newly initiate it. Assessing which normative concerns are authoritative over others and in which direction adjustments are called for is integral to the practices of discursive articulation and critical assessment.

The practice-based diversification of human ways of life also greatly multiplies the normative concerns at issue in their reproduction. New practices

enable new kinds of excellence, achievement, and mutual support, along with
capacities to recognize and respond to these concerns. These expressive freedoms are among the most distinctive features of human ways of life. Humans
have instituted new sciences, arts, technologies, relationships, identities,
pedagogies, political institutions, political criticisms, economic relations of
production and distribution, and ways of caring for others, along with more
dubious forms of criminality, exploitation, surveillance, alienation, and wan-
ton destruction. Recognition that practices embodying different normative
concerns make conflicting demands on one another couples with acknowledgment that more dubious or destructive practices also cultivate and rely on
skills, achievements, and appreciation. Together they highlight the urgency
of understanding the authority of critical assessments of normative concerns
and the constraints and opportunities brought about by their conjoined pursuit. How critical assessment and reconciliation of divergent concerns can be
objectively grounded rather than merely effectively dominant is sharpened
by those concerns' partial mutual opacity. Understanding and appreciating
normative diversity requires cultivating postures and skills to recognize and
be gripped by those concerns, which normally develop through participation
in the relevant practices.

The normative heterogeneity of human lives and its transformative shaping of the people who assess them critically has driven quests for a standpoint
of normative sovereignty. Widespread appeals to ahistorical conceptions of
reason exemplify efforts to transcend that normative heterogeneity in a critical standpoint with legitimate authority over other concerns. The suggestion
has been that we can cultivate independent critical capacity to assess those
concerns and interrelations, aspiring to a freedom from parochial influences
and contingent identifications, with authority for anyone because beholden
to no one in particular. It would provide an *objective* basis, independent of the
normative perspectives or standpoints at issue, for determining how those
divergent concerns should be accommodated together in interdependent
practices.[4]

A biologically based conception of the two-dimensional normativity of
human ways of life shows why this quest for a sovereign normative standpoint is misguided, and misconceives people's normative heterogeneity. Lance
(2015, 280) recently characterized this misconception of normative authority
as a "dual-aspect" analysis. On one side of this mistaken distinction are normative concerns people *happen* to accept or respond to—their having these

4. Allen, Forst, and Haugaard (2014) instructively discuss how this perennial issue arises
concerning the role of reason in criticizing and transforming social or political domination.

concerns may have been due to their biological or psychological makeup, conformity or resistance to their cultural context, or the persuasive influence of particular individuals. The other side encompasses the normative concerns they genuinely *ought* to have, with authority over what they do, whether they recognize and accept it or not. Those concerns may be authoritative due to their normative location—people have "genuine" responsibilities *as* parents, partners, coworkers, neighbors, citizens, and so forth—but independent of their de facto acceptance of that responsibility. Some responsibilities are conferred by explicit commitments—promises, contracts, etc.—but de facto acknowledgment of or motivation by those responsibilities is distinct from the grounds for their normative legitimacy, whether acknowledged or not. The quest for normative sovereignty seeks a basis for determining what concerns people ought to accept and live by, independent of whatever they "happen to" accept.

Understanding people's two-dimensional normative accountability as a biological phenomenon undermines this supposed separation between de facto and authoritative normative concerns. Consider the one-dimensional normativity of other organisms. Organisms have the goal of sustaining their lives and reproducing their lineage not simply as a de facto motivational tendency which they might or might not have. Organisms *are* the goal-directed process of sustaining life and lineage. They might succeed or fail to do so, but that goal's normative force is inseparable from the organism they are. An organism that fails to sustain itself or reproduce its lineage does not thereby give up its goal-directed normativity; it remains within that one-dimensional normative space while failing to fulfill its constitutive goal. Its normativity is nevertheless one-dimensional in not constraining *how* it responds to its constitutive goal-directedness apart from a finite dependence on its currently configured life process. It can only develop and evolve from the bodies, ways of life, and populational variation within that lineage, in the environments it inhabits, can move to, or partially reconstructs. Within those finite constraints, however, the only normative issue is whether it sustains life and lineage and not how it develops and evolves. What is at stake in the resolution of that issue is nevertheless literally a matter of life and death.

Human ways of life are more normatively complex and varied, but people similarly inhabit that normative space. The normative concerns shaping people's lives and the practices they take up are not contingent matters of fact they happen to accept. They are living organisms always already in a normative space. That space allows for alternative life trajectories and different possible configurations of practices developing over time, guided and governed by a more diverse array of normative concerns. None of those trajectories, practices, or concerns can be sustained in isolation, however. They depend on sustaining sufficiently supportive alignment with other people, other practices, and the discursive sense they can make of the lives they are living and the issues they confront. People are already situated within ongoing practices and relationships, including discursive practices whose terms partially express what is at issue for them and the practices they (continue to) participate in, and what is at stake in how those practices and their lives develop. They are already gripped by some normative concerns and ways to express or articulate them, while dealing with oversights, contradictions, and other unresolved issues. The practices making up their developmental environment encompass how people engage one another discursively, in relationships and practical settings, and in coordinating and guiding their own participation. People rely on others living their own vertically differentiated lives, who understand the situations they inhabit together in partly different ways with a different normative grip.

These practice-differentiated environments are temporally extended, because people are living bodies who must continue to sustain themselves through authoritative response to the predictively intelligible patterns of practice around them. Dual-aspect conceptions of normativity are dualistic—the distinction makes the relations among those aspects unintelligible—in two respects. First, they do not recognize that the relationships, practices, and normative concerns people have are not just contingent facts about them but are constitutive of them as humans living a practice-differentiated way of life.[5] Second, they fail to recognize how those normative concerns are not already determinate but instead open onto their authoritative continuation in ongoing patterns of interdependent practices and performances. Just as other organisms and lineages develop and evolve finitely, with and from particular bodies, ways of life, and populational variation in their environing circumstances, so with us. The normative concerns embodied in practices and their continuing interdependence are adequately specifiable only anaphorically, as issues and stakes in how people continue to live and sustain themselves. Those issues and stakes are vertically and horizontally differentiated, but interdependent and mutually accountable. Both individually and collectively, what is at issue is who those people have been and are becoming.

Kukla (2006, 2021b) develops a related conception of how people already inhabit an interdependent, temporally extended but normatively heterogeneous space. Kukla's discussion ostensibly concerns the objectivity of

---

5. Hegel (2018, chaps. 5–6) analyzes this inadequate understanding of normativity as the "alienation" characteristic of modern ways of life.

epistemic warrant but encompasses capacities to recognize and respond to legitimate normative authority more generally, including authoritative moral concerns. Kukla's account is instructive in retaining a vestigial role for the objective transcendence of people's normative heterogeneity as a regulative ideal. Understanding why the mutual normative accountability of human ways of life does not require that vestigial, regulative role for an objective standpoint culminates this first aspect of human normative finitude.

Kukla begins by emphasizing that normative heterogeneity is often construed too loosely.

> Contingent differences between our perceptual capacities . . . affect what we are able to notice. . . . This isn't enough, on its own, to provide a challenge to the belief that aperspectival objectivity is a condition for the legitimacy of reasons. In order to argue against aperspectivalism, then, we must be able to defend the subtle claim that the *epistemic* status of something as *warrant* can depend on the standpoint of the inquirer. Our contingent histories and resulting second natures must have the potential to make us not only *more or less rational,* but *able to perceive different reasons and access different warrants when being rational in response to the same causal inputs.* (Kukla 2006, 85)

Mere differences in prior experience or available evidence are not differences over which normative concerns are authoritative because they are corrigible once evidence is shared. That is not a difference in normative perspectives but only a contingently different access to relevant evidence bearing on shared normative concerns. Genuine normative heterogeneity nevertheless is pervasive. People have developed and taken up bodily postures and strategies shaped by the normative orientations of different practices and transformative relationships, even though those orientations are interdependent and thereby mutually accountable.

Kukla then agrees that people's abilities to recognize the normative significance of their shared surroundings do genuinely differ.

> If our ability to perceive inferentially fecund facts is a contingently inculcated second nature capacity, then there is no prima facie reason to think that we share it in all of its details. We should expect the rational deliverances of perception to vary depending on the experiences and practices that gave form to an inquirer's normative grasp of standard conditions and appropriate inferences. . . . Both [natural and moral perception] involve learned receptive sensitivity to normative relationships and inference-licensing reasons, whether these are reasons for actions or for beliefs. If this is right, then our perceptual capacities can provide variant warrant that fails the test of aperspectival objectivity. (Kukla 2006, 89)

Those differences nevertheless only concern the predictive sense of normativity. Normative accountability extends further, toward evaluative standpoints people *can* come to occupy in response to partially shared circumstances and to one another. Normative concerns and the postured receptivity to the goods or excellences they make appreciable are not given, self-enclosed positions. They are instead postured orientations toward partially shared circumstances, including responses to one another and one another's postured orientations. People can learn from others' ostensive guidance, testimony, and critical reasoning; learn to trust them where they lack time or skill to learn for themselves; and recognize the need to adjust their divergent orientations to sustain their practical interdependence. Although people's normative orientations are grounded in what they normally discern and appreciate, they are corrigible and educable by others' needs and concerns.

These sources of normative convergence and mutual transformation respond to worries that normative divergence could enable forceful imposition of the concerns shaping powerful practical alignments without legitimate or effective challenge (Kukla 2021b). Kukla, who shares my insistence on the developed normative diversity of a practice-differentiated mode of life, defends a further response. They claim that people's irreducibly divergent normative concerns answer to regulative ideals of epistemic democracy and an aperspectival normative sovereignty. People's mutual interdependence and accountability to a shared world that constrains their activities constitutes a need to treat normative differences as responsive to ostensive guidance and reasoned challenge, even in the face of recalcitrance.

> There is no reason to think that our rational capacities are in fact optimally plastic, nor that the perspective of the ideally educated perceiver is an attainable goal. . . . But simply consigning oneself or others to a partial and inaccessible standpoint has to count as a failure of responsibility. . . . The universal accessibility of warrant is a built-in ideal of inquiry and rational discourse. But we need to transform the logical place of that ideal . . . to a regulative principle governing our rational attempts to work toward a maximally complete and accurate grasp of the character of the empirical world. (Kukla 2006, 92–93)

Kukla's defense of this regulative conception of epistemic objectivity rightly emphasizes three important aspects of the practice-based diversification of people's biological normativity. First, a practice-based way of life depends on its environmental situatedness; people ought to align their activities and normative concerns with what their environments afford. Kukla presents a regulative ideal of the universality of warrant as defeasible means to a more basic concern for ontological objectivity—getting the world right in relevant ways and responding accordingly. Second, people's capacities for discernment and recognition are not given but are educable by ostensive guidance from others and their own further exploration and critical assessment. Normative accountability outruns people's current postures and concerns. The two-dimensional normative temporality of a practice-based way of life encompasses a transformative openness both to previously indiscernible or inconceivable empirical phenomena and to previously unrecognized or underappreciated normative concerns. The authoritative sense of normativity takes precedence over and outruns its predictive sense even though both are necessary and mutually informative. Third, Kukla rightly recognizes the constitutive interdependence of practices and the consequent need for mutual accommodation among practices and their normative concerns. This point converges with the underlying concern to understand the world rightly, since people's developmental environments are a nexus of interdependent practices. Not all practices or normative concerns should be recognized and accommodated—some are incompatible with others, and must be abandoned to accommodate them, or vice versa—but such judgments and outcomes should be responsive to the discernments and concerns of others affected by and interdependent with those understanding the actions and making judgments.

Despite these important insights, appeals to a universal normative standpoint even as a regulative ideal overlook other important aspects of the normative heterogeneity of a practice-differentiated way of life. That regulative ideal blocks adequate recognition that the conflicting normative concerns and people's capacities for their discursive articulation and critical assessment are part of the world whose transformation is at issue. Consider first the normative priority of ontological objectivity. Kukla (and others[6]) are right that people's ways of life and normative concerns depend on and answer to circumstances that are beyond their current comprehension or the target of mistaken beliefs. People's capacity for niche-constructive transformation of their surroundings is limited. Those limits cut both ways, however, because people inherit and must deal with a niche constructively transformed world.

---

6. Haugeland (1998) and Smith (2019, 1996) similarly emphasize the priority of ontological objectivity for moral and political as well as semantic and epistemic concerns; Kukla's account partly builds on theirs. Putnam (1975) argues that the semantic significance of what people say and do answers to objectively real, practice-independent modal relations among the objects and properties with which they interact, and Haslanger (2012) extends Putnam's modal semantic externalism to normative modalities.

Those transformations constrain subsequent changes in how people live and what normative concerns they can intelligibly pursue.

Consider some ontologically objective characteristics of the world we now inhabit: higher levels of carbon dioxide and other greenhouse gases in the earth's atmosphere; a substantial built infrastructure threatened by projected sea-level rise; massive agricultural land-clearing for and soil depletion by agricultural monocultures; extinction or habitat-reduction for many other species; and the associated population growth of humans and their codomesticated and commensal species. These niche-constructive transformations of people's biological environment are now integral to the world to which people's critical assessments and transformative responses are objectively accountable. Similar points apply to twentieth-century reconstructions of many built environments to enforce socioeconomic and racial stratification. That infrastructure of stratification and oppression is now part of the world in which a more just society would have to be built.[7] Consequential reconstructions of human environments have not come to a halt; what people do now changes the situations of subsequent generations, the consequences of their responses, and hence their normative significance.

As a second consideration, the niche-constructive transformation of human biological environments is not limited to physical reconstruction. I earlier objected to Wartenberg's (1990) ideal of individual rational autonomy as the telos of transformative power because people's interdependence with one another is not self-effacing. People's lives and capacities are embedded in relationships, practices, and patterns of development, and they develop differently through those educational practices, personal relationships, and institutional structures. These social considerations are intimately entangled with niche-constructive transformations of their physical circumstances and the conceptual resources available and intelligible to characterize and assess them. Critical assessment of institutions, practices, and people's developmental patterns consequently is not just undertaken *from* educable normative postures but is directed *toward* shifting forms of life.

People cannot acquire and assess on their own authority all they need to know to conduct their lives well. Our capacities for rational appreciation of

---

7. Highway systems in the United States exemplify material niche construction with long-term consequences for how people live. Many were designed in part to lock in patterns of enforced racial segregation, populational mobility, and energy use (Norton 2008; Caro 1974). Kukla (2021, chaps. 4–6) discusses how urban infrastructure can be partially repurposed despite such structural constraints, exemplified by Berlin after the Wall and Johannesburg after apartheid. Such repurposing is nevertheless still conditioned by its existing infrastructure, as Kukla's account makes clear.

normative concerns are not completely plastic. People's willingness to sustain dominating alignments that impose their own normative priorities on others will often not be overridden by critical assessment or countervailing alignments. Kukla takes an important step in this direction by recognizing that "a maximally inclusive perspective [would not be] that of an abstract 'ideal observer,' but rather that of an observer whose capacities are ideal given what actual humans with contingent, sensuous, receptive faculties can come to perceive" (2006, 92). Appeal to a universal normative standpoint even as a regulative ideal is nevertheless still overly idealizing. The regulative ideal of optimally rational capacities, optimally complete information, and optimally inclusive openness to others' normative concerns does not acknowledge the costs and difficulties involved in overcoming people's cognitive, affective, and moral limitations. Movement from people's actual capacities even to optimal rather than ideal capacities still involves overcoming normative "friction" (Tsing 2005), whose consequences also must be taken into account. Expanding people's moral and epistemic education and engaging in political activism against unrepentant domination, even to the extent of a supposedly optimal regulative ideal, requires developing and extending the practices to accomplish that end. Those transformations must be accommodated among other practices people maintain, including those that meet their biological needs in new ways or that sustain other practices and achievements that matter.

Put another way, what *would* be optimal rationality, information, and openness is at issue in the critical practices whose optimization is at stake. A regulative ideal of normative universality implicitly postulates that mutual adjustment of the normative concerns in practice-differentiated ways of life could at least ideally reach an equilibrium of optimal joint satisfaction. Such a regulative ideal would accommodate optimal rational responsiveness both to the concerns of particular practices and those that adjust those practices to accommodate their mutual interdependence. On a niche-constructive conception of practices, that adjudication of the two-dimensional normativity of a practice-differentiated way of life has a path-dependent, nonequilibrium normative dynamics. As interdependent, environmentally situated organisms, people share situations shaped by issues with different personal, epistemic, and political stakes. Any specific resolution of those issues transforms that situation, presenting them with new issues and reconfigured stakes amid realignments of power. These issues and stakes can genuinely grip situated agents without needing to project their eventual convergence in normative equilibrium. Kukla's regulative ideal of aperspectival objectivity is a last vestige of traditional aspirations to a sovereign normative standpoint above the fray of competing, transformative concerns. Such efforts instead would

elevate some concerns above others in the misleading guise of neutrality. Resolving the issues that arise amid those conflicts and their interdependence
instead always requires contestable but consequential judgments and transformative practices in which enabling and constraining power relations work
together.

Does rejecting any sovereign standpoint of conceptual universality or
objectivity, even as a regulative ideal, undermine normative transcendence
of the particularity and partiality of people's current ways of life? No. What
are often regarded as transcendent normative concerns—objectivity, justice,
truth, equality, or rejection of anthropocentrism—remain integral to critical
assessment of practices and their constitutive concerns. That role is always
situated and changing, however, and efforts to transcend current normative
limitations embed a history of specific issues and concerns. Reflection on histories of these concepts is instructive. Historians of science argue that the
concept of epistemic objectivity has a complex history with multiple overlapping and sometimes conflicting meanings.[8] What it means for inquiry,
judgment, or depiction to be objective cannot be specified abstractly. That
concept developed in response to issues arising in the course of inquiry, in response to different challenges, limitations, or recognizable forms of partiality.

The development and critical assessment of idealizing theories of justice
also suggestively indicates the situated significance of transcendent normative concerns. John Rawls's influential *A Theory of Justice* (1971), for example,
was advanced as an ideal theory. The theory was nevertheless substantively
framed by the contemporary prominence of utilitarian ethics and Marxian
political theories in asking what justice should mean in societies structured
by property rights and market exchange. Responding to communitarian
critics challenging the theory's prioritizing individual autonomy over concrete forms of communal and interpersonal dependence, Rawls (1993) re-
conceived the theory more narrowly. It became a "political" conception of
justice, shaped by modern societies' ineliminable pluralism of comprehensive conceptions of the good and seeking an overlapping consensus among
"reasonable" comprehensive conceptions. Rawls also recognized the theory's
historical framing not only by market economies but also the political history
of modern "liberal" societies.

Critics nevertheless rightly argued that Rawls's political conception of
justice remained overly idealized. Susan Okin (1989) argued that Rawls's

8. Instructive work on the multiple, situated meanings of objectivity include Daston (1992),
Daston and Galison (2007, 1992), Porter (1995), Douglas (2009). The concept also developed
somewhat differently in epistemology, ethics, aesthetics, and metaphysics.

earlier account systematically excluded critical assessment of heteronorma-tive, family-centered practices of child development and traditional gender-based domination and exploitation. These practices and gender-constitutive norms were built in to the theory's "ideal" conditions. Moving to a political conception of justice did not escape that limitation. Charles Mills (1997, 2017) and Elizabeth Anderson (2010) later argued that Rawlsian ideal theory also excluded the racial segregation built into American society from the scope of justice and obscured how racial oppression structures the theory's concep-tion of a just society. Pankaj Mishra (2013, 2020) argues that the tradition encompassing Rawlsian liberalism incorporates complicity with the broader legacies of racism and European colonialism.

The moral in light of the two-dimensionality and temporality of normativ-ity is that concepts such as justice or epistemic objectivity acquire genuinely transcendent normative significance, but they do not do so through a regula-tive ideal of a just society or objective knowledge. They instead respond to issues arising from discursive articulation and critical reflection within ongo-ing practices amid contested conceptions of what is at stake in resolving those issues. Normative concerns are constitutively temporal, pointing beyond cur-rent normative configurations toward open-ended possibilities for specific improvement and possible further considerations not yet recognized or taken into account. How people interact, align, dominate, resist, challenge, and rea-son with one another are never completed but always point toward new issues and reconceptions of what was already at stake in current circumstances. De-bates over ideal theories of justice further highlight that point. Okin's (1989) proposal that justice might require abolition of gender as a system of social domination determined by sex has since been repositioned by genderqueer and transgender practices and identifications. Recognition of and response to pervasive racial domination are similarly complicated by multiracial and other reconfigured racial identifications, and intersectional conceptions of race, ethnicity, sex, gender, and sexuality (Haslanger 2012, part 3; Glasgow et al. 2019). Moreover, what a just society could mean amid wrenching insti-tutional and infrastructural changes due to the effects and prospects of cli-mate change or novel pandemic disease may raise new issues that restructure what concern for justice calls for.

The character and role of discursive articulation, critical reflection, and empirical accountability in people's lives are themselves at issue in evolv-ing normative configurations. In advancing particular normative concerns and subordinating the claims to authority of others, people also develop and promote different practices and skills of critical assessment. Discursive articulation, critical reflection, and calling others to account are sufficiently integral to a practice-differentiated way of life that dystopian worries about their complete suppression are not credible. What adequately critical reflection and transformation of current ways of life require is part of what is at issue in sustaining a practice-differentiated way of life, along with which practices and concerns it encompasses. Rationality is not a given capacity but a suite of abilities and normative concerns initially developed and subsequently transformed in specific practical contexts.

The diversity of the contexts and capacities where rationality is at issue is readily visible in surveying issues for which development of those capacities has been at stake. Examples of contested forms of rationality include: encounters between mnemonic oral cultures and the critical juxtaposition of claims enabled by written texts and construals of those encounters as conflicts between rhetoric and logic or scripture and enlightened reason; new forms of discursive production, from printing presses to the internet and social media, which countered efforts at discursive regulation and control with a differently constrained proliferation of discourses and modes of reasoning; conflicts between cultivation of elite knowledge production and demands for epistemic democracy, including those emerging from epistemic empowering of workers, women, or colonial, postcolonial, and other racialized agents; the comparative significance of verbal reasoning and visual depiction, dramatic impersonation, musical expression, or video and film as loci of critical reflection and expression. Questions about education and development of people's critical capacities and access to those developmental activities also permeate these issues. Appeals to reason and worries about its suppression or ideological distortion always take shape in particular constellations of practices and the issues and stakes in their subsequent development. Adjudication of people's critical capacities, modes of reasoning, and ways of cultivating them is also a "messy retail business" not amenable to wholesale sovereign overview.

A consequence of the biologically grounded normative interdependence of practice-differentiated ways of life is that sustaining those ways of life requires building and rebuilding how people live together and engage one another critically in circumstances that outrun their control. Normative accountability is not about adjudicating different ways of life from a sovereign standpoint, but involves a situated transcendence that asks how to move beyond current ways of life and their issues or impasses to discern and build more adequate ways to live together. These considerations arise within current ways of life and incorporate discursive articulation of their futurally oriented normative concerns. Such critical articulations both depend predictively on current practices and normative articulations and aim to transcend them in ways situated by people's defeasible comprehension of their current possibilities and prospects.

The normative authority and force of that situated transcendence arises from how people's lives, practice-participation, capacities, and ways of life are at stake in the issues raised by the practices amid which they live. We each stake our lives and those of others on normative concerns we take up and respond to. Although the force of these issues ultimately invokes the life-and-death character of biological normativity, including possible extinction of some practices as subpatterns of human ways of life, they are two-dimensional. What is at stake in human evolution is not just whether but how the lineage continues, and people stake their lives on its continuation in some ways rather than others. Those stakes include sustaining specific capacities and concerns for critical transcendence of parochial limitations and their effectiveness in guiding that continuation. We cannot fully secure ourselves against irrationality, ideological domination, or suppression of dissent any more than we can eliminate epistemic error or conceptual incoherence. Interdependence with one another and the world inhabited together nevertheless also undermines dystopian concerns that normative adjudication could collapse into impositions by force or robust and inescapable ideological domination. People often do confront powerful, dominating alignments and ideological distortions of their critical capacities, but resistance to both remains an open possibility and a project.

Social-Ecological Interdependence

We can now take up an aspect of a naturecultural reconception of social practices whose import could not be adequately clarified earlier. A naturecultural conception of practices emphasizes evolutionary continuity between humans and other animals even in the complexity of people's practice-differentiated way of life. It also explains the manifest differences of human ways of life from those of other animals as the result of ratcheted, niche-constructive evolution. Those differences have provided a primary rationale for bifurcating human ways of life between social relations among persons and humanity's place in nature. People's *cooperative* interdependence, *mutual* normative accountability, and the *discursive* articulation of those cooperative projects and normative concerns together have often seemed to establish a distinctively human community that excludes other organisms. People depend on those organisms in myriad ways and often deliberately and forcefully utilize their capacities and their lives but do not enlist their active cooperation in common projects. People do live in symbiotic interdependence with many other species, most notably the codomesticated and commensal organisms whose evolved ways of life are intimately interdependent with those of humans,[9] but that dependence has usually been understood as different in kind from people's social and rational interdependence with one another.

This traditional bifurcation between natural and social forms of interdependence has been canonically expressed as a difference between instrumentally adapting actions to their causal consequences and cooperating with other rational agents in a common project. We have seen that conceptually articulated cooperation can take place even if the shape of common projects is differently conceived by those who take them up together. Humans reason with one another in extended discursive exchanges, but no other organisms encounter or join in the conversation. People also hold one another responsible for and accountable to normative concerns that have no meaning or authority for other organisms. Moreover, people take up practices and work together to change one another's performances and their shared surroundings to accord better with the normative concerns they take as authoritative. People may thereby seem to separate themselves more decisively from all other organisms as a community of rational agents sharing languages and a social world. Although humans are usually regarded as exclusive members of that community, membership is often identified communally and discursively rather than biologically. Brandom, for example, claims that, "We think of ourselves in broadest terms as the ones who say 'we.' It points to the one great Community comprising members of all particular communities—the Community of those who say 'we' with and to someone, whether the members of those different particular communities recognize each other or not" (1994, 4). The material conditions for membership in that community may have arisen from the evolution of the human lineage, but community identity and authority are understood as *normatively* constituted. On Brandom's account, the relevant norms are established by mutual recognition among members who can say, "we," but he thereby asserts a normative difference in who *should* be recognized as members of a community rather than a difference between natural entities subject to causation or natural law and social or rational beings responsive to norms (Brandom 1979).

A naturecultural conception of practices does not deny the differences between how people engage one another and their interactions with and

9. I speak of "codomestication" rather than "domestication," because the interdependence of human ways of life with codomesticated plants and animals required substantial modification of people's lives and practices to adjust to those of other species.

dependence on other organisms. It does, however, assimilate people's cooperative, discursive, two-dimensionally normative, *mutual* interdependence within
a more encompassing biological interdependence within and across taxa. The
relevant normative we is not confined to those who can explicitly say "we" to
one another. The reasons for this expansion begin with recognition that familiar contrasts between nature and the social world in which humans engage one
another as members of a discursive community assimilate other organisms to
nature conceived as an inexorably causal or lawlike realm.[10] Wartenberg (1990,
80–84) exemplifies such assimilation of biology to physics when he characterized people's "action-environments" by contrasting what is possible for a socially situated agent to what is physically possible.[11] He did not consider that the
relevant comparison might instead be what is biologically possible for people
as animals whose capacities and normative orientations are shaped by their developmental environments. That assimilation of biology to physical processes
overlooks a critical difference between people's abiotic circumstances and other
organisms whose bodies, life activities, and environmental habitats are integral
components of human biological environments.

   People interact causally with abiotic features of their environments, often in complex, nonlinear networks of causal interdependence exemplified
by the earth's climate system. Other organisms are integral to those causal
entanglements but as active, goal-directed participants. Those organisms
thereby complicate that causal interdependence. Organisms are not self-
contained entities but ongoing processes of environmental intra-action. Organismic life processes actively configure their circumstances as affordances
for their way of life by evolving and developing that way of life in response
to what circumstances afford for them. Through their phenotypically plastic development, cumulatively niche-constructive activity, reproduction, and
multidimensional evolution, all organisms are intra-active life processes that
are continually reconfigured in goal-directed ways. These environmentally
intra-active life processes incorporate, and are incorporated by, the niche-
constructive life processes of other organisms. These overlapping and mutually inclusive patterns of environmental intra-action cannot be understood
as independent causal influences merely added to one another. Their mutual
interdependence is constitutively indispensable to all, because other organisms' life processes counteract the cumulative effects of any single organismic

   10. Biological phenomena *are* appropriately understood as scientifically "lawful," but that
requires a different, more adequate understanding of scientific lawfulness. Lange (2000, 2007)
and I (Rouse 2015, chaps. 8–9) both discuss this sense of biological lawfulness.
   11. See discussion of Wartenberg's neglect of human biology in chap. 8 above.

lineage as unidirectional "biological pumps" (Laland, Odling-Smee, and Gilbert 2008, 557).

I have repeatedly cited Kim Sterelny's distinction between biological "detection agents" whose metabolism and behavior reliably respond to features of their environments, and more flexibly responsive organisms whose ways of life evolved to counter how other organisms evolved to interfere with simple detection patterns (2003, chap. 2). Even detection agents nevertheless respond to more complex patterns of interaction among organismic lineages. Their comparatively simple and reliable responsiveness to abiotic environmental "signals" may not mark a lack of interference or exploitation by other organisms, but only a dynamic evolutionary interplay between the fitness "costs" of that interference and the cognitive, energetic, and reproductive costs of more flexible responsiveness. Flexible cognitive/behavioral responsiveness to possibly conflicting environmental signals is not a difference in kind but only an adaptive shift in the dynamic evolutionary balance between capacities for flexible response and their metabolic costs.

A naturecultural conception of practices understands human ways of life as evolving a further shift in that dynamic balance. The evolution of cooperative interdependence; a practice-differentiated way of life and its discursive articulation; and the associated cognitive, behavioral, and metabolic challenges to which human bodies and developing bodily postures adjust are further flexible adaptations to more complex and variegated environments.[12] That complexity is not just an exogenous environmental condition but has been made and remade by people's niche-constructive ways of life. Humans have migrated and adapted to diverse physical surroundings and opened alternative ways of life together in part by materially reconstructing those settings and developing different bodily postures and skills in response. This book has argued that a practice-differentiated way of life and its discursively articulated two-dimensional normativity are integral to that iteratively niche-constructive process.

Other organisms—animals, plants, fungi, and the three microbial kingdoms—have been integral to the environments to which humans have been skillfully responsive, reconstructive, and adaptive amid overlapping abiotic surroundings. Those organisms have in turn been responsive, resistant, adaptive, and sometimes significantly unadaptive to aspects of people's niche-constructive reconfigurations that figured in their own developmental and

12. Recall Potts's (1996) argument from archaeological and paleoclimate data that the hominin lineage initially evolved amid dramatic, rapid, and often oscillating climatic shifts. Humans arguably are adapted not to stable environmental conditions but to frequent environmental change, including those wrought by their own activities.

selective environments. Humans have dramatically altered many abiotic circumstances and adapted to others, but we have *co*evolved and *co*develop with
other organisms. The phenotypically plastic development, niche-constructive
activity, and population evolution of those organisms amid people's practice-
differentiated activities continually provoke further realignments in the
practices that encompass those organisms. The cooperative interdependence through which humans diversified their ways of life, in part through
reasoned, discursive involvement with one another, thus belongs to a more
encompassing evolutionary and developmental interdependence with other
organisms. The effects of human activity on those other lineages have dominated recent discussions of human ecology under the headings of an ongoing
mass extinction heralding a geological epoch, the Anthropocene, demarcated
by the extent of human geophysical influence on the earth. A naturecultural
conception of practices as niche constructive recognizes that the influences
among organismic lineages have been reciprocal, even when the impacts in
different directions are disproportionate.

Thoroughly surveying the extent and character of human ecological interdependence with other organismic populations and lineages or its changing shape due to humans' global migration and socioecological transitions
is beyond the scope of this book. These topics have also been well addressed
elsewhere. Four groups of examples will instead indicate the importance of
recognizing people's *discursive* negotiation of normative concerns at issue in
our practices as continuous with our more extensive interdependence with
other organisms' ways of life.

Consider first the social foraging that many accounts of hominin evolution
take as the lineage's initial shift toward what became a practice-differentiated
way of life. As Sterelny (2012a, chap. 1; 2021, chap. 1) forcefully argued, this
way of life required extensive cooperative activity and intergenerationally accumulated expertise to find and exploit diverse sources of food, shelter, and
materials for cooperatively refined toolkits, and to avoid or defend against
other predators. These activities were interdependent with the many species
of plants and animals that became important foraged resources. That way of
life requires extensive knowledge of how to find or track, identify, catch or
extract, prepare and protect *these* food resources and respond effectively to
competitors for *them*. That was not a static set of capacities, but a continuing
realignment of skills and behavior with changing patterns of resource availability and behavior. As Sterelny noted:

> Foragers do have prepared eyes and minds; they are expert natural histori-
> ans of their local patch. . . . The ancient tuber and corn harvesters depended

on much hard-won information, if underground storage organs really were important resources from *erectus* on. Fruits are designed to be eaten. But plants do not welcome herbivore consumption of their storage organs, and hence they are protected both mechanically and chemically. . . . [Moreover] the populations of target species are depleted. Predators become increasingly rare, wary, or both. These environmental effects also create co-evolutionary opportunities for [other] species . . . (2012b 12–13, 16–17)

A foraging way of life *is* an intimate ecological, informational, and metabolic interdependence with many other organisms and their habitats. Cooperative foragers' social and cognitive interrelations with one another are closely aligned with how they interact with diverse other species to sustain that way of life.

The significance of that interdependence is clearly illustrated by the extinction of the comparably complex social-ecologically interdependent way of life of Neanderthals. The most compelling accounts of Neanderthal extinction emphasize at least three factors in combination. Ecological change that replaced temperate forests with colder, open tundra was surely central. The resulting demographic depletion of Neanderthal populations made it harder to sustain relevant expertise and other cognitive capital across generations, and possibly to maintain hunting bands of sufficient size. Both issues were nevertheless also conjoined with how Neanderthal skills and knowledge were interdependent with a gradually disappearing fauna and flora. Sterelny again highlights relevant considerations.

[Neanderthals] were adapted to a sinking local optimum, ambush hunting. This local fitness peak was separated from higher optima by a fitness trench. To survive in the steppe-tundra, they would have to switch from ambush hunting to pursuit-endurance hunting. That switch would have extremely high transition costs. The steppe-tundra grazers (reindeer, musk oxen, and the like) are a distinct fauna to that of the temperate forests, so information about natural history and habitat, tracking, target identification, and target behavior would all have to be relearned. Likewise, the weapons skills are different. . . . At least initially, the learning costs would be high, and there would be few, or no locally available models. (2012b, 68)

The moral is that social cooperation, transgenerational growth of cognitive capital, and the metabolic costs of cognitive complexity and neotenous development are not self-contained features of a lineage but closely integrate with the prevalence and behavior of other organisms. Those cognitive skills and cultural traditions had to align with other ways of life as reliable resources and manageable threats.

Early hominins' foraging lifeways have now mostly been displaced by practice-differentiated ways of life ultimately based on agricultural cultivation and herding, which provide my second group of examples. The domestication of a few plant and animal species, massive expansion of their cultivation, and consequent diminution of habitat and populations of other organisms have often been described as producing human-dominated ecosystems (Potts 1996, chap. 2). That description underestimates the active evolutionary role of codomesticated and commensal organisms in shaping their own interdependence with human ways of life. That intimate involvement is not a human achievement, but the outcome of extended coevolution among humans and codomesticated and commensal animals and plants (Budiansky 1999; Noske 1989). Ethnobotanist Mark Plotkin nevertheless notes that, "In the movement toward a global economy, there has been a trend to concentrate on fewer and fewer species. Today, less than 20 plant species produce most of the world's food. Furthermore, the four major carbohydrate crop species—wheat, corn, rice, and potatoes—feed more people than the next 26 most important crops combined" (Plotkin 1988, 107). Many other plant species cannot be directly consumed by humans but can be digested by other animals. Human efforts to take advantage of the concentrated metabolic affordances of their domesticated animals have now had even more prodigious impact. A recent survey of the proportionate disposition of the earth's biomass indicated that "Today, the biomass of humans (≈0.06 Gt C) and the biomass of livestock (≈0.1 Gt C, dominated by cattle and pigs) far surpass that of wild mammals, which has a mass of ≈0.007 Gt C. This is also true for wild and domesticated birds, for which the biomass of domesticated poultry (≈0.005 Gt C, dominated by chickens) is about threefold higher than that of wild birds (≈0.002 Gt C). In fact, humans and livestock outweigh all vertebrates combined, with the exception of fish" (Bar-On et al. 2018, 6508). The growth and expansion of human populations are directly connected to the evolved codomestication and both extensive and intensive cultivation or herding of a limited number of plant and animal species. Those practices also led to coevolutionary commensal relations with other species that adapted to practices ranging from agricultural cultivation to concentrated urban societies and their waste products (Boivin et al. 2016).

The process of domestication has often been imagined as instrumental control enabled by humans' epistemic insights and technologically enhanced skills as a sociocultural legacy of iterated niche construction. That conception is not altogether mistaken, but it neglects how the process of codomestication depends on developmental patterns, social relationships, affective temperaments, and genetic reservoirs of the organisms involved (Budiansky

1999; Noske 1989). Some organisms cannot be effectively domesticated, partly for lack of congenial temperaments or social dominance that humans can redirect developmentally. Other populations lack relevant genetic or developmental variation on which selective breeding can operate to express human-desired traits. Successful codomestication nevertheless also required extensive adjustment of *people's* activities, skills, and interpersonal associations to accord with the habits and capacities of codomesticated species. Codomestication or commensal relations were often not deliberate outcomes but instead the result of gradual, mutually coevolutionary adaptation. Sometimes that coevolution led to speciation events such as the divergence of dogs and wolves. Other cases involved practice-differentiation in human ways of life, as some people's lives became more tightly bound to agricultural cultivation or herding, and others depended on those practices in more limited ways. The preeminently cited example of "gene-culture co-evolution" (Feldman and Cavalli-Sforza 1976; Laland and Brown 2011, chap. 7; Lewens 2015, sect. 5.5) has been the spread of genes for adult lactose tolerance in coevolution with dairying practices (Holden and Mace 1997). Those genetic changes were nevertheless a relatively small consequence of a much more extensive interdependence among human herders and cattle, as herders' practices and ways of life became intimately responsive to the needs, habits, capacities, and hazards of the cattle that provided sustenance for them and many others.

Efforts to maintain conceptual separation of humans' biological involvement in nature from sociocultural, discursive participation in human communities must also confront the integral significance of codomesticated and commensal organisms for how people engage one another. That begins with recognition that the thermodynamic nonequilibrium of all living systems on earth is almost entirely sustained by the solar energy flux, whose direct biological absorption is accomplished by plants and microbes which together also compose the vast majority of earth's biomass (Bar-On, Phillips, and Milo 2018). That recognition extends to the diverse social practices through which the human appropriation of biologically mediated solar energy is accomplished, notably in food production and distribution, but also including delayed appropriation of prior solar absorption as combustible timber, fossil fuels, and "biofuels."[13] The initial formations of continuing, centralized political authority were closely connected to cultivation of grain crops that enable both a substantial surplus harvest and its surveillance and expropriation

13. Apart from nuclear fission and the limited achievement of artificial nuclear fusion, all fuels are biofuels, despite conventional use of that term for combustible chemicals derived from agricultural products.

(Scott 2017).[14] The growth and spread of urban centers and agricultural hinterlands and the scope of European settler colonialism have been significantly shaped by coevolution of microbial pathogens and the immune systems of people living in close proximity, with devastating effects on previously unexposed human populations. These epidemic disasters often occurred well in advance of European political or military engagement with indigenous populations (McNeill 1976; Crosby 1986; Diamond 1998; Wolf 1982).

Encounters between European settler colonists and indigenous populations in the Americas and Australasia also involved domesticated and commensal organisms accompanying the Europeans in addition to their coevolved microbes. As Alfred Crosby argued, "If the Europeans had arrived in the New World and Australasia with twentieth-century technology in hand, but no animals, they would not have made as great a change as they did by arriving with horses, cattle, pigs, goats, sheep, asses, chickens, cats, and so forth. Because these animals are self-replicators, the efficiency and speed with which they can alter environments, even continental environments, are superior to those for any machine we have thus far devised" (1986, 173). The role of animals in aiding human mobility and migration, in land clearance and other augmentations of bodily capacities, in warfare, in pollinating plants for human consumption, and in human sociality as pets, among many other involvements, indicate the extent to which human practices and communities are multispecies intra-actions.

The continuing significance of people's evolving interdependence with commensal microbial populations provides a third group of examples of how our practice-differentiated ways of life are a multispecies phenomenon. Understanding that interdependence begins with recognition that human bodies and those of all other multicellular eukaryotes are symbiotic holobiont communities in which microbes have functionally indispensable roles in metabolism, development, immunity, and other bodily processes (Bordenstein and Theis 2015; Gilbert 2017; Gilbert, Sapp, and Tauber 2012). Human populations and practices have nevertheless also coevolved with and adapted to many microbial pathogens, long the primary remaining predators of humans. Even before people recognized bacteria or viruses as microbial life forms, they adjusted many practices to accommodate and mitigate the

14. Graeber and Wengrow (2021) provocatively argue for more complex historical relations between social foraging and agricultural cultivation in human history and resist straightforward correlations among agricultural practices, urban societies, and centralized political authority. They nevertheless do not dispute that larger societies with centralized political authority do depend on agricultural codomestication of plants and animals.

effects of their unrecognized microbial participants. Food storage, waste disposal, health care, sanitation, interpersonal association, animal husbandry, sexuality, warfare, and more changed in response to diseases now recognized as of microbial origin, including those affecting people's companion species. Frequent epidemic disease outbreaks have pervaded human ways of life with consequent realignments of many practices in their wake.[15] Understanding microbes as disease vectors prompted new practices and adjustments to old ones, but also had further coevolutionary effects, including bacterial evolution and lateral transfer of antibiotic resistance, gradual reduction of microbial virulence, and possibly the increasing prevalence of human autoimmune disorders.

Two recent global pandemics, wrought by the newly emergent retroviruses HIV and SARS-CoV-2, are illustrative. Both viruses first evolved through close interactions among humans and other animals, as viruses mutated to infect new hosts and spread rapidly via uptake in diverse practices. The viruses' own reproduction, developmental patterns, and evolution reveal that many social practices are also constitutively shaped by microbial exchange. The circulation, reproduction, and evolution of HIV not only both displayed and established its multifaceted involvement in many social practices, including sexual activities, identities, and politics; medical treatments and public health; drug use; regulation of scientific research; intranational and international economic and health disparities; and intersectional power alignments around race, sexuality, gender, nationality, and economic class. The evolution of HIV also revealed and intervened in people's prior mutual accommodation with other microbes that became "opportunistic infections" in concert with HIV infection. Within some power-differentiated practice complexes, the politically contested development of treatment and mitigation strategies transformed the HIV epidemic from an acute, epidemic disease to a chronic, treatable medical condition. In other settings, it remains epidemic and acute, closely coupled with economic conditions, culturally-inflected sexual and other social power relations, human migration, and state and transnational policies and authority. People have realigned their participation in many practices to mitigate, resist, or sometimes exacerbate circulation and

15. Herlihy (1997) and Cantor (2001) discuss the sociopolitical significance of the most prominent historical example of epidemic disease, medieval and early modern Eurasia's bubonic plague epidemics. Kenny (2021) accessibly surveys the long course of human interaction with microbial infectious agents, while Garrett (1994) prominently addresses the prospect of enhanced vulnerability to epidemic disease.

reproduction of HIV, and in response HIV developed, evolved, and engaged in its own forms of niche construction.

The SARS-CoV-2 pandemic is in early stages of coevolution with humans. Its circulation and reproduction have nevertheless brought out more clearly how practices widely regarded as social phenomena have always incorporated and accommodated the circulation and changing virulence of various microbial companion species. Mundane and familiar practices of economic production and exchange, familial relations, education and other dependent care, transportation, recreation, and social-ritual activities are also practices of microbial exchange. In retrospect, we can see that those practices had gradually realigned to accommodate and mitigate the effects of already-prevalent microbes. People's interactions with SARS-CoV-2 have extensively disrupted or exacerbated those realignments. The virus's spread and its virulence were also differentiated by socioeconomic stratification and political authority and culture, as with many other human disease conditions. Its future course is unclear, but longer-term reconfiguration of many practices disrupted by SARS-CoV-2 will likely be closely entangled with how vaccination and prophylactic practices align with the virus's own reproduction, evolution, and niche construction, and with many other practices intersecting with mass vaccination. The shape of those other practices will depend significantly on the variability and mutability of the virus in response to selection pressures imposed by partially vaccinated host populations and effective alignment or misalignment with vaccination and medical treatment for SARS-CoV-2 infection.

A fourth and final group of examples of the constitutive entanglement of our practice-differentiated ways of life with the life processes of other organisms come from the many lineages that have been *unable* to develop and evolve in sustainable alignment with those practices. Global migration of human populations was accompanied by the disappearance or retreat of many other species unable to sustain themselves within the ecosystems partially absorbed within people's practice-differentiated environments. Whether directly removed by hunting or land clearance, or indirectly affected by habitat loss, resource removal, toxic by-products, or interactions with human-commensal species, these lineages' capacities for developmental plasticity and evolvability could not sustain them in these novel environmental conditions. The absence of those flora and fauna then significantly affected other species that had previously interacted with them. Humans have thereby dramatically reconfigured our own ecological, practice-articulated interdependence with other organisms, from the loss of top predators and consequent population growth among mesopredators to general loss of biodiversity in the wake of human populations' expansion and practice-diversification. If the earth is indeed undergoing an "insect apocalypse" (Goulson 2019) or a sixth mass extinction (Kolbert 2014), these characteristically destructive entanglements of people's practice-differentiated environments with the selective environments of other species will become all the more evident and consequential.

These misalignments of human practices with other organisms include the repeated, unsustainable "harvesting" or other destruction of species or populations that were for a time integral components of some practices and ways of life. Nor were these effects merely replaceable in turn by other resources usable to similar ends, even though sequentially destructive exploitations are familiar (e.g., Cronon 1991). The resulting transformations in how people relate to one another and understand themselves can have profound conceptual significance. Jonathan Lear has called attention to an especially striking but hardly unique case in the reminiscences of the Crow warrior Plenty Coups after his people were confined to reservations by advancing North American settler colonialism: "Plenty Coups refused to speak of his life after the passing of the buffalo, so that his story seems to have been broken off, leaving many years unaccounted for. . . . 'But when the buffalo went away the hearts of my people fell to the ground, and they could not lift them up again. After this nothing happened'" (Lear 2006, 2, quoting Linderman 1962, vii). The discursive practices with supposedly "meta-biological meanings [that] concern goals, purposes, and discriminations of better or worse" (Taylor 2016) incorporate people's interdependence with many companion species, including those species resisting that incorporation to the end.

Skeptical readers might accept the extent of human interdependence with other organisms and still be inclined to treat discursive practices, rational normativity, and critical self-understanding as a nearly decomposable component of human ways of life. In response, we should first recall from earlier chapters that discursive practices are only partially autonomous. They are integral to almost everything we do in our practice-differentiated way of life, including its two-dimensional normativity and our discursive immersion in everyday practical settings. The discursive articulation of the practices composing people's biological environments is itself also a form of iterated material and behavioral niche construction. People's bodies and cognitive processes have coevolved with practice-differentiated, discursively articulated ways of life, and those embodied capacities develop anew in each generation within an historically shaped field of practices. Separating a social world or even just our discursive practices from our bodily sustenance and development as living organisms, and our cognitive capacities and accountability to reasons in lives vertically differentiated amid concrete discursive practices, suggests a conception of the normative concerns shaping discursive life as misleadingly frictionless.[16] Normative concerns central to people's cognitive and discursive practices—sense making and its rationality, objectivity, clarity, perspicuousness, and more—are not already-determinate concepts shaped only by abstracted inferential relations to other equally disembodied meanings or truth conditions expressed disquotationally.

A "space of reasons" (Sellars 1997, 2007) has instead been opened and sustained by the niche-constructive development and ongoing reproduction of specific patterns of utterances and other discursive performances.[17] Reasoning, understanding, and their place in human ways of life always take concrete material form in people's ongoing lives amid those discursive practices. That concreteness incorporates the historically evolved languages people speak; the conceptual relations recognized, responded to, and built on in particular discursive contexts in those languages; the materiality of what is said, written, depicted, enacted, or sung, when, where, to whom, and how; and how those utterances, inferences, and more complex discursive constructions are taken up in people's vertically differentiated lives and the horizontally differentiated practices where they encounter and engage with one another. Anna Lowenhaupt Tsing eloquently expressed this aspect of people's discursive lives in talking about concepts as cultural "universals": "The universals that mobilize people, then, do not fulfill their own dreams to travel anywhere at any time. But this does not make them wrong-headed or irrelevant. . . . The knowledge that makes a difference in changing the world is knowledge that travels and mobilizes, shifting and creating new forces and agents of history in its path. However, those who claim to be in touch with the universal are notoriously bad at seeing the limits and exclusions of their knowledge" (2005, 8). Separating social life and discursive practices from evolving ways of life obscures the extent to which the character and scope of reasoning and epistemic assessment are at issue in the ongoing realignment and reproduction of discursive practices. In thinking and reasoning with one another, people sustain and develop supportive alignments among practices that enable how they live and shape their own development. Who "I" am and "we" are, and what roles social coordination and critical reasoning play in "our" lives are at stake for any of us in the ongoing development of critical engagement with one another. The working out of that issue cannot be separated from its place amid other issues arising in the sociopolitical ecology of human ways of life.

16. Tsing (2005) informatively analogizes "movements" of concepts in and through discursive practices to how friction both enables and slows movement of material bodies and fluids.

17. I have argued (Rouse 2015) that the Sellarsian "space of reasons" *is* humans' selective and developmental environment.

This aspect of people's social-ecological interdependence with one another and with other organisms and their biological environments shows up with greater clarity amid the now-rapidly-changing conditions of climate, species migration and extinction, and adaptive responses that will shape the future development and evolution of human ways of life. People's niche-constructive activities had significant effects on the earth's climate in advance of any understanding of the heat-trapping effects of greenhouse gases, just as we coevolved with microbial populations prior to recognizing their presence. In the case of climate effects, that ignorance was partially overcome by building an extensive scientific infrastructure for measuring climate-relevant variables, extracting historical climate signals from ice and sediments, running computational simulations of climate dynamics, and modeling the economic impact of climate change and the costs and benefits of its mitigation (Edwards 2010; Weart 2008; Malone and Yohe 1992). That scientific infrastructure and its epistemic achievements were themselves enabled by the very industrial economies and global connections whose climatological effects they made intelligible. Growing conceptual articulation of the interdependence of the climate system's physical dynamics with global political economies and demographics provides a telling and consequential case for understanding how discursive practices and critical assessments align with changes in other practice-differentiated environments. The limits to people's power to alter niche-constructive activities in light of conceptual understanding and critical assessment are now confronting a severe test, one in which our interdependence with other lineages will play a central role.

The scope and extent of ongoing global climatic changes have yet to be fully manifest. The lack of effective mitigation and adaptation since those possibilities first became more clearly understood has already led to substantial disruption of people's socio-ecologically interdependent, practice-differentiated ways of life. The time lag between carbon emissions and climate equilibria ensures more significant effects to come regardless of people's subsequent response. All of these effects are and will be mediated by shifting power relations shaping people's social-ecological action-environments and the normative concerns at issue there, including those arising from geographic and demographic changes in populations of other species. What then happens will exemplify especially clearly the two-dimensional normativity of practice-differentiated ways of life and their ineliminably social-ecological interdependence. Questions about how various practices will change or die out and what normative concerns will shape human ways of life cannot be adequately addressed in isolation. These questions will be all the more explicitly intertwined with the issue of what could be done to maintain (some of) those practices and concerns amid demographic changes and geographic shifts in populations of other organisms. Sufficiently cooperative practical realignments to work out sustainable ways of living together with one another and with other species, under often-destructive shifts in material conditions, may well depend in turn on mitigating recognized forms of political injustice, oppression, and exclusion. Those political concerns are nevertheless also exacerbated by initial responses to climate change. The resulting changes will surely affect the material and conceptual relations among the discourses and practices of social-ecological adaptation to climate change; the place of conceptually articulated reasoning and rationally guided niche reconstruction in human ways of life; the fate of power-inflected considerations of justice and moral or political responsibility; and the power alignments that differentially enable or constrain people's capacities and prospects. In those conjoined conceptual, political, and climatological conditions, however, the notion of a human social world founded on but nearly autonomous from physical and biological nature will seem at best a quaint anachronism and at worst a destructive delusion.

Acknowledgments

I dedicate this book with enduring gratitude to Mark Okrent and the late John Haugeland, whose friendship and sustained philosophical conversations have been indispensable to my work and have greatly enriched my life.

Research, writing, and publication of the book have been generously supported by a three-year grant from the Natural Sciences Division of the John Templeton Foundation and a sabbatical and leave from Wesleyan University. I gratefully acknowledge this support, with special thanks to Matthew Walhout, Templeton's vice president for the natural sciences, for encouragement and guidance, and to foundation staff for their assistance. Wesleyan remains a wonderfully supportive home for my teaching and research.

The project was prompted by conversations at the 2017 International Summer Workshop on Practice-Based Studies at the University of Warwick Business School and a workshop on Practice, Practices, and Pragmatism at University of Aalborg-Copenhagen. Thanks to Davide Nicolini for the invitation to Warwick, to Anders Buch and Ted Schatzki for organizing the Copenhagen workshop, and to participants in both events for stimulating discussions and encouragement.

I owe special gratitude to friends and colleagues whose thoughtful comments on the manuscript saved me from errors, confusions, and expressive infelicities, while also encouraging the project. Thanks to Shari Clough, Matthew Hancocks, Steven Levine, Torsten Menge, Andy Norman, Mark Okrent, and Preston Stovall. Two anonymous reviewers for the University of Chicago Press contributed thoughtful, unusually detailed reports. Their extensive, constructive comments guided substantial reconstruction of the manuscript, greatly improving its clarity, accessibility, and framing.

My understanding of human ways of life as a nexus of interdependent practices developed over a long time, incurring many intellectual debts. Annual meetings of the International Society for Phenomenological Studies were indispensable opportunities to discuss and refine many themes of the book. I thank all participants for the congeniality and philosophical richness of these meetings, with special gratitude for conversations there and elsewhere to Bill Blattner, Steve Crowell, Sacha Golob, John Haugeland, Stephan Käufer, Quill Kukla, Mark Lance, Jeff Malpas, Mark Okrent, Joan Wellman, and Kate Withy.

My interest in the relevance of evolutionary biology for social theory and science studies began in conversation with Vassiliki Betty Smocovitis and Hans-Jörg Rheinberger, and has been guided and sustained by Wesleyan Biology colleagues Sonia Sultan, Annie Burke, Fred Cohan, and Barry Chernoff, with special thanks to Annie Burke, whose graduate seminar on developmental evolution I audited in 2006.

Many other scholars contributed notably to my conception of this project, while also deserving absolution for its conclusions. Thanks to Linda Martín Alcoff, Steve Angle, Rachel Ankeny, Karen Barad, Bob Brandom, Alessandra Buccella, Shari Clough, Christina Crosby, Elena Cuffari, Dan Dennett, Bill deVries, Bert Dreyfus, John Dupre, Catherine Elgin, Brian Fay, Clifford Geertz, Gillian Goslinga, Ian Hacking, Sally Haslanger, Tony Hatch, Philipp Haueis, Deborah Heath, Steve Horst, Henry Jackman, Tom Kuhn, Quill Kukla, Lisa Lloyd, Alan Love, Jill Morawski, Lynn Hankinson Nelson, William Paris, Victoria Pitts-Taylor, Mark Risjord, Dick Rorty, Paul Roth, Ted Schatzki, Joan Scott, Sanford Shieh, Jan Slaby, Gil Skillman, Brian Cantwell Smith, Elise Springer, Alessandra Tanesini, Charles Taylor, Stephen Turner, Tom Wartenberg, Sam Wheeler, Charlotte Witt, Andrea Woody, Alison Wylie, and Jack Zammito.

Colleagues in the Philosophy Department and the Science in Society Program at Wesleyan University provide ongoing intellectual sustenance and a congenial community of scholars. Many issues and themes here were explored in courses at Wesleyan, where I benefited as always from enthusiastic and thoughtful engagement by my students.

Many audiences constructively engaged my presentations on these issues. In addition to those mentioned, I thank those attending these workshops or colloquia: Wittgenstein and Practice, University of Bergen, Norway; the Markfest honoring Mark Okrent, and an earlier talk to the Philosophy Department at Bates College, Lewiston, Maine; the Department of Philosophy at University of California at Santa Cruz; the joint meeting of European Network for Philosophy of the Social Sciences and Philosophy of Social Science

Roundtable, Hannover, Germany; the Sociology Department at Colby College, Waterville, Maine; Action, Agency, and Posthumanism, University of Aalborg-Copenhagen, Denmark; Why Rules Matter, Prague, Czech Republic; Rice/Leipzig Seminar, The Normativity of Nature/The Nature of Normativity, Houston, Texas; a three-day workshop on *Articulating the World*, Freie Universität, Berlin, Germany; a presidential research talk and talks to the Social Sciences Division and Center for the Humanities at Wesleyan University; the Department of Philosophy, Miami University; Society for the Philosophy of Science in Practice, Aarhus University, Denmark; the Philosophy Department, Rice University; Naturalism and Normativity in the Social Sciences, University of Hradec Kralove, Czech Republic; Center for Study of Mind in Nature, University of Oslo, Norway; Society for Analytical Feminism at Eastern Division, American Philosophical Association, Washington, D.C.; Temporal Externalism, York University, Toronto, Ontario; Philosophy of Social Science Roundtable, University of California, Santa Cruz; Loemker Conference on Philosophy of Sociology and Anthropology, Emory University, Atlanta, Georgia; and Practices and Social Order, Universität Bielefeld, Germany.

I continue to benefit from the advice, encouragement, and patience of Karen Merikangas Darling, the science and technology studies editor at the University of Chicago Press. I could not ask for more thoughtful and judicious editorial support.

I reserve deepest gratitude for my family. Sally Grucan's unwavering love, patience, and commitment sustained me throughout work on the book, and long before. My sons, Brian Grucan Rouse and Martin Grucan Rouse, continue to inspire me and give me hope.

References

Adas, Michael. 2014. *Machines as the Measure of Men*. 2nd ed. Ithaca: Cornell University Press.

Alder, Ken 2002. *The Measure of All Things*. New York: Free Press.

Allen, Amy, Rainer Forst, and Mark Haugaard. 2014. "Power and Reason, Justice and Domination: A Conversation." *Journal of Political Power* 7, no. 1 (March): 7–33.

Amundson, Ron. 2005. *The Changing Role of the Embryo in Evolutionary Thought*. Cambridge: Cambridge University Press.

Anderson, Elizabeth. 2010. *The Imperative of Integration*. Princeton: Princeton University Press.

Anderson, Michael. 2014. *After Phrenology*. Cambridge, MA: MIT Press.

Anderson, Warwick. 2003. *The Cultivation of Whiteness*. New York: Basic Books.

Ankeny, Rachel, and Sabina Leonelli. 2016. Repertoires: A Post-Kuhnian Perspective on Scientific Change and Collaborative Research. *Studies in History and Philosophy of Science, Part A* 60:18–28.

Anscombe, Elizabeth. 1957. *Intention*. Ithaca: Cornell University Press.

Arendt, Hannah. 1958. *The Human Condition*. 2nd ed. Chicago: University of Chicago Press.

———. 1965. *On Revolution*. New York: Viking.

———. 1970. *On Violence*. New York: Harcourt Brace Jovanovich.

Aristotle. 1941. *Metaphysics*. In *The Basic Works of Aristotle*, edited by Richard McKeon and translated by W. D. Ross, 689–934. New York: Random House.

Ásta. 2018. *Categories We Live By*. Oxford: Oxford University Press.

Avital, Eytan, and Eva Jablonka. 2000. *Animal Traditions*. Cambridge: Cambridge University Press.

Bachrach, Peter, and Morton Baratz. 1962. "Two Faces of Power." *American Political Science Review* 56, no. 4 (December): 947–52.

Bakke, Gretchen. 2016. *The Grid*. New York: Bloomsbury.

Baptist, Edward. 2014. *The Half Has Never Been Told*. New York: Basic Books.

Barad, Karen. 2007. *Meeting the Universe Halfway*. Durham: Duke University Press.

Barkow, Jerome, Leah Cosmides, and John Tooby, eds. 1992. *The Adapted Mind*. Oxford: Oxford University Press.

Barnes, Barry. 1988. *The Nature of Power*. Cambridge: Polity Press.

Bar-On, Yinon, Rob Phillips, and Ron Milo. 2018. "The Biomass Distribution on Earth." *PNAS* 115 (25): 6506–11. doi.org/10.1073/pnas.1711842115.

Bechtel, William. 1993. "Integrating Sciences by Creating New Disciplines." *Biology and Philosophy* 8:277–99.

———. 2005. *Discovering Cell Mechanisms.* Cambridge: Cambridge University Press.

Bickerton, Derek. 2008. *Bastard Tongues.* New York: Hill and Wang.

———. 2009. *Adam's Tongue.* New York: Hill and Wang.

———. 2014. *More Than Nature Needs.* Cambridge: Harvard University Press.

———. 2016. *Roots of Language.* New Edition. Berlin: Language Science Press.

Biddle, Justin, and Rebecca Kukla. 2017. "The Geography of Epistemic Risk." In *Exploring Inductive Risk*, edited by Kevin Elliott and Ted Richards, 215–37. Oxford: Oxford University Press.

Bloor, David. 1976. *Knowledge and Social Imagery.* London: Routledge.

Blue, Stanley, and Nicola Spurling. 2017. "Connective Tissue in Hospital Life." In *The Nexus of Practices*, edited by Allison Hui, Theodore Schatzki, and Elizabeth Shove, 24–37. New York: Routledge.

Boivin, Nicole, Melinda Zeder, Dorian Fuller, Alison Crowther, Gregor Larson, Jon Erlandson, Tim Denham, and Michael Petraglia. 2016. "Ecological Consequences of Human Niche Construction: Examining Long-Term Anthropogenic Shaping of Global Species Distributions." *PNAS* 113, no. 23 (June): 6388–96.

Bordenstein, Seth, and Kevin Theis. 2015. "Host Biology in Light of the Microbiome: Ten Principles of Holobionts and Hologenomes." *PLOS Biology.* https://doi.org/10.1371/journal.pbi0.1002226.

Bourdieu, Pierre. 1977. *Outline of a Theory of Practice.* Translated by Richard Nice. Cambridge: Cambridge University Press.

———. 1990. *The Logic of Practice.* Translated by Richard Nice. Stanford: Stanford University Press.

———. 1991. *Language and Symbolic Power.* Edited by John Thompson. Translated by Gino Raymond. Cambridge, MA: Harvard University Press.

Bowker, Geoffrey, and Susan Leigh Star. 1999. *Sorting Things Out.* Cambridge, MA: MIT Press.

Boyd, Robert and Peter Richerson. 1985. *Culture and the Evolutionary Process.* Chicago: University of Chicago Press.

Brandom, Robert. 1976. "Truth and Assertibility." *Journal of Philosophy* 73: 137–49.

———. 1979. Freedom and Constraint by Norms. *American Philosophical Quarterly* 16: 187–96.

———. 1983. "Asserting." *Journal of Philosophy* 17, no. 4 (November): 637–50.

———. 1994. *Making It Explicit.* Cambridge: Harvard University Press.

———. 2000. *Articulating Reasons.* Cambridge, MA: Harvard University Press.

———. 2002. *Tales of the Mighty Dead.* Cambridge, MA: Harvard University Press.

———. 2019. *A Spirit of Trust.* Cambridge, MA: Harvard University Press.

Brandon, Robert, and Janis Antonovics. 1996. "The Coevolution of Organism and Environment." In *Concepts and Methods in Evolutionary Biology*, edited by Robert Brandon, 161–78. Cambridge: Cambridge University Press.

Brown, Matthew. 2017. "Placing Power in Practice Theory." In *The Nexus of Practices*, edited by Allison Hui, Theodore Schatzki, and Elizabeth Shove, 169–82. New York: Routledge.

Budiansky, Steven. 1999. *The Covenant of the Wild.* New Haven: Yale University Press.

Buch, Anders, and Theodore Schatzki, eds. 2019. *Questions of Practice in Philosophy and Social Theory.* New York: Routledge.

Buller, David. 2005. *Adapting Minds*. Cambridge, MA: MIT Press.

Burge, Tyler. 1979. "Individualism and the Mental." *Midwest Studies in Philosophy* 4, no. 1 (September): 73–122.

Buss, David. 1999. *Evolutionary Psychology*. Boston: Allyn and Bacon.

Butler, Judith. 1990. *Gender Trouble*. New York: Routledge.

———. 1993. *Bodies that Matter*. New York: Routledge.

Callon, Michel. 1986. "Elements of a Sociology of Translation: Domestication of the Scallops and the Fishermen of St Brieuc Bay." In *Power, Action and Belief*, edited by John Law, 196–233. London: Routledge & Kegan Paul.

———. 1987. "Society in the Making: the Study of Technology as a Tool for Sociological Analysis." In *The Social Construction of Technological Systems*, edited by Wiebe Bijker, Thomas Hughes, and Trevor Pinch, 83–103. Cambridge, MA: MIT Press.

———. 1991. "Techno-economic Networks and Irreversibility." In *A Sociology of Monsters*, edited by John Law, 132–165. London: Routledge.

Cantor, Norman. 2001. *In the Wake of the Plague*. New York: Simon and Schuster.

Caro, Robert. 1974. *The Power Broker*. New York: Alfred Knopf.

Carr, David. 1986. *Time, Narrative, and History*. Bloomington: Indiana University Press.

Cartwright, Nancy. 1983. *How the Laws of Physics Lie*. Oxford: Clarendon Press.

———. 2003. "From Causation to Explanation and Back." In *The Future for Philosophy*, edited by Brian Leiter, 230–45. Oxford: Oxford University Press.

———. 2019. *Nature, the Artful Modeler*. Chicago: Open Court.

Cavalli-Sforza, Luca, and Marcus Feldman. 1981. *Cultural Transmission and Evolution*. Princeton: Princeton University Press.

Chemero, Anthony. 2009. *Radical Embodied Cognitive Science*. Cambridge, MA: MIT Press.

Cheney, Dorothy, and Robert Seyfarth. 1990. *How Monkeys See the World*. Chicago: University of Chicago Press.

Cheney, Dorothy, and Robert Seyfarth. 2007. *Baboon Metaphysics*. Chicago: University of Chicago Press.

Clifford, James, and George Marcus, eds. 1986. *Writing Culture*. Berkeley, CA: University of California Press.

Coates, Ta-Nehisi. 2015. *Between the World and Me*. New York: Random House.

Collins, Harry, and Steven Yearley. 1992. "Epistemological Chicken." In *Science as Practice and Culture*, edited by Andrew Pickering, 301–26. Chicago: University of Chicago Press.

Cosmides, Leah, and John Tooby. 1987. "From Evolution to Behavior: Evolutionary Psychology as the Missing Link." In *The Latest on the Best*, edited by John Dupre, 277–307. Cambridge, MA: MIT Press.

Crane, Tim. 1992. "The Nonconceptual Content of Experience." In *The Contents of Experience*, edited by Tim Crane, 136–57. Cambridge: Cambridge University Press.

Creary, Nicholas, ed. 2012. *African Intellectuals and Decolonization*. Athens, OH: Ohio University Press.

Cronk, Lee, Napoleon Chagnon, and William Irons, eds. 2000. *Adaptation and Behavior*. New York: de Gruyter.

Cronk, Lee, and Beth Leech. 2013. *Meeting at Grand Central*. Princeton: Princeton University Press.

Cronon, William. 1991. *Nature's Metropolis*. New York: W. W. Norton.

Crosby, Alfred. 1986. *Ecological Imperialism*. Cambridge: Cambridge University Press.

Crosby, Christina. 2016. *A Body, Undone*. New York: NYU Press.

Cummins, Robert. 2002. "Haugeland on Representation and Intentionality." In *Philosophy of Mental Representation*, edited by Hugh Clapin, 122–37. Oxford: Oxford University Press.

Dahl, Robert. 1957. "The Concept of Power." *Behavioral Science* 2 (July): 201–15.

———. 1958. "A Critique of the Ruling Elite Model." *American Political Science Review* 52, no. 2 (June) 463–69.

———. 1961. *Who Governs?* New Haven: Yale University Press.

Daston, Lorraine. 1992. "Objectivity and the Escape from Perspective." *Social Studies of Science* 22, no. 4 (November): 597–618.

Daston, Lorraine, and Peter Galison. 1992. "The Image of Objectivity." *Representations* 40:81–128.

———. 2007. *Objectivity*. New York: Zone Books.

Davidson, Donald. 1980. *Essays on Action and Events*. Oxford: Oxford University Press.

———. 1984. *Inquiries into Truth and Interpretation*. Oxford: Oxford University Press.

———. 1986. "A Nice Derangement of Epitaphs." In *Truth and Interpretation*, edited by Ernest LePore, 433–46. Oxford: Blackwell.

———. 2005a. *Truth and Predication*. Cambridge, MA: Harvard University Press.

———. 2005b. *Truth, Language, and History*. Oxford: Oxford University Press.

Dawkins, Richard. 1982. *The Extended Phenotype*. Oxford: Oxford University Press.

Deacon, Terrence. 1997. *The Symbolic Species*. New York: Norton.

De Certeau, Michel. 1984. *The Practice of Everyday Life*. Translated by Steven Rendall. Berkeley, CA: University of California Press.

Dennett, Daniel. 1987. *The Intentional Stance*. Cambridge, MA: MIT Press.

———. 1991a. *Consciousness Explained*. New York: Little, Brown.

———. 1991b. "Real Patterns." *Journal of Philosophy* 88, no. 1 (January): 27–51.

———. 1995. *Darwin's Dangerous Idea*. New York: Simon and Schuster.

———. 2013. "Kinds of Things—Toward a Bestiary of the Manifest Image." In *Scientific Metaphysics*, edited by Don Ross, James Ladyman, and Harold Kincaid, 96–107. Oxford: Oxford University Press.

———. 2014. "The Evolution of Reasons." In *Contemporary Philosophical Naturalism and its Implications*, edited by B. Bashour and H. Muller, 47–62. New York: Routledge.

———. 2020. "Herding Cats and Free Will Inflation." *Proceedings of the American Philosophical Association* 94:149–63.

de Waal, Frans. 2016. *Are We Smart Enough to Know How Smart Animals Are?* New York: W. W. Norton.

Diamond, Jared. 1998. *Guns, Germs, and Steel*. New York: W. W. Norton.

Di Paolo, Ezequiel. 2005. "Autopoiesis, Adaptivity, Teleology, Agency." *Phenomenology and the Cognitive Sciences* 4:429–52.

Di Paolo, Ezequiel, Elena Cuffari, and Hanne De Jaegher. 2018. *Linguistic Bodies*. Cambridge, MA: MIT Press.

Dor, Daniel. 2015. *The Instruction of Imagination*. Oxford: Oxford University Press.

Dor, Daniel, and Eva Jablonka. 2000. "From Cultural Selection to Genetic Selection: A Framework for the Evolution of Language." *Selection* 1, nos. 1–3 (January): 33–55.

———. 2001. "How Language Changed the Genes." In *New Essays on the Origin of Language*, edited by Jürgen Trabant and Sean Ward, 149–75. Berlin: de Gruyter.

———. 2004. "Culture and Genes in the Evolution of Human Language." In *Human Paleoecology in the Levantine Corridor*, edited by N. Goren-Inbar and J. Speth, 105–15. Oxford: Oxbow Press.

———. 2010. "Plasticity and Canalization in the Evolution of Linguistic Communication: An Evolutionary-Developmental Approach." In *The Evolution of Human Language*, edited by Viviane Déprez, Richard Larson, and Hiroko Yamakido, 135–47. Cambridge: Cambridge University Press.

Douglas, Heather. 2000. "Inductive Risk and Values in Science." *Philosophy of Science* 67, no. 4 (December): 559–79.

———. 2009. *Science, Policy, and the Value-Free Ideal*. Pittsburgh: University of Pittsburgh Press.

Dreyfus, Hubert. 1979. *What Computers Can't Do*. New York: Harper & Row.

———. 1991. *Being-in-the-World*. Cambridge: MIT Press.

———. 2002. "Responses." In *Heidegger, Coping, and Cognitive Science*, edited by Mark Wrathall and Jeff Malpas, 313–49. Cambridge: MIT Press.

———. 2014. *Skillful Coping*, edited by Mark Wrathall. Oxford: Oxford University Press.

Dreyfus, Hubert, and Stuart Dreyfus 1986. *Mind Over Machine*. New York: Free Press.

Driver, Julia. 2012. *Consequentialism*. New York: Routledge.

Dupré, John. 2004. "Understanding Contemporary Genomics." *Perspectives on Science* 12, no. 3 (Fall): 320–38.

———. 2012. *Processes of Life*. Oxford: Oxford University Press.

———. 2021. *The Metaphysics of Biology*. Cambridge: Cambridge University Press.

Dupré, John, and Daniel Nicholson, eds. 2018. *Everything Flows*. Oxford: Oxford University Press.

Duster, Troy. 2003. *Backdoor to Eugenics*. 2nd ed. New York: Routledge.

Ebbs, Gary. 2000. "The Very Idea of Sameness of Extension Across Time." *American Philosophical Quarterly* 37, no. 3 (July): 245–68.

———. 2009. *Truth and Words*. Oxford: Oxford University Press.

Edelman, Lee. 2004. *No Future*. Durham, NC: Duke University Press.

Edwards, Paul. 2010. *A Vast Machine*. Cambridge, MA: MIT Press.

Edwards, Paul, Steven Jackson, Geoffrey Bowker, and Cory Knobel, eds. 2007. *Understanding Infrastructure*. Ann Arbor: Deep Blue.

Eisenstein, Elizabeth. 1979. *The Printing Press as Agent of Change*. Cambridge: Cambridge University Press.

Epstein, Brian 2015. *The Ant Trap*. Oxford: Oxford University Press.

Evans, Gareth 1982. *The Varieties of Reference*. Edited by J. McDowell. Oxford: Oxford University Press.

Fausto-Sterling, Anne. 1992. *Myths of Gender*. 2nd ed. New York: Basic Books.

Fay, Brian. 1987. *Critical Social Science*. Ithaca: Cornell University Press.

———. 1996. *Contemporary Philosophy of Social Science*. Oxford: Blackwell.

Feldman, Marcus, and Luca Cavalli-Sforza. 1976. "On the Theory of Evolution under Genetic and Cultural Transmission with Application to the Lactose Absorption Problem." In *Mathematical Evolutionary Theory*, edited by Marcus Feldman, 145–73. Princeton: Princeton University Press.

Fine, Arthur. 1986. *The Shaky Game*. Chicago: University of Chicago Press.

Fine, Cordelia. 2010. *Delusions of Gender*. New York: W. W. Norton.

Foucault, Michel. 1971. *The Order of Things*. Translated by Alan Sheridan. New York: Pantheon.

———. 1977. *Discipline and Punish*. Translated by Alan Sheridan. New York: Random House.

———. 1978. *The History of Sexuality*, vol. 1, *An Introduction*. Translated by Robert Hurley. New York: Random House.

———. 1982. "The Subject and Power." In *Michel Foucault*, edited by Hubert Dreyfus and Paul Rabinow, 208–26. Chicago: University of Chicago Press.

———. 2003. *Society Must Be Defended*. Translated by David Macey. New York: St. Martin's Press.

Fraser, Nancy. 1981. "Foucault on Modern Power: Empirical Insights and Normative Confusions." *Praxis International* 1, no. 3 (October): 272–87.

Freire, Paolo. 1970. *Pedagogy of the Oppressed*. New York: Herder and Herder.

Fricker, Miranda. 2007. *Epistemic Injustice*. Oxford: Oxford University Press.

Frye, Marilyn. 1983. *The Politics of Reality*. New York: Crossing Press.

Galison, Peter. 1997. *Image and Logic*. Chicago: University of Chicago Press.

———. 1998. "Feynman's War: Modeling Weapons, Modeling Nature." *Studies in History and Philosophy of Modern Physics* 29, no. 3 (September): 391–434.

———. 2003. *Einstein's Clocks, Poincaré's Maps*. New York: W. W. Norton.

Gallagher, Shaun. 2017. *Enactivist Interventions*. Oxford: Oxford University Press.

———. 2020. *Action and Interaction*. Oxford: Oxford University Press.

Garfinkel, Harold. 1967. *Studies in Ethnomethodology*. Atlantic Highlands, NJ: Prentice-Hall.

Garrett, Laurie. 1994. *The Coming Plague*. New York: Farrar, Straus & Giroux.

Gaventa, John. 1980. *Power and Powerlessness*. Urbana, IL: University of Illinois Press.

Geertz, Clifford. 1973. *The Interpretation of Cultures*. New York: Basic Books.

Gibbard, Allan. 1990. *Wise Choices, Apt Feelings*. Oxford: Oxford University Press.

Gibson, James. 1979. *The Ecological Approach to Visual Perception*. Boston: Houghton Mifflin.

Giddens, Anthony. 1984. *The Constitution of Society*. Berkeley: University of California Press.

Gilbert, Scott. 2002. "The Genome in its Ecological Context: Philosophical Perspectives on Interspecies Epigenesis." *Annals of the New York Academy of Science* 981 (December): 202–18.

———. 2016. "Developmental Plasticity and Developmental Symbiosis: The Return of Eco-Devo." *Current Topics in Developmental Biology* 116:415–33.

———. 2017. "Holobiont by Birth: Multilineage Individuals as the Concretion of Cooperative Practices." In *Arts of Living on a Damaged Planet*, edited by Anna Tsing, Heather Swanson, Elaine Gan, and Nils Bobanoff, M73–M89. Minneapolis: University of Minnesota Press.

Gilbert, Scott, and David Epel. 2009. *Ecological Developmental Biology*. Sunderland, MA: Sinauer Associates.

Gilbert, Scott, John Opitz, and Rudolf Rapp.1996. "Resynthesizing Evolutionary and Developmental Biology." *Developmental Biology* 173, no. 2 (February): 357–72.

Gilbert, Scott, Jan Sapp, and Alfred Tauber. 2012. "A Symbiotic View of Life: We Have Never Been Individuals." *Quarterly Review of Biology* 87, no. 4 (December): 325–41.

Giroux, Henry. 2020. *On Critical Pedagogy*. 2nd ed. London: Bloomsbury Academic.

Glasgow, Joshua, Sally Haslanger, Chike Jeffers, and Quayshawn Spencer. 2019. *What is Race?* Oxford: Oxford University Press.

Godfrey-Smith, Peter. 1996. *Complexity and the Function of Mind in Nature*. Cambridge, MA: MIT Press.

———. 2001. "On the Status and Explanatory Structure of Developmental Systems Theory." In *Cycles of Contingency*, edited by Susan Oyama, Paul Griffiths, and Russell Gray, 283–97. Cambridge, MA: MIT Press.

———. 2002. "On the Evolution of Representational and Interpretive Capacities." *The Monist* 85, no. 1 (January): 50–69.

Goffman, Irving. 1959. *The Presentation of Self in Everyday Life*. New York: Doubleday.

———. 1983. "The Interaction Order." *American Sociological Review* 48, no. 1 (February): 1–17.

Goodman, Nelson. 1954. *Fact, Fiction, and Forecast*. Cambridge, MA: Harvard University Press.

Goody, Jack. 1977. *The Domestication of the Savage Mind*. Cambridge: Cambridge University Press.

Gould, Stephen Jay. 1996. *The Mismeasure of Man*. 2nd ed. New York: W. W. Norton.

Goulson, Dave. 2019. "The Insect Apocalypse, and Why it Matters." *Current Biology* 29, no. 19 (October): R967–R971. https://doi.org/10.1016/j.cub.2019.06.069.

Graeber, David, and David Wengrow. 2021. *The Dawn of Everything*. New York: Farrar, Straus & Giroux.

Gray, Russell 1992. "Death of the Gene." In *Trees of Life*, edited by Paul Griffiths, 165–209. Dordrecht: Kluwer.

Grice, H. Paul. 1988. *Studies in the Ways of Words*. Cambridge: Harvard University Press.

Griffiths, Paul, and Russell Gray. 2001. "Darwinism and Developmental Systems." In *Cycles of Contingency*, edited by Susan Oyama, Paul Griffiths, and Russell Gray, 195–218. Cambridge, MA: MIT Press.

Griffiths, Paul, and Stotz, Karola. 2018. "Developmental Systems Theory as a Process Theory." In *Everything Flows*, edited by John Dupré and Daniel Nicholson, 225–45. Oxford: Oxford University Press.

Grover, Dorothy, Joseph Camp, and Nuel Belnap. 1975. "A Prosentential Theory of Truth." *Philosophical Studies* 27, no. 1 (February): 73–125.

Guala, Francesco. 2016. *Understanding Institutions*. Princeton: Princeton University Press.

Habermas, Jürgen. 1970. *Toward a Rational Society*. Boston: Beacon Press.

———. 1986. "The Genealogical Writing of History: On Some Aporias in Foucault's Theory of Power." *Canadian Journal of Political and Social Theory* 10:1–9.

———. 1987. *The Philosophical Discourse of Modernity*. Translated by F. Lawrence. Cambridge, MA: MIT Press.

Hacking, Ian. 1983. *Representing and Intervening*. Cambridge: Cambridge University Press.

———. 2002. *Historical Ontology*. Cambridge, MA: Harvard University Press.

Halberstam, Judith. 2005. *In a Queer Time and Place*. New York: New York University Press.

Hammerstein, Peter, ed. 2003. *Genetic and Cultural Evolution of Cooperation*. Cambridge, MA: MIT Press.

Hansell, Michael. 1984. *Animal Architecture*. Oxford: Oxford University Press.

———. 2007. *Built by Animals*. Oxford: Oxford University Press.

Haraway, Donna. 1989. *Primate Visions*. New York: Routledge.

———. 1997. *Modest_Witness@Second_Millenium.FemaleMan©_Meets_ Oncomouse™: Feminism and Technoscience*. New York: Routledge.

———. 2008. *When Species Meet*. Minneapolis, MN: University of Minnesota Press.

Hartman, Saidya. 1997. *Scenes of Subjection*. Oxford: Oxford University Press.

Haslanger, Sally. 2012. *Resisting Reality*. Oxford: Oxford University Press.

———. 2018. "What Is a Social Practice?" *Philosophy* 82 (July): 231–47.

———. 2021. *Ideology in Practice*. Milwaukee: Marquette University Press.

Hatch, Anthony. 2016. *Blood Sugar*. Minneapolis: University of Minnesota Press.

———. 2019. *Silent Cells*. Minneapolis: University of Minnesota Press.

Haugaard, Mark. 2010. "Power: A Family Resemblance Concept." *European Journal of Cultural Studies* 13 (4): 419–38.

Haugeland, John. 1998. *Having Thought*. Cambridge, MA: Harvard University Press.

———. 2002. "Reply to Cummins on Representation and Intentionality." In *Philosophy of Mental Representation*, edited by Hugh Clapin, 138–44. Oxford: Oxford University Press.

———. 2013. *Dasein Disclosed*. Cambridge, MA: Harvard University Press.

Hauser, Marc. 1996. *The Evolution of Communication*. Cambridge, MA: MIT Press.

Hayward, Clarissa. 2000. *De-Facing Power*. Cambridge: Cambridge University Press.

Hayward, Clarissa, and Steven Lukes. 2008. "Nobody to Shoot? Power, Structure, and Agency: A Dialogue." *Journal of Power* 1:5–20.

Hegel, Georg W. F. 2018. *The Phenomenology of Spirit*. Translated by Terry Pinkard. Cambridge: Cambridge University Press.

Heidegger, Martin. 1962. *Being and Time*. Translated by John Macquarrie and Edward Robinson. New York: Harper & Row.

———. 1975. *Basic Problems of Phenomenology*. Translated by Albert Hofstadter. Bloomington: Indiana University Press.

Herlihy, David. 1997. *The Black Death and the Transformation of the West*. Cambridge, MA: Harvard University Press.

Herman, Barbara. 2022. *Kantian Commitments*. Oxford: Oxford University Press.

Heyes, Cecilia. 2012. "New Thinking: The Evolution of Human Cognition." *Philosophical Transactions of the Royal Society B* 367, no. 1599 (August): 2191–96. doi:10.1098/rstb.2012.01111.

———. 2018. *Cognitive Gadgets*. Cambridge, MA: Harvard University Press.

Hinde, Robert. 1974. *Biological Bases of Human Social Behavior*. New York: McGraw-Hill.

Hitchcock, Christopher. 2003. "Of Humean Bondage." *British Journal for the Philosophy of Science* 54, no. 1 (March): 1–25.

Hobbes, Thomas. 1994. *Leviathan*. Edited by Edward Curley. Indianapolis: Hackett.

Holden, Clare, and Ruth Mace. 1997. "Phylogenetic Analysis of the Evolution of Lactose Tolerance in Adults." *Human Biology* 69:605–28.

Honneth, Axel. 1991. *The Critique of Power*. Translated by Kenneth Baynes. Cambridge, MA: MIT Press.

Hrdy, Sarah Blaffer. 2009. *Mothers and Others*. Cambridge, MA: Harvard University Press.

Hubbard, Ruth. 1990. *The Politics of Women's Biology*. New Brunswick, NJ: Rutgers University Press.

Hughes, Thomas. 1983. *Networks of Power*. Baltimore, MD: Johns Hopkins University Press.

Hui, Allison, Theodore Schatzki, and Elizabeth Shove, eds. 2017. *The Nexus of Practices*. New York: Routledge.

Hume, David. 1993. *An Enquiry Concerning Human Understanding*. 2nd ed. Indianapolis: Hackett.

Hunter, Floyd. 1953. *Community Power Structure*. Chapel Hill: University of North Carolina Press.

Ingold, Tim. 1995. "'People Like Us': The Concept of the Anatomically Modern Human." *Cultural Dynamics* 7, no. 2 (July): 187–214.

———. 2001. "From Complementarity to Obviation: On Dissolving the Boundaries Between Social and Biological Anthropology, Archaeology, and Psychology." In *Cycles of Contingency*, edited by Susan Oyama, Paul Griffiths, and Russell Gray, 255–79. Cambridge, MA: MIT Press.

———. 2022. "Evolution Without Inheritance: Steps to an Ecology of Learning." *Current Anthropology* 63 (S25): S32–S55.

Ingold, Tim, and Gisli Palsson. 2013. *Biosocial Becomings*. Cambridge: Cambridge University Press.

Isaac, Jeffrey. 1987. "Beyond the Three Faces of Power: A Realist Critique." *Polity* 20, no. 1 (Autumn): 4–30.

Jablonka, Eva, and Marion Lamb. 2005. *Evolution in Four Dimensions*. Cambridge, MA: MIT Press.

———. 2020. *Inheritance Systems and the Extended Evolutionary Synthesis*. Cambridge: Cambridge University Press.

Jackman, Henry. 1999. "We Live Forwards But Understand Backwards: Linguistic Practices and Future Behavior." *Pacific Philosophical Quarterly* 80:157–77.

Johns, Adrian. 1998. *The Nature of the Book*. Chicago: University of Chicago Press.

Kafer, Alison. 2013. *Feminist, Queer, Crip*. Bloomington: Indiana University Press.

Kant, Immanuel. 1993. *Grounding for the Metaphysics of Morals*. Translated by J. Ellington. Indianapolis: Hackett Publishing.

———. 1998. *Critique of Pure Reason*. Translated by Paul Guyer and Allan Wood. Cambridge: Cambridge University Press.

———. 2000. *Critique of the Power of Judgment*. Translated by Paul Guyer and Eric Mathews. Cambridge: Cambridge University Press.

Kenny, Charles. 2021. *The Plague Cycle*. New York: Scribner.

Kidd, Ian, José Medina, and Gaile Pohlhause Jr., eds. 2019. *Routledge Handbook of Epistemic Injustice*. New York: Routledge.

Kitcher, Philip. 1985. *Vaulting Ambition: Sociobiology and the Quest for Human Nature*. Cambridge, MA: MIT Press.

Klein, Ursula. 2003. *Experiments, Models, Paper Tools*. Stanford: Stanford University Press.

Kohler, Robert. 1994. *Lords of the Fly*. Chicago: University of Chicago Press.

Kolbert, Elizabeth. 2014. *The Sixth Extinction*. New York: Henry Holt.

Korsgaard, Christine. 2009. *Self-Constitution*. Oxford: Oxford University Press.

Kripke, Saul. 1980. *Naming and Necessity*. Cambridge, MA: Harvard University Press.

Kripke, Saul. 1982. *Wittgenstein on Rules and Private Language*. Cambridge, MA: Harvard University Press.

Kroeber, A. L., and Clyde Kluckhohn. 1963. *Culture*. New York: Vintage.

Kuhn, Thomas. 1970. *The Structure of Scientific Revolutions*. 2nd ed. Chicago: University of Chicago Press.

Kukla, Quill Rebecca, 2000. "Myth, Memory and Misrecognition in Sellars' 'Empiricism and the Philosophy of Mind.'" *Philosophical Studies* 101, nos. 2–3 (December): 161–211.

———. 2002. "The Ontology and Temporality of Conscience." *Continental Philosophy Review* 35 (March): 1–34.

———. 2006. "Objectivity and Perspective in Empirical Knowledge." *Episteme* 3, nos. 1–2 (January): 80–95.

———. 2008. "Medicalization, 'Normal Function,' and the Definition of Health." In *Routledge Companion to Bioethics*, edited by John Arras, E. Fenton, and Rebecca Kukla, 513–30. New York: Routledge.

———. 2012. "Performative Force, Convention, and Discursive Injustice." *Hypatia* 29, no. 2 (Spring): 1–17.

———. 2017. "Ostension and Assertion." In *Giving a Damn*, edited by Zed Adams and Jacob Browning, 103–30. Cambridge, MA: MIT Press.

———. 2018. "Embodied Stances: Realism without Literalism." In *The Philosophy of Daniel Dennett*, edited by Bryce Huebner, 3–31. Oxford: Oxford University Press.

———. 2021a. *City Living*. Oxford: Oxford University Press.

———. 2021b. "Situated Knowledge, Purity, and Moral Panic." In *Applied Epistemology*, edited by Jennifer Lackey, 37–66. Oxford: Oxford University Press.

Kukla, Quill Rebecca, and Mark Lance. 2009. *Yo! and Lo!* Cambridge, MA: Harvard University Press.

———. 2014. "Intersubjectivity and Receptive Experience." *Southern Journal of Philosophy* 52, no. 1 (March): 22–42.

Lakoff, George, and Mark Johnson. 1980. *Metaphors We Live By*. Chicago: University of Chicago Press.

Laland, Kevin. 2017. *Darwin's Unfinished Symphony*. Princeton: Princeton University Press.

Laland, Kevin, and Gillian Brown. 2011. *Sense and Nonsense: Evolutionary Perspectives on Human Behavior*. 2nd ed. Oxford: Oxford University Press.

Laland, Kevin, John Odling-Smee, and Marcus Feldman. 2001. "Niche Construction, Ecological Inheritance, and Cycles of Contingency in Evolution." In *Cycles of Contingency*, edited by Susan Oyama, Paul Griffiths, and Russell Gray, 117–26. Cambridge, MA: MIT Press.

Laland, Kevin, John Odling-Smee, and Scott Gilbert. 2008. "EvoDevo and Niche Construction: Building Bridges." *Journal of Experimental Zoology (Mol Dev Evol)* 310, no. 7 (August): 549–66.

Laland, Kevin, Tobias Uller, Marcus Feldman, Kim Sterelny, Gerd Müller, Armin Moczek, Eva Jablonka, and John Odling-Smee, 2014. "Does Evolutionary Theory Need a Rethink? Yes, Urgently." *Nature* 514 (October): 161–64.

———. 2015. "The Extended Evolutionary Synthesis: Its Structure, Assumptions, and Predictions." *Proceedings of the Royal Society B* 282, no. 1813 (August): 1019.

Lance, Mark. 1997. "The Significance of Anaphoric Theories of Truth and Reference." *Philosophical Issues* 8:181–98.

———. 1998. "Some Reflections on the Sport of Language." *Philosophical Perspectives* 12:219–40.

———. 2000. "The Word Made Flesh: Toward a Neo-Sellarsian View of Concepts and Their Analysis." *Acta Analytica* 15, no. 25: 117–135.

———. 2015. "Life is Not a Box Score: Lived Normativity, Abstract Evaluation, and the Is/Ought Distinction." In *Meaning without Representation*, edited by Steven Gross, Nicholas Tebben, and Michael Williams, 279–305. Oxford: Oxford University Press.

———. 2017. "Language Embodied and Embedded: Walking the Talk." In *Giving a Damn*, edited by Zed Adams and Jacob Browning, 161–86. Cambridge, MA: MIT Press.

Lance, Mark, and Andrew Blitzer. 2012. "Phenomenology without Ontology." Unpublished paper presented at the American Philosophical Association Eastern Division Meeting, December 28, 2012.

Landecker, Hannah, and Aaron Panofsky. 2013. "From Social Structure to Gene Regulation and Back: A Critical Introduction to Environmental Epigenetics for Sociology." *Annual Review of Sociology* 39:333–57.

Landes, David. 1983. *Revolution in Time*. Cambridge, MA: Harvard University Press.

Lane, Nick. 2015. *The Vital Question*. New York: W. W. Norton.

Lange, Marc. 2002. *Natural Laws in Scientific Practice*. Oxford: Oxford University Press.

———. 2007. "Laws and Theories." In *A Companion to Philosophy of Biology*, edited by Sahotra Sarkar and Anya Plutinski, 489–505. Oxford: Blackwell.

Latour, Bruno. 1983. "Give Me a Laboratory and I Will Raise the World." In *Science Observed*, edited by Karin Knorr-Cetina and Michael Mulkay, 141–70. London: Sage.

———. 1987. *Science in Action*. Cambridge, MA: Harvard University Press.

———. 2005. *Reassembling the Social*. Oxford: Oxford University Press.

Lawson, Tony. 2019. *The Nature of Social Reality*. New York: Routledge.

Lear, Jonathan. 2006. *Radical Hope*. Cambridge, MA: Harvard University Press.

Lenoir, Timothy. 1997. *Instituting Science*. Stanford: Stanford University Press.

Lewens, Tim. *Organisms and Artifacts*. Cambridge, MA: MIT Press.

———. 2015. *Cultural Evolution*. Oxford: Oxford University Press.

Lewontin, Richard. 1974. "The Analysis of Variance and the Analysis of Causes." *American Journal of Human Genetics* 26, no. 3 (May): 400–11.

———. 1982. "Organism and Environment." In *Learning, Development and Culture*, edited by Henry C. Plotkin, 151–70. New York: Wiley.

———. 1983. "Gene, Organism, and Environment." In *Evolution*, edited by D. S. Bendall, 273–85. Cambridge: Cambridge University Press.

———. 2001a. "Gene, Organism, and Environment: A New Introduction." In *Cycles of Contingency*, edited by Susan Oyama, Paul Griffiths, and Russell Gray, 55–57. Cambridge, MA: MIT Press.

———. 2001b. *The Triple Helix*. Cambridge, MA: Harvard University Press.

Lewontin, Richard, Steven Rose, and Leon Kamin. 1984. *Not in Our Genes*. New York: Pantheon.

Linderman, Frank B. 1962. *Plenty-Coups: Chief of the Crows*, Lincoln: University of Nebraska Press.

Lloyd, Elisabeth. 2004. "Kanzi, Evolution and Language." *Biology and Philosophy* 19 (September): 577–88.

Lloyd, Elisabeth, and Marcus Feldman. 2002. "Evolutionary Psychology: A View from Evolutionary Biology." *Psychological Inquiry* 13, no. 2 (April): 150–56.

Longino, Helen. 2013. *Studying Human Behavior*. Chicago: University of Chicago Press.

Love, Alan. 2006. "Evolutionary Morphology and Evo-Devo: Hierarchy and Novelty." *Theory in Biosciences* 124 (3–4): 317–33.

Love, Alan, and William Wimsatt, eds. 2019. *Beyond the Meme*. Minneapolis: University of Minnesota Press.

Lovelock, James. 1979. *Gaia*. Oxford: Oxford University Press.

Lukes, Steven. 2005. *Power*. 2nd ed. New York: Palgrave MacMillan.

Machiavelli, Niccolò. 1988. *The Prince*. Edited by Quentin Skinner. Cambridge: Cambridge University Press.

MacIntyre, Alasdair. 1981. *After Virtue*. Notre Dame: University of Notre Dame Press.

———. 1988. *Whose Justice? Which Rationality?* Notre Dame: University of Notre Dame Press.

Malone, Thomas, and Gary Yohe. 1992. "Toward a General Methodology with which to Evaluate the Social and Economic Impact of Global Change." *Global Environmental Change* 3:101–10.

Marcus, George, and Michael Fischer. 1986. *Anthropology as Cultural Critique*. Chicago: University of Chicago Press.

Margulis, Lynn. 1998. *Symbiotic Planet*. New York: Basic Books.

Martin, Emily. 1991. "The Egg and the Sperm: How Science Has Constructed a Romance Based on Stereotypical Male-Female Roles." *Signs* 16, no. 3 (Spring): 485–501.

———. 1998. "Anthropology and the Cultural Study of Science." *Science, Technology, and Human Values* 23, no. 1 (Winter): 24–44.

Marx, Karl. 1973. *Grundrisse*. Translated by Martin Nicolaus. New York: Random House.

Mauss, Marcel. 1979. "Body Techniques." In *Sociology and Psychology*, translated by Ben Brewster, 97–135. London: Routledge & Kegan Paul.

May, Todd. 1994. *The Political Philosophy of Poststructuralist Anarchism*. State College, PA: Pennsylvania State University Press.

Maynard Smith, John. 1978. "Optimization Theory in Evolution." *Annual Review of Ecology and Systematics* 9: 31–56.

Maynard Smith, John. 1982. *Evolution and the Theory of Games*. Cambridge: Cambridge University Press.

McDowell, John. 1994. *Mind and World*. Cambridge, MA: Harvard University Press.

McDowell, John. 2010. "Brandom on Observation." In *Reading Brandom*, edited by Jeremy Wanderer and Bernard Weiss, 129–44. New York: Routledge.

McFall-Ngai, Margaret, Michael G. Hadfield, Thomas C. G. Bosch, and Jennifer J. Wernegreen. 2013. "Animals in a Bacterial World, a New Imperative for the Life Sciences." *PNAS* 110, no. 9 (January): 3229–36.

McMenamin, Mark, and Dianna McMenamin. 1994. *Hypersea*. Columbia University Press.

McNeill, William. 1976. *Plagues and Peoples*. Garden City, NJ: Anchor.

Meloni, Maurizio, John Cromby, Des Fitzgerald, and Stephanie Lloyd, eds. 2018. *The Palgrave Handbook of Biology and Society*. London: Palgrave MacMillan.

Menge, Torsten. 2015. *The Power of Genealogy*. PhD diss., Georgetown University.

———. 2018. "The Role of Power in Social Explanation." *European Journal of Social Theory* 21, no. 1 (July): 22–38.

———. 2019. "Violence and the Materiality of Power." *Critical Review of International Social and Political Philosophy*. https://doi/full/10.1080/13698230.2019.1700344.

———. 2020. "Fictional Expectations and the Ontology of Power." *Philosophers' Imprint* 20 (29): 1–22.

Merleau-Ponty, Maurice. 2012. *Phenomenology of Perception*. Translated by Donald Landes. New York: Routledge.

Millikan, Ruth Garrett. 1995. "Pushmi-Pullyu Representations." *Philosophical Perspectives* 9: 185–200.

Mills, C. Wright. 1956. *The Power Elite*. Oxford: Oxford University Press.

Mills, Charles. 1997. *The Racial Contract*. Ithaca: Cornell University Press.

———. 2017. *Black Rights/White Wrongs*. Oxford: Oxford University Press.

Mills, Susan, and John Beatty. 1979. "The Propensity Interpretation of Fitness." *Philosophy of Science* 46, no. 2 (June): 263–86.

Mintz, Sydney. 1985. *Sweetness and Power*. New York: Viking.

Mishra, Pankaj. 2013. *From the Ruins of Empire*. New York: Picador.

———. 2020. *Bland Fanatics*. London: Verso.

Mitchell, Timothy. 2011. *Carbon Democracy*. London: Verso.

Mol, Anne-Marie. 2002. *The Body Multiple*. Durham, NC: Duke University Press.

Morgan, T. J. H., N. T. Uomini, L. E. Rendell, L. Chouinard-Thuly, S. E. Street, H. M. Lewis, C. P. Cross, C. Evans, R. Kearney, I. de la Torre, A. Whiten, and K. N. Laland. 2015. "Experimental Evidence for the Co-Evolution of Hominin Tool-Making, Teaching, and Language." *Nature Communications* 6. doi:10.1038/ncomms7029.

Morriss, Peter. 2002. *Power*. 2nd ed. Manchester: Manchester University Press.

Müller-Wille, Staffan, and Hans-Jörg Rheinberger. 2012. *A Cultural History of Heredity*. Chicago: University of Chicago Press.

Mumford, Lewis. 1934. *Technics and Civilization*. New York: Harcourt.

Nelson, Alondra. 2016. *The Social Life of DNA*. Boston: Beacon Press.

Ng, Karen. 2020. *Hegel's Concept of Life*. Oxford: Oxford University Press.

Nicolini, Davide. 2012. *Practice Theory, Work, and Organization*. Oxford: Oxford University Press.

Nietzsche, Friedrich. 1967. *On the Genealogy of Morals and Ecce Homo*. Translated by Walter Kauffman and Robert Hollingdale. New York: Random House.

Noë, Alva. 1999. "Thought and Experience." *American Philosophical Quarterly* 36, no. 3 (July): 257–65.

———. 2004. *Action in Perception*. Cambridge, MA: MIT Press.

———. 2009. *Out of Our Heads*. New York: Hill and Wang.

Noske, Barbara. 1989. *Humans and Other Animals*. London: Pluto Press.

Oakeshott, Michael. 1975. *On Human Conduct*. Oxford: Clarendon Press.

O'Callaghan, Casey. 2017. *Beyond Vision*. Oxford: Oxford University Press.

Odling-Smee, John, Kevin Laland, and Marcus Feldman. 2001. "Niche Construction, Ecological Inheritance, and Cycles of Contingency in Evolution." In *Cycles of Contingency*, edited by Susan Oyama, Paul Griffiths, and Russell Gray, 117–26. Cambridge, MA: MIT Press.

———. 2003. *Niche Construction*. Princeton: Princeton University Press.

Okin, Susan. 1989. *Justice, Gender and the Family*. New York: Basic Books.

Okrent, Mark. 2007. *Rational Animals*. Athens, OH: Ohio University Press.

———. 2018. *Nature and Normativity*. New York: Routledge.

O'Regan, J. Kevin. 2011. *Why Red Doesn't Sound Like a Bell*. Oxford: Oxford University Press.

Ortner, Sherry. 1984. "Theory in Anthropology since the Sixties." *Comparative Studies in Society and History* 26, no. 1 (January): 126–66.

Ortner, Sherry. 2006. *Anthropology and Social Theory*. Durham, NC: Duke University Press.

Oyama, Susan. 1985. *The Ontogeny of Information*. Cambridge: Cambridge University Press.

Oyama, Susan, Paul Griffiths, and Russell Gray, eds. 2001. *Cycles of Contingency*. Cambridge, MA: MIT Press.

Peacocke, Christopher. 2001. "Does Perception Have a Nonconceptual Content?" *Journal of Philosophy* 98:239–64.

Peregrin, Jaroslav. 2014. *Inferentialism*. New York: Palgrave MacMillan.

Pickering, Andrew. 1984. *Constructing Quarks*. Chicago: University of Chicago Press.

Pickering, Andrew, ed. 1992. *Science as Practice and Culture*. Chicago: University of Chicago Press.

Pigliucci, Massimo. 2010. "Phenotypic Plasticity." In *Evolution: The Extended Synthesis*, edited by Massimo Pigliucci and Gerd Müller, 355–78. Cambridge, MA: MIT Press.

Pigliucci, Massimo, and Gerd Müller, eds. 2010. *Evolution: The Extended Synthesis*. Cambridge, MA: MIT Press.

Pippin, Robert. 2008. *Hegel's Practical Philosophy*. Cambridge: Cambridge University Press.

Pitkin, Hannah. 1972. *Wittgenstein and Justice*. Berkeley: University of California Press.

Pitts-Taylor, Victoria. 2016. *The Brain's Body*. Durham, NC: Duke University Press.

Plotkin, Mark. 1988. "The Outlook for New Agricultural and Industrial Products from the Tropics." In *Biodiversity*, edited by Edward Wilson, 106–16. Washington, DC: National Academy Press.

Polanyi, Michael. 1958. *Personal Knowledge*. Chicago: University of Chicago Press.

Polsby, Nelson. 1960. "How to Study Community Power: The Pluralist Alternative." *Journal of Politics* 22, no. 3 (August): 474–84.

———. 1968. "Study of Community Power." *International Encyclopedia of the Social Sciences*, vol. 3, 157–63. New York: Macmillan and The Free Press.

———. 1980. *Community Power and Political Theory*. New Haven: Yale University Press.

Polyakov, Michael. 2012. "Practice Theories: The Latest Turn in Historiography?" *Journal of the Philosophy of History* 6, no. 2 (January): 218–35.

Pomeranz, Kenneth. 2000. *The Great Divergence*. Princeton: Princeton University Press.

Porter, Theodore. 1995. *Trust in Numbers*. Princeton: Princeton University Press.

Potts, Richard. 1996. *Humanity's Descent*. New York: Avon.

———. 2013. "Hominin Evolution in Settings of Strong Environmental Variability." *Quaternary Science Reviews* 73, no. 1 (August): 1–13.

Prinz, Jesse. 2004. *Gut Reactions*. Oxford: Oxford University Press.

Putnam, Hilary. 1975. "The Meaning of 'Meaning.'" In *Mind, Language, and Reality*, 215–71. Cambridge: Cambridge University Press.

Quine, W. v. O. 1960. *Word and Object*. Cambridge, MA: MIT Press.

———. 1969. *Ontological Relativity and Other Essays*. New York: Columbia University Press.

Rabinow, Paul. 1996. *Essays on the Anthropology of Reason*. Princeton: Princeton University Press.

Rawls, John. 1955. "Two Concepts of Rules." *Philosophical Review* 64, no. 1 (January): 3–32.

———. 1971. *A Theory of Justice*. Cambridge, MA: Harvard University Press.

———. 1980. "Kantian Constructivism in Moral Theory." *Journal of Philosophy* 77, no. 9 (1980): 515–72.

———. 1993. *Political Liberalism*. New York: Columbia University Press.

Reardon, Jenny. 2005. *Race to the Finish*. Princeton: Princeton University Press.

Reckwitz, Andreas. 2002. "Toward a Theory of Social Practices: A Development in Culturalist Theorizing." *European Journal of Social Theory* 5, no. 2 (May): 243–63.

Rheinberger, Hans-Jörg. 1995. "From Microsomes to Ribosomes: 'Strategies' of Representation, 1935–55." *Journal of the History of Biology* 28:49–89.

Richardson, Sarah. 2013. *Sex Itself*. Chicago: University of Chicago Press.

Richerson, Peter, and Robert Boyd. 2005. *Not By Genes Alone*. Chicago: University of Chicago Press.

Risjord, Mark. 2014. "Structure, Agency, and Improvisation." In *Rethinking the Individualism-Holism Debate*, edited by Julie Zahle and Finn Collin, 219–36. Dordrecht: Springer.

Rorty, Richard. 1991. "Representation, Social Practice, and Truth." In *Objectivity, Relativism, and Truth*, 151–61. Cambridge: Cambridge University Press.

Rosenblueth, Arturo, Norbert Wiener, and Julian Bigelow. 1943. "Behavior, Purpose, and Teleology." *Philosophy of Science* 10, no. 1: 18–24.

Roth, Paul. 2020. *The Philosophical Structure of Historical Explanation*. Evanston: Northwestern University Press.

Rouse, Joseph. 1987. *Knowledge and Power*. Ithaca: Cornell University Press.

———. 1990. "The Narrative Reconstruction of Science." *Inquiry* 33, no. 2 (August): 179–96.

———. 1996a. "Beyond Epistemic Sovereignty." In *The Disunity of Science*, edited by Peter Galison and David Stump, 398–416. Stanford: Stanford University Press.

———. 1996b. *Engaging Science*. Ithaca: Cornell University Press.

———. 1999. "Understanding Scientific Practices: Cultural Studies of Science as a Philosophical Program." In *The Science Studies Reader*, edited by Mario Biagioli, 442–56. New York: Routledge.

———. 2000. "Coping and its Contrasts." In *Heidegger, Coping, and Cognitive Science*, edited by Mark Wrathall and Jeff Malpas, 7–28. Cambridge, MA: MIT Press.

———. 2002. *How Scientific Practices Matter*. Chicago: University of Chicago Press.

———. 2005a. "Mind, Body and World: Todes and McDowell on Bodies and Language." *Inquiry* 48, no. 1 (August): 38–61.

———. 2005b. "Power/Knowledge." *The Cambridge Companion to Foucault*, edited by Gary Gutting, 2nd edition, 95–122. Cambridge: Cambridge University Press.

———. 2007. "Social Practices and Normativity." *Philosophy of the Social Sciences* 37:46–56.

———. 2009. "Standpoint Theories Reconsidered." *Hypatia* 24, no. 4 (Fall): 244–52.

———. 2013. "What Is Conceptually Articulated Understanding?" In *Mind, Reason, and Being-in-the-World: The McDowell-Dreyfus Debate*, edited by Joseph Schear, 250–71. London: Routledge.

———. 2015. *Articulating the World*. Chicago: University of Chicago Press.

———. 2018. "Naturecultural Inferentialism." In *From Rules to Meanings: New Essays on Inferentialism*, edited By Ondřej Beran, Vojtěch Kolman, Ladislav Koreň, 239–48. London: Taylor and Francis.

———. 2021. "Stance and Being." *Journal of the American Philosophical Association* 7, no. 1 (Spring): 20–39.

———. 2022. "Liberal or Radical Naturalism?" In *Routledge Handbook of Liberal Naturalism*, edited by Mario de Caro and David Macarthur, 177–89. New York: Routledge.

———. Forthcoming. "Theories of Practices." In *Oxford Handbook of Social Ontology*, edited by Stephanie Collins, Brian Epstein, Sally Haslanger, and Hans-Bernhard Schmid. Oxford: Oxford University Press.

Ryle, Gilbert. 1971. "The Thinking of Thoughts: What is *Le Penseur* Doing?" In *Collected Essays 1929–1968*, 494–510. New York: Barnes and Noble.

Sacks, Harvey. 1995. *Lectures on Conversation*. Oxford: Blackwell.

Savage-Rumbaugh, Sue, and Roger Lewin. 1994. *Kanzi*. New York: Wiley and Sons.

Savage-Rumbaugh, Sue, Stuart Shanker, and Talbot Taylor. 1998. *Apes, Language, and the Human Mind*. Oxford: Oxford University Press.

Schatzki, Theodore. 1996. *Social Practices*. Cambridge: Cambridge University Press.

———. 2002. *The Site of the Social*. University Park: Pennsylvania State University Press.

———. 2010. *The Timespace of Human Activity*. Lanham, MD: Rowman & Littlefield.

———. 2019. *Social Change in a Material World*. New York: Routledge.

Schatzki, Theodore, Karin Knorr-Cetina, and Eike von Savigny, eds. 2001. *The Practice Turn in Contemporary Theory*. New York: Routledge.

Schear, Joseph, ed. 2013. *Mind, Reason, and Being-in-the-World*. New York: Routledge.

Scheman, Naomi. 2011. *Shifting Ground*. Oxford: Oxford University Press.

Schoen, Brian. 2009. *The Fragile Fabric of Union*. Baltimore: Johns Hopkins University Press.

Schrödinger, Erwin. 1992. *What Is Life?* Cambridge: Cambridge University Press.

Scott, James. 2017. *Against the Grain*. New Haven, CT: Yale University Press.

Searle, John. 1969. *Speech Acts*. Cambridge: Cambridge University Press.

———. 1983. *Intentionality*. Cambridge: Cambridge University Press.

————. 1995. *The Construction of Social Reality*. New York: Free Press.

————. 2010. *Making the Social World*. Oxford: Oxford University Press.

————. 2018. Constitutive Rules. *Argumenta* 4, no. 1: 51–54.

Sellars, Wilfrid. 1997. *Empiricism and the Philosophy of Mind*. Cambridge, MA: Harvard University Press.

————. 2007. *In the Space of Reasons*. Cambridge, MA: Harvard University Press.

Shapin, Steven, and Simon Schaffer. 1985. *Leviathan and the Air Pump*. Princeton: Princeton University Press.

Shove, Elizabeth, Mika Pantzar, and Matt Watson. 2012. *The Dynamics of Social Practice*. London: Sage.

Simon, Herbert. 1981. *The Sciences of the Artificial*. 2nd ed. Cambridge, MA: MIT Press.

Smith, Brian Cantwell. 1988. "The Semantics of Clocks." In *Aspects of Artificial Intelligence*, edited by James Fetzer, 3–31. Dordrecht: Springer.

————. 1996. *On the Origin of Objects*. Cambridge, MA: MIT Press.

————. 2015. "So Boundary as Not to Be an Object at All." In *Boundary Objects and Beyond*, edited by Geoffrey Bowker, 219–27. Cambridge, MA: MIT Press.

————. 2019. *The Promise of Artificial Intelligence*. Cambridge, MA: MIT Press.

Smocovitis, Vassiliki Betty. 2020. "Historicizing the Synthesis: Critical Insights and Pivotal Moments in the Long History of Evolutionary Theory." In *The Theory of Evolution*, edited by David Mindell and Samuel Scheiner, 25–45. Chicago: University of Chicago Press.

Soler, Léna, Sjoerd Zwart, Michael Lynch, and Vincent Israel-Jost, eds. 2014. *Science After the Practice Turn*. New York: Routledge.

Solnit, Rebecca. 2010. *Infinite City*. Berkeley: University of California Press.

Soluri, John. 2005. *Banana Cultures*. Austin: University of Texas Press.

Springer, Elise. 2013. *Communicating Moral Concern*. Cambridge, MA: MIT Press.

Sterelny, Kim. 2003. *Thought in a Hostile World*. Oxford: Blackwell.

————. 2012a. *The Evolved Apprentice*. Cambridge, MA: MIT Press.

————. 2012b. "Language, Gesture, Skill: The Coevolutionary Foundations of Language." *Philosophical Transactions of the Royal Society B* 367, no. 1599 (August): 2141–51. doi:10.1098/rstb.2012.0116.

————. 2021. *The Pleistocene Social Contract*. Oxford: Oxford University Press.

Stringer, Chris and Peter Andrews. 2012. *The Complete World of Human Evolution*. 2nd ed. London: Thames and Hudson.

Sultan, Sonia. 2015. *Organism and Environment*. Oxford: Oxford University Press.

Symons, Donald. 1989. "A Critique of Darwinian Anthropology." *Ethology and Sociobiology* 10, nos. 1–3 (January): 131–44.

Tanesini, Alessandra. 2014. "Temporal Externalism: A Taxonomy, an Articulation, and a Defense." *Journal for the Philosophy of History* 8:1–19.

Taussig, Michael. 1980. *The Devil and Commodity Fetishism in South America*. Chapel Hill: University of North Carolina Press.

Taylor, Charles. 1964. *The Explanation of Behavior*. London: Routledge & Kegan Paul.

————. 1984. "Foucault on Freedom and Truth." *Political Theory* 12, no. 2 (May): 152–83.

————. 1985a. *Human Agency and Language*. Cambridge: Cambridge University Press.

————. 1985b. *Philosophy and the Human Sciences*, Cambridge: Cambridge University Press.

————. 1991. "The Dialogical Self." In *The Interpretive Turn*, edited by David Hiley, James Bohman, and Richard Shusterman, 304–14. Ithaca: Cornell University Press.

———. 1995. *Philosophical Arguments*. Cambridge, MA: Harvard University Press.

———. 2016. *The Language Animal*. Cambridge, MA: Harvard University Press.

Tengblad, Stefan, ed. 2012. *The Work of Managers*. Oxford: Oxford University Press.

Thompson, Evan. 2007. *Mind in Life*. Cambridge, MA: Harvard University Press.

Todes, Samuel. 2001. *Body and World*. Cambridge: MIT Press.

Tomasello, Michael. 2008. *Origins of Human Communication*. Cambridge, MA: MIT Press.

———. 2014. *A Natural History of Human Thinking*. Cambridge, MA: Harvard University Press.

———. 2019. *Becoming Human*. Cambridge, MA: Harvard University Press.

Tomasello, Michael, Ann Kruger, and Hilary Ratner. 1993. "Cultural Learning." *Behavioral and Brain Sciences* 16, no. 3 (September): 495–552.

Tomasello, Michael, Alicia Melis, Claudio Tennie, Emily Wyman, and Esther Herrmann. 2012. "Two Key Steps in the Evolution of Human Cooperation." *Current Anthropology* 53, no. 6 (December): 673–92.

Tomlinson, Gary. 2015. *A Million Years of Music*. New York: Zone Books.

———. 2018. *Culture and the Course of Human Evolution*. Chicago: University of Chicago Press.

Traweek, Sharon. 1988. *Beamtimes and Lifetimes*. Cambridge, MA: Harvard University Press.

———. 1992. "Border Crossings: Narrative Strategies in Science Studies and among Physicists in Tsukuba Science City, Japan." In *Science as Practice and Culture*, edited by Andrew Pickering, 429–465. Chicago: University of Chicago Press.

———. 1996. "*Kokusaika, Gaiatsu*, and *Bachigai*: Japanese Physicists' Strategies for Moving into the International Political Economy of Science." In *Naked Science*, edited by Laura Nader, 174–97. New York: Routledge.

Tsing, Anna Lowenhaupt. 2005. *Friction*. Princeton: Princeton University Press.

Turner, J. Scott. 2000. *The Extended Organism*. Cambridge, MA: Harvard University Press.

Turner, Stephen. 1994. *The Social Theory of Practices*. Chicago: University of Chicago Press.

———. 2007. "Practice Then and Now." *Human Affairs* 17, no. 2 (December): 111–25.

———. 2008. "Practices as the New Fundamental Social Formation in the Knowledge Society." *Družboslovne Razprave* 24:49–64.

Wagner, Günter 2000. "What is the Promise of Developmental Evolution? Part I." *Journal of Experimental Zoology (Molecular and Developmental Evolution)* 288:95–98.

———. 2001. "What is the Promise of Developmental Evolution? Part II." *Journal of Experimental Zoology (Molecular and Developmental Evolution)* 291:305–9.

———. 2003. "What is the Promise of Developmental Evolution? Part III." *Journal of Experimental Zoology (Molecular and Developmental Evolution)* 300B:1–4.

Walzer, Michael. 1983. "The Politics of Michel Foucault." *Dissent* 30 (4): 481–90.

Warin, Megan, Vivienne Moore, and Michael Davies. 2016. "Epigenetics and Obesity: The Reproduction of *Habitus* Through Intracellular and Social Environments." *Body and Society* 22, no. 4 (July): 52–78.

Wartenberg, Thomas. 1990. *The Forms of Power*. Philadelphia: Temple University Press.

———, ed. 1992. *Rethinking Power*. Albany, NY: State University of New York Press.

Weart, Spencer. 2008. *The Discovery of Global Warming*. Cambridge, MA: Harvard University Press.

West-Eberhard, Mary Jane. 2003. *Developmental Plasticity and Evolution*. Oxford: Oxford University Press.

Wheeler, Samuel. 2000. *Deconstruction as Analytic Philosophy*. Stanford: Stanford University Press.

Whiten, Andrew, Jane Goodall, William McGrew, Toshiada Nishida, Vernon Reynolds, Yuki-
    maru Sugiyama, Caroline Tutin, Richard Wrangham, and Christophe Boesch. 1999. "Cul-
    tures in Chimpanzees." *Nature* 399 (June): 682–85.
Williams, Eric. 1994. *Capitalism and Slavery*. 2nd ed. Chapel Hill: University of North Carolina
    Press.
Wilson, Alan. 1985. "The Molecular Basis of Evolution." *Scientific American* 253 (October):
    148–57.
————. 1991. "From Molecular Evolution to Body and Brain Evolution." In *Perspectives on Cel-
    lular Regulation*, edited by Judith Campisi and A. B. Pardee, 331–40. New York: John Wiley/
    A.R. Liss.
Wilson, Edward. 1975. *Sociobiology*. Cambridge, MA: Harvard University Press.
Wilson, George, and Samuel Shpall. 2016. "Action." *Stanford Encyclopedia of Philosophy*, edited
    by Edward N. Zalta (Winter 2016), https://plato.stanford.edu/archives/win2016/entries
    /action/.
Wimsatt, William. 2019. "Articulating Babel: A Conceptual Geography for Cultural Evolution."
    In *Beyond the Meme*, edited by Alan Love and William Wimsatt, 1–41. Minneapolis: Univer-
    sity of Minnesota Press.
Winterhalder, Bruce, and Eric Alden Smith. 2000. "Analyzing Adaptive Strategies: Human Be-
    havioral Ecology at Twenty-Five." *Evolutionary Anthropology* 9:51–72.
Withy, Katharine. 2011. "Making Sense of Heidegger on Thrownness." *European Journal of Phi-
    losophy* 22, no. 1 (June): 61–81.
Wittgenstein, Ludwig. 1953. *Philosophical Investigations*. Translated by Elizabeth Anscombe.
    Oxford: Blackwell. Parenthetical references in the text are to part and paragraph numbers.
Wolf, Eric. 1982. *Europe and the People without History*. Berkeley: University of California Press.
Woody, Andrea. 2004. "Telltale Signs: What Common Explanatory Strategies in Chemistry Re-
    veal about Explanation Itself." *Foundations of Chemistry* 6:13–43.
Wrangham, Richard. 1999. "Evolution of Coalitionary Killing." *Yearbook of Physical Anthropol-
    ogy* 42:1–30.
Wrangham, Richard. 2009. *Catching Fire*. New York: Basic Books.
Wray, Gregory, Douglas Futuyma, Richard Lenski, Trudy MacKay, Dolph Schluter, Joan Strass-
    mann, and Hopi Hoekstra. 2014. "Does Evolutionary Theory Need a Rethink? No, All is
    Well." *Nature* 514:161–64.
Zahle, Julie, and Finn Collin, eds. 2014 *Rethinking the Individualism-Holism Debate*. Cham:
    Springer.
Zammito, John. 2004. *A Nice Derangement of Epistemes*. Chicago: University of Chicago Press.

Index

accountability, normative, 4, 6–7, 22, 54–56, 80, 103, 106, 118, 155–66, 176, 181–82, 202, 219, 255–56, 259, 264–65, 268, 270, 276, 281–84. *See also* normativity

Acheulian stone tools, 78

acknowledgement. *See* recognitives

action, 16, 30, 32–33, 40, 44, 51, 90, 111, 133–34, 175–77, 212, 230, 233–35, 239, 241–42, 246–48, 274

action-environments, 234–35, 284, 295

adaptation, 7, 63, 69, 72, 75, 129, 131, 222, 249, 285

affects, affective, 20, 40–41, 52, 56, 8, 94, 103, 160, 162, 170,180, 201, 204, 215, 227, 233–34, 243, 245, 253, 267, 270, 278, 288–89

affordances, environmental, 11, 43, 68, 93, 100, 187, 284

agents, agency, 61–62, 84–85, 128, 163, 197, 206, 231, 234–49, 258, 265, 270. *See also* persons

agriculture, 6, 81, 236–37, 277, 288–91

Alder, Ken, 221n21

alignment, 151–54, 194–95, 233–42, 251, 254–56, 268, 270, 273, 277, 294. *See also* misalignments

alleles, 67

altruism, 55, 78–79

anaphora, anaphoric: dependence on prior processes, 165, 202n9, 217–19, 224, 227, 251, 260n27; goals as, 133, 273; issues and stakes as, 79n24, 158–60, 251, 273

anatomy, human, 1, 5, 99–100

Anderson, Elizabeth, 238, 280

animal behavior, nonhuman, 5, 58, 74–77, 91–94, 98n19, 119; with flexible environmental response, 77, 94, 188–89, 235, 262, 269, 285. *See also* cognition: animal (nonhuman); detection agents; organisms

animal cognition. *See* cognition: animal (nonhuman)

animal communication. *See* communication, animal

animal cultures. *See* cultures, animal

animals, social, 1, 16, 94, 190–91, 234, 257

Anthropocene, 81, 285

anthropocentrism, 13–14, 128, 279

anthropology, 3, 29–31, 47, 54

anthropomorphism, 56

Antonovics, Janis, 63, 74, 110

Arendt, Hannah, 231

Aristotle, 59–61, 64, 133–34

articulation, conceptual. *See* conceptual articulation

artifacts. *See* equipment

artificial intelligence (AI), 3, 186

assertion, asserting, 174–185, 214

Ásta, 207–9, 233

autonomy, 231, 244–48, 277; of disciplines or domains, 4, 26

autopoiesis, 61–62, 199

Avital, Eytan, 74–75

baboons, 190–91

Barad, Karen, 45, 62, 90n10, 110n2, 186n22, 192, 239n10

Beatty, John, 59

Bechtel, William, 223

behavioral ecology, human, 8, 57, 131

behaviorism, 7, 28–29, 86, 246

Bickerton, Derek, 73, 75, 76n19, 80, 101, 135, 169n1, 171nn6–8, 198n3, 214

Biddle, Justin, 252–53

Bigelow, Julian, 60

biochemistry, 1, 5–6

biology, 1–9; developmental, 2, 11, 66; ecological-developmental, 10, 14–16, 58, 64, 68–81, 84, 99,

Printed and bound by CPI Group (UK) Ltd, Croydon, CR0 4YY

05/03/2024

14465210-0005